Ethnobotany and Conservation of Biocultural Diversity

edited by

Thomas J. S. Carlson

and

Luisa Maffi

Advances in Economic Botany, Volume 15

Series Editor: Charles M. Peters

THE NEW YORK BOTANICAL GARDEN PRESS

Published by
The New York Botanical Garden Press
Bronx, New York

Composition and interior design by Sarah MacLennan Kerr, Oberlin, Ohio.
Cover design by Charles M. Peters.
Manufacturing by Maple-Vail Book Manufacturing Group.

Front cover: *Left.* Basket weaver collecting sedges in a saline wetland, Rift Valley, western Uganda (T. Cunningham). *Middle.* Salako Dayak in West Kalimantan, Indonesia extracting damar (oleo-resin) from a young *Shorea lamellata* tree (C. Peters). *Right.* Hotï boy taking aim with children's blowgun (*Arthrostylidium schomburgkii*), Kayamá, Bolívar State (S. Zent).

Back cover: Traditional dance by village women near Bwindi Impenetrable National Park, western Uganda (C. Peters).

The paper used in this publication meets the requirements of the American National Standard for Information Sciences—Permanence of Paper for Publications and Documents in Libraries and Archives (ANSI/NISO Z39.48–1992)

Printed in the United States of America using soy-based ink on recycled paper.

MetLife Foundation is a Leadership Funder of The New York Botanical Garden Press.

Library of Congress Cataloging-in-Publication Data

Ethnobotany and conservation of biocultural diversity / edited by Thomas J. S. Carlson and Luisa Maffi.
p. cm. — (Advances in economic botany : v. 15)
Includes bibliographical references (p.).
ISBN 0-89327-453-4 (pbk. : alk. paper)
1. Ethnobotany. 2. Ethnoscience. 3. Ethnobiology. 4. Biological diversity. 5. Indigenous peoples—Ecology. 6. Conservation of natural resources. 7. Sustainable development. I. Carlson, Thomas J. S. II. Maffi, Luisa. III. Series.
GN476.73.E834 2003
581.6′3—DC21

2003009163

Contents

Introduction: Ethnobotany and Conservation of Biocultural Diversity
Thomas J. S. Carlson and Luisa Maffi 1

Part 1. Indigenous Knowledge and the Creation and Conservation of Biodiversity

1. Maintaining and Restoring Biocultural Diversity: The Evolution of a Role for Ethnobiology
Luisa Maffi 9

2. Ethnobotanical Convergence, Divergence, and Change among the Hotï of the Venezuelan Guayana
Stanford Zent and Egleé López-Zent 37

3. Amazonian Indians as Ecological Disturbance Agents: The Hotï of the Sierra de Maigualida, Venezuelan Guayana
Egleé López-Zent and Stanford Zent 79

4. Indigenous Knowledge and Conservation of the Rain Forest: Ethnobotany of the Barí of Venezuela
Manuel Lizarralde 113

5. Ethnobotanical Ground-Truthing and Forest Diversity in the Western Amazon
Glenn H. Shepard Jr., Douglas W. Yu, and Bruce W. Nelson 133

Part 2. Knowledge and Sustainable Use of Traditional Plant Resources

6. Indigenous Nomenclature, Classification, and Utilization of African Rattans
Terry C. H. Sunderland 175

7. Crafts and Conservation: The Ecological Footprint of International Markets on an African Resource
A. B. Cunningham and S. K. Choge 215

8. Effects of Economic Liberalization on Dao Women's Traditional Knowledge, Ecology, and Trade of Medicinal Plants in Northern Vietnam
Jennifer C. Sowerwine 235

9. Medical Ethnobotany and Therapeutic Uses of *Croton lechleri* Müell.-Arg. for Human Health
Thomas J. S. Carlson 263

Part 3. Ethical Issues in Ethnobiology

10. Ethnobiology and Globalization: Science and Ethics at the Turn of the Century
Miguel N. Alexiades 283

11. Weighing the Proverbial "Ounce of Prevention" against the "Pound of Cure" in a Biocultural Context: A Role for the Precautionary Principle in Ethnobiological Research
Kelly Bannister and Katherine Barrett 307

List of Contributors **341**

Index **343**

Introduction: Ethnobotany and Conservation of Biocultural Diversity

Thomas J. S. Carlson and Luisa Maffi

The field of ethnobotany (and more generally ethnobiology) traces its roots to two distinct research traditions: a long-standing interest in how human societies around the world make use of plants (and animals) in their local environments; and a more recent (mid-1950s onward) interest in how humans perceive, classify, and name the natural world (for additional details on this history, see Berlin 1992: 3–5). Research in the former tradition has historically been carried out mostly, though not exclusively, by scholars trained in biology. The latter research tradition, on the other hand, has largely developed out of the specialized anthropological niche of ethnoscience and comparative cognitive studies.

Over the last few decades these two traditions have progressively gone from proceeding along separate (and sometimes uneasily coexisting) tracks to increasing, and mutually enriching, integration. A decisive step in that direction was signified in the 1990s by the emergence of a more encompassing ethnoecological perspective on human–environment relationships (Toledo 1992). Ethnoecological research integrates biological, cultural, and linguistic studies to understand the knowledge and beliefs that indigenous and traditional communities have about their biological environments and their practices of use and management of the natural resources found in their ecosystems. Toledo (2000) synthesizes this approach by defining ethnoecology as a discipline that explores three interrelated levels of human–environment relationships: the *corpus* (knowledge), the *cosmos* (beliefs), and the *praxis* (practices) displayed by human societies in their interactions with local ecosystems. Through this integrated approach, ethnobiology has also been notably gaining in methodological sophistication, moving toward the establishment of its own distinctive methods tool kit (Martin 1995; Alexiades 1996; Cunningham 2001).

In the course of the 1990s, another major development was the growing recognition that one emergent property of ecosystems having experienced significant human presence and human resource use over time is that they are biocultural systems—systems that have been jointly shaped by biological and cultural dynamics—and that much biodiversity found in place (including in habitats commonly thought of as pristine, such as tropical rain forests) is actually anthropogenic (Posey 1999). This in turn led to the realization that local ecological knowledge, beliefs, and practices have much to contribute to the conservation of biodiversity as well as to sustainable use of natural resources (Maffi et al. 2000). There is also now a growing recognition of the contributions of ethnobotany to the human health needs of local rural communities (Carlson 2001). During the same period, the application of indigenous and traditional knowledge increasingly came to be seen as valuable also in the context of development of new commercial products, such as in the pharmaceutical, dietary-supplement, and agricultural domains (ten Kate & Laird 1999).

The key role of ethnobiological and ethnoecological research at all of these levels has contributed significantly to the "coming of age" of the discipline and to raising its profile in both academic and nonacademic circles. This role has also contributed to putting a particularly bright spotlight on ethnobiology in relation to a variety of ethical concerns surrounding the documentation and use of traditional knowledge by outside scholars and more generally in connection to the relationship between scholars and the local communities they work with. Paradoxically, ethnobiology is currently a discipline that has both made some of the most progressive strides toward addressing these ethical issues (see, e.g., the Code of Ethics and Draft Standards of Practice of the International Society of Ethnobiology at <http://guallart.dac.uga.edu/ISE>; also Posey 1996; Posey & Dutfield 1996) and been the object of (occasionally vehement) criticism for not doing enough to address them. Far from shying away from these complexities and contradictions, most ethnobiologists continue their efforts to grapple with them and with the problems posed by research and applied work with indigenous and traditional peoples in an increasingly globalized and politicized world, while pursuing their goal to produce solid theory, reliable methods, and valid knowledge.

The chapters in this book, based in part on a symposium by the same title held at the Sixteenth International Botanical Congress in Saint Louis, Missouri, 1–7 August 1999, showcase recent ethnobotanical research conducted within this general framework by members of a new generation of ethnobiologists, including case studies from the tropical environments of the Amazon Basin, Africa, and Asia. The volume is divided into three parts. Part 1 focuses on the contributions of traditional ecological knowledge to the creation and conservation of biodiversity. Part 2 is devoted to the knowledge and sustainable use of traditional plant resources. Part 3 deals with ethical issues in ethnobiology.

In the first chapter of part 1, Luisa Maffi traces the history of the emergence of a biocultural perspective on the diversity of life as a coevolutionary process that, ever since humans have been on the planet, has shaped the environment through the joint action of both natural and cultural forces. She frames this history within the context of the enduring debate about the relationships between "nature" and "culture" and reviews a variety of positions that have been expressed in favor of or against the biocultural perspective, both theoretically and in relation to environmental conservation. She also discusses recent controversies about the nature and role of indigenous knowledge. Against this background, Maffi concludes with an analysis of the role of ethnobiology in this connection and of the possible synergies with fields of the natural sciences, such as ecosystem health and sustainability science, that centrally embrace a humans-in-environment perspective.

The remaining chapters of Part 1 explore various ways in which indigenous knowledge and practices influence biodiversity and its conservation. Stanford Zent and Egleé López Zent apply quantitative research methods to researching the factors that influence ethnobotanical knowledge convergence, divergence, and change among four different Hotï communities in the Amazon Basin of Venezuela. They analyze a variety of factors, including the biological environment, patterns of settlement, social structure, interethnic contact, religious practices, educational institutions, material culture, and access to modern technology. The authors suggest how their findings can be used to develop culturally and biologically appropriate conservation initiatives for the Hotï area. In a related study, Egleé López Zent and Stanford Zent demonstrate how the Hotï people act as "agents of creative disturbance" in the local environment through their subsistence behaviors. They infer that through long-term occupation of their habitat, the Hotï have probably had a considerable impact on the composition, diversity, structure, and function of their forest ecosystem and may have enhanced its biotic richness and complexity. Zent and Zent point to the significance of these findings for biocultural conservation, in that they show that the subsistence activities of indigenous peoples constitute a pervasive and integral component in the dynamic constitution of Amazonian ecosystems.

The chapter by Manuel Lizarralde describes his research on the ethnobotanical and ethnoecological knowledge of the Barí indigenous people in the northwest lowlands of Venezuela. Focusing on forest trees, the author finds that the Barí not only possess an impressive knowledge of tree species but also have a high potential use for these species, both directly (for food, medicine, fuel, construction material, technology, and commercial timber) and indirectly (as indicators of the possible presence of animals the Barí hunt). Lizarralde shows how this knowledge can be applied in support of Barí claims over their lands and for the conservation of biodiversity in their forests.

Glenn H. Shepard Jr., Douglas W. Yu, and Bruce W. Nelson present a method for assessing rain-forest habitat diversity through "ethnobotanical ground-truthing": a triangulation among LANDSAT imaging, ecological field research, and indigenous ethnobotanical and ethnoecological knowledge. This research, conducted among the Matsigenka indigenous group of the Peruvian Amazon, finds that the Matsigenka have a complex system of habitat classification based on vegetative and other biotic features as well as topographic, hydrological, and edaphic characteristics and disturbance regimes. The Matsigenka recognize all forest types currently known to ecologists working in the region and also distinguish some types not apparently known to ecologists. Half of the forest diversity recognized by the Matsigenka is beyond LANDSAT's powers of resolution. This highlights the importance of collaborating with local indigenous people when conducting studies on biotope classification.

The first three chapters of Part 2 are devoted to traditional uses of local plant resources and analyze the processes of both social and ecological change involved in the extension of such uses from subsistence-based to market-based economies. Terry

C. H. Sunderland's research on the indigenous nomenclature and classification of African rattans in West and Central Africa suggests that use of the rattan resources is subtler and more complex than previously thought. This complexity is reflected in local indigenous nomenclatural systems. Integration of the local indigenous taxonomy into Western taxonomic and ecological studies is crucial for a fuller understanding of this plant resource, one of the most important non-timber forest resources of the lowland African forests, and thus for its appropriate and sustainable development.

The chapter by Anthony B. Cunningham and S. K. Choge describes their research in East Africa, where small-scale subsistence production of wooden crafts and basketry rarely has negative social or ecological impacts. However, the authors find that large-scale commercial trade in baskets and carved woodcrafts, especially those of slow-growing species, can have negative impacts on both the ecology and sociology of the region. Cunningham and Choge point out that the commercial craft trade can generate valued local revenue through the use of traditional skills but that the long-term viability of rural craft production projects and the multiple-use value of craft resources must also be taken into account. They highlight the need for more sustainable harvesting, production, and marketing practices related to these crafts.

Jennifer C. Sowerwine introduces a gender perspective in her research on Dao women's traditional knowledge, production, and trade of medicinal plants in the mountains of northern Vietnam. She finds that economic liberalization is having mixed effects on these women's traditional medicinal knowledge and practices. On one hand, the decline in government subsidies for health care has resulted in a rise in demand for traditional botanical medicine due to its affordability. On the other hand, there is increasing evidence of overharvesting, while policies for highland development such as reforestation of so-called barren land and intensification of public land use create barriers to women's free movement across the landscapes where medicinal plants are gathered. Furthermore, the introduction of intensive agricultural production reduces the amount of time younger women can spend learning about medicinal plants from their female elders. All of these processes raise questions about the long-term sustainability of both medicinal plant resources and Dao women's knowledge of them.

The last chapter of part 2 describes aspects of research on traditional medicinal plant resources. Thomas J. S. Carlson stresses how tropical countries contain the most biologically and culturally diverse traditional medicinal systems in the world and how ethnobotanical diversity from tropical ecosystems supplies affordable medicines and foods to rural communities, significantly contributing to public health. He points out that, as ecosystems are degraded and languages are lost, these traditional botanical systems and the health care they provide become significantly diminished. This highlights the need to conserve not only ecosystems but also the vitality of the ethnolinguistic groups who live within these ecosystems. The chapter makes particular reference to research conducted on *Croton lechleri* (*sangre de drago*), a traditional medicinal plant in the western Amazon Basin. He describes the experimental biology and human clinical studies that confirmed the therapeutic activity of the ethnobotanical uses of this plant to treat diarrhea, respiratory infections, herpes simplex, and skin wounds.

The two chapters of part 3 delve into the ethical complexities and dilemmas raised by ethnobiologists' work on traditional knowledge. Miguel N. Alexiades points to the commodification of genetic and cultural resources, the privatization of education, science, and development, and the unprecedented articulation between the local and the global that characterize the dawn of the twenty-first century. He argues that ethnobiology as a discipline is especially affected by these global transformations, because it lies at the interface between nature and culture, between genetic resources and environmental knowledge, between local and international communities, and between science and society. He describes the challenges and opportunities facing ethnobotanists with regard to these dynamic processes, as they seek to develop sustainable and ethical research agendas.

Kelly Bannister and Katherine Barrett focus on the ethical issues surrounding the publication of findings from ethnobiological research. They consider uncertainties and potential harms that may result from publication and third-party appropriation of such research results. Through a proposed extension of the Precautionary Principle, the authors suggest that alternative avenues for the dissemination of research results should be actively sought. They also suggest that current academic merit systems be redesigned to encourage and reward direct support to local indigenous peoples and to foster closer partnerships between indigenous and academic communities.

The overall conclusions that emerge from the chapters in this book are: first, that conservation projects would benefit from integrating indigenous and traditional communities into the development and implementation of conservation plans and that, in turn, such projects should seek to benefit these communities; and second, that the conservation of biological diversity and the maintenance of healthy ecosystems should proceed hand in hand with support for the survival, health, and continued development of the indigenous and traditional societies that live within and directly depend on these ecosystems.

Literature Cited

Alexiades, M. N. (ed.). 1996. Selected Guidelines for Ethnobotanical Research: A Field Manual. Advances in Economic Botany, 10. Bronx: New York Botanical Garden.

Berlin, B. 1992. Ethnobiological Classification: Principles of Categorization of Plants and Animals in Traditional Societies. Princeton, N.J.: Princeton University Press.

Carlson, T. J. S. 2001. Language, ethnobotanical knowledge, and tropical public health. Pp. 489–502 *in* L. Maffi (ed.), On Biocultural Diversity: Linking Language, Knowledge, and the Environment. Washington, D.C.: Smithsonian Institution Press.

Cunningham, A. B. 2001. Applied Ethnobotany: People, Wild Plant Use and Conservation. London and Sterling, Va.: Earthscan Publications.

Maffi, L., G. Oviedo & P. B. Larsen. 2000. Indigenous and Traditional Peoples of the World and Ecoregion Conservation: An Integrated Approach to Conserving the World's Biological and Cultural Diversity. Gland, Switzerland: WWF and Terralingua.

Martin, G. J. 1995. Ethnobotany: A Methods Manual. London: Chapman & Hall.

Posey, D. A. (ed.). 1996. Traditional Resource Rights: International Instruments for Protection and Compensation for Indigenous Peoples and Local Communities. Gland, Switzerland, and Cambridge: IUCN.

——— (ed.). 1999. Cultural and Spiritual Values of Biodiversity. London and Nairobi: Intermediate Technology Publications and UNEP.

——— & G. Dutfield. 1996. Beyond Intellectual Property: Toward Traditional Resource Rights for Indigenous Peoples and Local Communities. Ottawa: International Development Research Centre.

Ten Kate, K. & S. A. Laird. 1999. The Commercial Use of Biodiversity: Access to Genetic Resources and Benefit-Sharing. London: Earthscan.

Toledo, V. M. 1992. What is ethnoecology? Origins, scope and implications of a rising discipline. Etnoecológica 1(1): 5–21.

———. 2000. Ethnoecology: A conceptual framework for the study of indigenous knowledge on nature. Plenary lecture, Seventh International Congress of Ethnobiology, Athens, Ga., 22–27 October 2000.

Part 1

Indigenous Knowledge and the Creation and Conservation of Biodiversity

1

Maintaining and Restoring Biocultural Diversity: The Evolution of a Role for Ethnobiology

Luisa Maffi

Few intellectual debates have had a longer, and more controversial, history than that about the relationships between "nature" and "culture." There are numerous versions of this debate, relevant to different contexts and domains of thought. In the incarnation that more closely bears on current environmental issues, the debate focuses at its most theoretical on the place of humans, as bearers of the capacity for learned behavior (culture), vis-à-vis the rest of life on earth (nature), as well as on the very definition and scope of the two key concepts. A variety of philosophical, religious, cosmological, and scholarly views have been expressed in this regard. At one end are those views that see humans as separate from, and often dominant over, other life forms. Here, culture is commonly conceptualized as technological advance (and the material products thereof) and the tool for increasing control over nature. At the other end are views that place humans squarely within an interdependent web of life, as an intrinsic component of nature. From this perspective, culture may be understood as a part of humans' natural endowment: an innate adaptive faculty that confers on humans' extraordinary plasticity in relation to their natural and social environments, thus giving rise, within parameters, to a dizzying diversity of forms of knowledge, beliefs, values, and behaviors (and their material and nonmaterial products).

One might want to label the former view as "anthropocentric" and the latter as "biocentric," but the ideological landscape conjured by these views is more complex than that. This becomes especially apparent in looking at how such theoretical positions play out when it comes to the practice of environmental conservation. Some of the most radical proponents of biocentric perspectives on conservation, such as deep ecologists, while viewing humans as part of an interdependent web of life, also tend to espouse a "human-hands-off" stance, thus de facto joining hands with those conservation biologists who go so far in taking a "biocentric last stand for nature" as to conceptualize ecosystems as separate from, and devoid of, humans.[1] These two positions thus present different interpretations of "biocentric," while sharing what might be termed a paradoxical "misanthropic anthropocentrism." Another perspective, while embracing a biocentric "humans-in-the-web-of-life" perspective, attempts to (anthropocentrically) uphold human values in conservation. Representatives of this view find themselves in contrast with both positions above but would also see the "technological-fix" anthropocentrists as very strange bedfellows indeed. And so on, with more permutations.

An important aspect of this complex debate centers on whether humans as a species are innate maximizers of short-term gain, who, if given free rein, will compulsively wreck the environment, or there are specific and specifiable historical conditions under which such destructive tendencies may manifest themselves, while in other circumstances human–environment relationships may be more sustainable. The role of indigenous and traditional peoples[2] and their ecological knowledge (TEK)[3] and behavior is also significantly at issue here, and so is the relationship between indigenous (or traditional) knowledge and scientific knowledge.

The interplay and tension among these multiple views has shaped the environmental field for at least the past twenty years. Conservation institutions' agendas have informed as well as been informed by it, and the flow of funding to conservation initiatives has directed as well as been directed by it. As a consequence, the pendulum has repeatedly swung between more human-inclusive and more human-exclusive approaches to conservation. This chapter reviews some of the key aspects of this recent history and attempts to identify the role that ethnobiology (along with related areas, such as ecological anthropology) has played and can play in this context, through the kind of in-depth and multifaceted work on traditional knowledge and use of flora and fauna that is exemplified in this book.

From the Biodiversity Crisis to a Biocultural Approach to the Diversity of Life

When, during the 1980s, the current biodiversity extinction crisis became a widespread focus of scientific and public attention (Wilson 1988), biologists identified it as the first in the history of the planet to be, by and large, a human-made phenomenon, caused by the negative impacts of human activities on the environment. That assessment was undoubtedly correct. The main causes that have been identified for the degradation of the world's ecosystems and the depletion of the biodiversity in them—from changes in land use to overharvesting of natural resources, waste disposal, alien species invasions, extreme natural events, and climate change—all stem from human activities, or have a major human activity component to them. The evidence of such macroscopic human-made effects led to a tendency to conceptualize humans as a whole as part of the problem—if not *the* problem—and argue that humans could only be part of the solution by "taking their hands off" the environment. In their alarmed calls to arms in defense of the environment, biologists generally overlooked the possibility that not all humans might have the same negative impacts on the environment—or that some human societies might not have a negative impact, or might even have a positive impact, on biodiversity. Action they promoted mostly favored human-exclusionary approaches to conservation.

Around the same time, however, a different perspective, one that linked environmental protection to sustainable development by stressing the integration of human needs in conservation, was emerging. This perspective was already apparent in prominent international documents such as the proceedings of the 1982 World Parks Congress (McNeely & Miller 1984) and the report *Caring for the Earth* (IUCN et al. 1991). It then came to the forefront at the 1992 United Nations Conference on Environment and Development (UNCED, the Rio Summit) and was variously reflected in the ensuing international action plans and instruments (Agenda 21, Rio Declaration, Convention on Biological Diversity [CBD], and subsequent others), as well as in major global biodiversity conservation plans and policies (WRI et al. 1992; Heywood 1995; IUCN 1997; IUCN & WWF 1998; WWF 1998). The main thrust of these processes and activities was that stemming the biodiversity crisis was not feasible without ensuring viable economic development of human societies, particularly those located in the parts of the world where biodiversity is both most abundant and most threatened (such as the Tropics).

Here, the conservation-cum-sustainable-development agenda met the growing body of work in the social sciences on cross-cultural variation in the environmental consequences of human behavior. Such research analyzed the circumstances, currently and historically, of both negative and positive impacts of human activities on biodiversity and ecosystem health, in small-scale societies as well as with the rise of complex civilizations. It provided evidence that human relationships with the environment are a highly complex and diverse phenomenon and that they should be understood on the basis of a more nuanced evaluation of the interactions among a wide range of social, cultural, economic, political, and ecological variables.

The results of this research generally suggested that small-scale societies with a history of continued and undisturbed occupation of given territories would, over time, tend to develop and maintain detailed and accurate knowledge about their ecological niches, as well as sustainable ways of using and managing natural resources. Ethnobiologists and ecological anthropologists extensively documented TEK, often finding it to be more in-depth than Western scientific knowledge of local environments and pointing to its value for environmental conservation and sustainable use and for the analysis and monitoring of long-term ecological changes (Berlin et al. 1974; Hunn 1977; Brokensha et al. 1980; Nations 1980; Norgaard 1981; Bulmer 1982; Williams & Hunn 1982; Hames & Vickers 1983; Alcorn 1984; Nabhan 1989; Posey & Balée 1989; Atran 1990, 1993; Taylor 1990; Berlin 1992; Blackburn & Anderson 1993; Williams & Baines 1993; Balée 1994; Warren et al. 1995; Berkes 1999; Blount & Gragson 1999; Medin & Atran 1999; Minnis & Elisens 2000). Research also identified historical cases of unsustainable human–environment relationships, which could often be linked to migrations or to the rise of complex civilizations living beyond the capacity of local ecosystems, as well as cases in which environmental degradation resulted from rapid socioeconomic and cultural change and other circumstances affecting the lifestyles and livelihoods of indigenous and traditional populations (Diamond 1986, 1987, 1991; Johnson 1989; Redford 1990; Hames 1991; Ponting 1991; Bahn & Flenley 1992; Denevan 1992a, 1992b; Ellen 1994; Eldredge 1995; Flannery 1995; Grinde & Johansen 1995; Kirch & Hunt 1996; Atran & Medin 1997; Kirch 1997).

During the 1990s these developments led to the rise of an integrated biocultural perspective at the intersection of the environmental sciences and the social sciences[4] (Oldfield & Alcorn 1991; Shiva et al. 1991; Chapin 1992; Nietschmann 1992; Clay 1993; Colchester 1994; Toledo 1994; Wilcox & Duin 1995; Alcorn 1997; McNeely 1997; Nabhan 1997; Stevens 1997; Posey 1999; Maffi 2001a). This perspective identified cultural and linguistic diversity[5] as manifestations of the diversity of life, analogous to biodiversity in their nature and dynamics. As with ecosystems, diversity was seen as a key factor in the vitality and resilience of cultural and linguistic systems. According to this view, all these systems are also interrelated, in that the diversity of human life is what provides "the human intellectual 'gene pool'" from which human populations draw to adapt to local environmental conditions and change (McNeely 1997: 192). These adaptations, constantly evolving in response to the changing parameters of the environment, are institutionalized in social organization, cultural knowledge, beliefs and values, technology, and language. They include knowledge of local flora and fauna and ecological relations and dynamics, as well as practices of natural resource use and regulation.

Within this framework, the hypothesis was proposed that ecosystems that have experienced significant human presence and human resource use over time are biocultural systems—systems that have been shaped jointly by biological and cultural dynamics—not only in terms of humans causing environmental degradation and resource depletion but also in terms of humans contributing, directly or indirectly, to the maintenance and even to the creation of biodiversity. Research pointed to the anthropogenic nature of many areas previously thought of as "wildernesses" and to many "pristine" landscapes as, in fact, cultural landscapes; that is, landscapes created or modified by human activities such as natural forest management, cultivation, and the use of fire (Bulmer 1982; Nabhan 1989; Blackburn & Anderson 1993; Posey 2001), as well as by humans acting as biodiversity-enhancing "agents of creative disturbance" (E. L. Zent & S. Zent 2004), although the modifications may be so subtle and sometimes operate on such a large scale that they are easily confused with the natural evolution of the landscape. It was suggested that cultural landscapes can also be said to be "linguistically anthropogenic" (Maffi 1998), in that human relationships with the environment are encoded in and conveyed by words, expressions, stories, legends, and songs that project the history of those relationships onto the land. Likewise, a notion of "linguistic ecologies" was proposed (Mühlhäusler 1996) to describe local networks of relationships encompassing not only a human group's linguistic and social environment but also the physical environment, all seen as interrelated parts of a whole. The idea of "logosphere," analogous to that of biosphere, was also put forth (Krauss 1996) to hypothesize the existence of another web of life: a web of human languages and knowledge that would have coevolved with the environment over the course of human history.

Studies comparing the global geographical patterns of the world's species and languages (Harmon 1996, 1998b; Maffi 1998; Oviedo et al. 2000; Harmon & Maffi 2002)—where species richness and language richness (numbers of species and languages) were taken as proxies for the full gamut of variation implied in the concepts of biological and cultural diversity[6]—showed a significant overlap in distribution, particularly in the case of endemic species and languages. According to estimates (Durning

1992), 60–80% of the more than 6000 existing oral languages are spoken by indigenous peoples, implying that such peoples constitute most of the world's cultural diversity, in spite of their overall small population numbers (approximately 300 million, or 5% of the world's population; Gray 1999). In these cross-mappings, the most extensive overlap was found in tropical rainforest regions—the most biodiversity-rich areas and also the most culturally diverse in terms of different indigenous groups living in them. But the mappings showed presence of indigenous peoples in most habitat types in areas of the Americas, Europe, Asia, Africa and Oceania that cover perhaps 20% of the world's land surface (or three times the total surface covered by protected areas; Martin 1993).

Several large-scale biogeographical factors were suggested as possibly accounting for these correlations by comparably affecting both biological and linguistic diversity, especially endemism (Harmon 1996). Such factors include: extensive land masses with a variety of terrains, climates, and ecosystems; island territories, especially with internal geophysical barriers; and tropical climates, fostering higher numbers and densities of species. All of these factors might increase linguistic diversity by increasing mutual isolation between human populations and thus favoring linguistic diversification. An additional possible factor for biodiversity–linguistic diversity correlations might be the coevolution of small-scale societies with their local ecosystems, in which, over time, humans interacted closely with the environment, modifying it as they adapted to it and developing specialized knowledge of it (Harmon 1996). In order to convey this vital knowledge and ways of acting upon it for individual and group survival, humans would develop specialized ways of talking about it (Harmon 1996; Maffi 1998). Thus the local languages, through which this knowledge was encoded and transmitted, would in turn have become molded by and specifically adapted to their socioecological environments (cf. Mühlhäusler 1995, 1996; Nettle 1996; Balée 2001).

These findings—along with the observation that in numerous instances indigenous and traditional peoples have had a light ecological footprint on the environment in which they have lived for generations and that they may even be the creators of some of the biodiversity found in place—led to the claim that indigenous peoples can be considered as "the most important stewards of the earth" (Martin 1993: xvi) and to the suggestion that fostering the persistence of cultural (and linguistic) diversity will support biodiversity conservation, and vice versa. This was seen as a pressing task, in view of the fact that a major crisis of diversity loss is also occurring in the world's many cultures and languages (Burger 1987; Gray 1991; Robins & Uhlenbeck 1991; Durning 1992, 1993; Hale 1992; Krauss 1992, 1996; Goehring 1993; Miller 1993; Harmon 1996, 1998a, 1998b; Maffi 1998, 2001a, 2001b; Maffi et al. 1999; Oviedo et al. 2000). Estimates of the impending demise of native languages (and, by implication, much of the cultural diversity expressed by them) indicated that as many as 90% of the 6,000+ currently spoken languages may be extinct or face extinction before the end of the twenty-first century (Krauss 1992, 1996; Harmon 1995).

Research stressed that, although linguistic and cultural change has taken place throughout human history, human societies have been able to maintain vitality and continuity through change as long as they could maintain their autonomy, and could thus retain control over the process of change itself (Chapin 1994; Hill 1997, 2001). Rapid and sweeping change, instead, can severely hamper the ability of a human group to adapt on its own terms to changing social and ecological environment and thus maintain its cultural and linguistic distinctiveness. As societies are increasingly channeled into a cash economy driven and sustained by high levels of consumption and short-term overexploitation of natural resources, the common consequences are displacement of human populations, restriction on their mobility, alteration of subsistence economies due to loss of traditional land rights and to declines in biodiversity, breakdown of social structures, and acculturation, all of which threaten the survival of many indigenous and other distinctive cultures.[7] Language shift often ensues, along with loss of TEK and of the ability to interact sustainably with the environment (Diamond 1993; Nabhan & St. Antoine 1993; Wilkins 1993; Woodbury 1993; Mühlhäusler 1995, 1996; Fishman 1996; Maffi 1998, 2001a, 2001b; Maffi et al. 1999; Zent 1999, 2001; Florey 2001; Hill 2001; Hunn 2001; Lizarralde 2001; Nabhan 2001). Preliminary assessments of the status of indicators of the world's cultural diversity—from the use of local languages to ethnic affiliation, forms of social organization, subsistence practices, land management, diet, medicine, art, and religion—revealed a downward trend in diversity in all cases (Harmon 1992), adding to the sense of urgency vis-à-vis the conservation of biocultural diversity (see also Harmon & Loh 2002).

Biocultural Diversity on Conservation and Human Rights Agendas

The growing awareness of the potential value of indigenous and traditional peoples' relationships with local ecosystems for conservation efforts was reflected in the main international environmental documents elaborated during the 1990s. For example, of indigenous peoples *Caring for the Earth* stated:

> Their cultures, economies, and identities are inextricably tied to their traditional lands and resources. Hunting, fishing, trapping, gathering or herding continue to be major sources of food, raw materials and income. Moreover, they provide native communities with a perception of themselves as distinct, confirming continuity with the past and unity with the natural world. Such activities reinforce spiritual values, an ethic of sharing, and a commitment to stewardship of the land, based on a perspective of many generations. (IUCN et al. 1991: 61, Box 11)

The CBD and related documents acknowledged the role of traditional knowledge in biodiversity conservation. Article 8(j) of the CBD, <http://www.biodiv.org/convention/articles.asp>, provides that each contracting party must:

> Subject to its national legislation, respect, preserve and maintain knowledge, innovations and practices of indigenous and local communities embodying traditional lifestyles relevant for the conservation and sustainable use of biological diversity and promote their wider application with the approval and involvement of the holders of such knowledge, innovations and practices and encourage the equitable sharing of the benefits arising from the utilization of such knowledge, innovations and practices.

Article 10(c) of the CBD requires parties to "protect and encourage customary use of biological resources in accordance with traditional cultural practices that are compatible with conservation or sustainable use requirements." This implies that conservation organizations should support customary regulations relating to natural resource use and management to the extent that they enhance biodiversity conservation. In other words, this and other related documents (e.g., Agenda 21, chapter 26.3) stressed that, in addressing environmental threats and the underlying causes of biodiversity loss, a coincidence of needs and interests may arise between conservation organizations and indigenous and traditional peoples, thus increasing the potential for a joint commitment to and partnership in conservation plans.

On their part, indigenous and traditional peoples active in the main international environmental and human rights forums argued that they would be more likely to engage in such partnerships if they were to be directly involved in determining the criteria for compatibility of traditional cultural practices with conservation and sustainable use of biological resources. They saw environment and development issues from a human rights perspective and placed their role vis-à-vis biodiversity conservation in the context of the recognition of their fundamental human rights—such as the right to self-determination, land/territorial and traditional resource rights, property rights, cultural and linguistic rights, and environmental rights; i.e., the rights of present and future generations to enjoy a healthy life in a healthy environment. They pursued these goals through participation in relevant processes, from the CBD's Conferences of the Parties to the U.N. Working Group on Indigenous Populations in charge of elaborating the Draft U.N. Declaration on the Rights of Indigenous Peoples and, more recently, the newly formed U.N. Permanent Forum on Indigenous Issues (Greaves 1994; Skutnabb-Kangas & Phillipson 1994; Brush & Stabinsky 1996; Posey 1996; Posey & Dutfield 1996; Daes 1997; Niec 1998; ten Kate & Laird 1999; Skutnabb-Kangas 2000; Maffi 2001c; Laird 2002).

In response to these developments, major conservation organizations, such as WWF and IUCN, elaborated principles and policies meant to guide partnerships with indigenous communities and organizations in conserving biodiversity within indigenous peoples' lands and territories and in promoting sustainable use of natural resources (WWF 1996; IUCN 1997; WCPA et al. 1999; Oviedo et al. 2000). International financial institutions such as the World Bank had already elaborated directives in relation to development activities affecting indigenous peoples (World Bank 1991).[8] Political entities involved in development cooperation, such as the Organization of American States and the European Union, followed suit (OAS 1997; European Commission 1998; European Union 1998). A number of conservation projects began to be implemented in partnership with indigenous peoples and traditional communities, building on recognition of the need to respect indigenous peoples' and traditional communities' rights and to support security of tenure and access to lands and resources, as a prerequisite to their commitment to conservation (see, e.g., Beltran 2000; Weber et al. 2000). International organizations also began to devote attention to the joint erosion of biological, cultural, and linguistic diversity and to the loss in intellectual resources represented by the loss of indigenous knowledge (Maffi et al. 1999; The Nation 1999; Posey 1999).

Action began to be called for in this connection—witness, e.g., the Recommendations on Scientific Research to be Undertaken for Implementation of Article 8(j), issued by the Fourth Meeting of the Conference of the Parties to the CBD, which stated: "Efforts should be made to identify those areas in which loss of traditional knowledge is occurring rapidly, so that appropriate steps can be taken to prevent this loss" (Document UNEP/CBD/COP/4/Inf.18, quoted in Bannister & Barrett 2004: 327).

The Pendulum Swings Back

Toward the end of the 1990s these developments seemed to spell a significant change in the field of conservation biology. As one conservationist put it, it appeared that "conservation biology has come to accept that humans will always be part of ecological systems, both natural and altered." Theoretically, then, we are closer to the point where outside scientists may become not only more accepting of human presence and the need to collaborate with local people but also more aware of the positive aspects of local people's presence and their knowledge (Steinmetz 1998: 2–3, citing Meffe & Carroll 1997).

Accepting this view did not mean actually breaking entirely new ground. Rather, it meant going forward by going back to the very origins of ecology (Odum 1953) and embracing a conceptualization of humans as an intrinsic part of the landscape and as linked to the rest of the biotic and abiotic world by means of complex patterns of ecological relationships (as outlined, e.g., in Nevah and Lieberman's [1984] Total Human Ecosystem concept). It also meant acknowledging human presence and activities (and thus human needs) as an intrinsic component in the development of visions for biodiversity conservation.

If signs of movement in that direction could be detected at that point, however, they were counterbalanced by expressions of growing discontent in the conservation community. Conservation-cum-sustainable-development, argued some prominent conservation biologists, just was not working (Terborgh 1999; Soulé 2000). The argument was that, all conservation hype notwithstanding, during the 1990s the rates of deforestation (and other signs of environmental depletion) had increased, and the main culprit was to be identified in the kind of initiatives that had been promoted during that period: from natural forest management to controlled harvesting of forest products to bioprospecting and ecotourism. In these discontents' analysis, all of these initiatives had attempted to wed conservation to sustainable development but ultimately had done nothing to stop overexploitation driven by "our collective obsession with maximizing economic growth" (Terborgh 1999). Indigenous peoples, now with exploding populations, shotguns and plows at hand, and access to the market economy, were seen by these critics as prey to the same "obsession." The discontents' drastic conclusions were: as much land as possible in the Tropics (and other priority areas worldwide) should be placed under tight government control—with funding shifted back from economic development outside to strict protection within these areas; and no people should be allowed in as residents or resource users, following the model of the U.S. national park system.

One might question the view of people (indigenous or otherwise) as invariably and inevitably short-term economic maximizers.[9] One might also observe that an "obsession with maximizing economic growth" configures anything but sustainable development and that perhaps part of the problem might reside in the progressive skewing of the sustainable development agenda—from resting firmly and evenly on the three "pillars" of environment, society, and economy, to perilously balancing on only one, economy—thus reverting to business-as-usual, top-down unsustainable development driven by global economic forces.[10] Some have also noted the irony implied in advocating a strict preservationist approach that would relocate people outside the forested lands in which they reside (in the same way as Native Americans were relocated outside Yellowstone and Yosemite National Parks), while calling for the reintroduction of top predators—which is precisely what, according to other views, indigenous people have been (Shepard 2002; Kay 1998).[11]

Yet these preservationist positions have been enjoying renewed favor in conservation circles, both fueled and sustained by new injections of funding. In turn, this is promoting a move of conservation organizations away from community-based projects. These developments have not gone unnoticed in anthropological circles. Concerned ecological and environmental anthropologists fear that whatever gains were made toward the integration of social needs and aspirations in conservation agendas are now being eroded, as the use of community participatory approaches begins to be abandoned or becomes increasingly tokenistic, reduced to a "managerial technique" for producing "consensus" rather than for building local capacity (Conservation and Community 2001). These anthropologists express frustration at what they see as conservation organizations' enduring "top-down conservation planning, the failure to understand equity and benefit issues, the lack of social history in project design and implementation, and the poor quality of social research" (Conservation and Community 2001: 1). At the same time, they acknowledge that anthropologists share part of the blame—in that they have not been very visible and effective in the conservation arena for lack of institutional structures to communicate and promote the contribution of anthropology's distinctive ethnographic approach—and seek to create such channels for communication as well as for fruitful collaboration.

On their part, some conservation biologists do acknowledge that success in protecting nature preserves will depend largely on the participation of local people. Taking this into account leads them to advocate, rather than community-based conservation, "conservation-based community" (Soulé 2000: 60). This notion focuses on conservation, not development, and on participatory processes that promote conservation by encouraging "pride in nature protection and a sense of ownership by local people" (p. 60) through programs that address "the vital needs both of nature and of local people who must become allies in nature protection" (p. 62). From this camp, calls for collaboration with social scientists are being heard, at least from the point of view of helping identify the mechanisms that will lead people to perform the needed behavioral changes. Social scientists are likely to sympathize in principle with this idea of conservation, yet they may find this approach still too top-down in practice, in terms of who is defining conservation for whom, and this proposed ground for interdisciplinary collaboration too limited and somewhat self-serving.

A lively debate thus continues. It also intersects a related one that, though having been around for over a decade, has been picking up again lately: over the myth or reality of the "ecologically noble savage" (or, in a more specific incarnation, the "ecological Indian"). The latter debate came notably to the fore in the early 1990s (Redford 1990), when biologists began to criticize what they considered to be a romanticized and essentializing view of indigenous people as "conservationists" used and abused by anthropologists to promote an ideological and political agenda of indigenous peoples as stewards of nature. (Interestingly, some anthropologists have expressed similar concerns about the misuse of the concept—in that it might end up being counterproductive if indigenous peoples should fail to conform to that image—but have attributed the misuse to environmentalists; Brosius 1997.) Since then, the argument has been variously pursued between biologists and anthropologists (e.g., Alcorn 1993, 1996; Redford & Stearman 1993), and it flared up recently on the pages of *Conservation Biology* (Chicchón 2000; Marsh et al. 2000; Redford & Sanderson 2000; Schwartzman et al. 2000a, 2000b; Terborgh 2000).

The main bone of contention has been whether indigenous people display conscious conservation-oriented behavior, or conservation has been the unintended combined by-product of low population densities, low-impact technologies, and long-term adaptation to natural cycles for resource extraction, along with the indirect effect of cultural sanctions (such as taboos). Also at issue has been whether any such conservationist or conservation-like behavior is the prevalent pattern among indigenous people, or the scale is tipped on the side of destructive exploitation (with oft-quoted examples going from the extinction of megafauna in North America and in the Pacific islands to clear-cutting and burning in tropical forests). Similar discussions have arisen around recent historical analyses, charging that the "ecologically noble savage" is indeed a myth in the case of Native Americans (Krech 1999). This version of the argument appears to take a middle ground vis-à-vis the historical record of human presence on the North American continent, seeing it as a complex one that reveals both environmentally friendly and destructive tendencies. At the same time, however, it arguably tips the scale by eschewing any interpretation of the more eco-friendly kinds of behaviors as possibly suggesting that Native Americans may have been intuitively cognizant of ecological limits—or at least by implying that native population numbers, more than any such knowledge, had to do with the state of the environment in precolonial times.

Here, the "ecological Indian" debate engages with the complementary one (reviewed in Mann 2002) that has arisen around what has been dubbed the "pristine myth" (Denevan 1992a, 1992b; Cronon 1995; Lentz 2000). At issue is the extent to which what European colonists saw as the American continent's "pristine wildernesses" actually were anthropogenic environments (i.e., as previously discussed, environments significantly altered and, indeed, created over long periods of time by the continent's original inhabitants, through a variety of means including cultivation, soil modification, selective enhancement of plant and animal species, and the use of fire).[12]

Resolution of this controversy depends both on anthropological and archaeological field data and on estimates of precolonization population numbers. Proponents of the "anti-pristine-myth" theory vary in their calculations of these numbers but agree in claiming that the numbers were large enough to bring about extensive modifications of ecosystems and that, in turn, these modifications (as documented in numerous sites) allowed the land to support high population densities (including in areas of notoriously soil-nutrient-poor Amazonia). This all would have changed with the arrival of Europeans, due to the devastating proportions of the epidemics they triggered in the non-immune native populations. Both the dramatic depopulation and the consequent diminished human intervention in the environment would have lent support to colonists' reading of the landscape as largely untrammeled by humans (even more so because they had a vested interest in seeing the land as *terra nullius*). Opponents of this view react vehemently, even charging, for instance, that to suggest that, with appropriate modification, the Amazon Basin may have sustained much higher populations than it now does is to give free rein to more human influx and to increased deforestation (Mann 2002).

Few would disagree in principle with the importance of carefully documenting the nature and dynamics of traditional knowledge, its historical as well as current impacts, both positive and negative, on the environment, the history of human occupation and modification of ecosystems, and the causes for the success or failure of community-based projects. In practice, however, debates over these issues, as reviewed above, tend to engender polarized positions. Such polemics often seem to have more to do with matters of institutional or ideological turf than with dispassionate attempts to improve our understanding of human–environment relationships. Furthermore, one might suggest that ultimately the main point is not to argue over indigenous peoples' heads as to whether they are "conservationists" in a sense that satisfies biologists' (or anyone else's) notions—an approach that might contribute to indigenous peoples' further disenfranchisement and marginalization. In other words, the debates that are accompanying the current swinging back of the pendulum may end up hurting those who have already been most seriously affected by social, political, economic, and environmental problems for which they largely are not responsible and do not control. Rather, the point is "how [indigenous peoples] can better participate in and benefit from conservation policies and projects for their lands," both making the most of traditionally well adapted resource use practices and adopting new, conscious conservation measures when a changing situation may require doing so in order to maintain their subsistence base (Shepard 2002: 127).

Traditional Knowledge and Its Discontents

Better understanding of traditional knowledge and its integration with other forms of knowledge (including Western scientific knowledge) in the quest for new solutions would therefore seem to be one key element for achieving both conservation and sustainability. As indicated above, there is an already fairly long history of research on traditional knowledge in the writings of ethnobiologists, ethnoecologists, and other ethnoscientists and through the formulations of indigenous and other traditional peoples themselves (especially since traditional knowledge became enshrined in the Rio Summit documents such as the CBD). Within these circles, however, a recent contentious topic has been precisely the nature of traditional (or indigenous) knowledge,[13] its use in development and conservation projects, and its relationships with scientific knowledge. One aspect of this controversy is exemplified by a debate sparked by an article in the journal *Indigenous Knowledge and Development Monitor* (IKDM)[14] (Agrawal 1995a; see also Agrawal 1995b), which was followed by several commentaries (IKDM 1996a, 1996b, 1996c).[15]

The main purpose of Agrawal's article is to question the distinctions between indigenous (or traditional) knowledge and Western scientific knowledge sometimes made by scholars and practitioners (and by indigenous people themselves). This purpose seems to stem from a concern that making such distinctions harks back to now-discredited theories of "primitive mentality" (Lévi-Bruhl 1966) and of the "savage mind" (Lévi-Strauss 1966). Agrawal's criticism of the differences commonly mentioned in relation to indigenous versus Western scientific knowledge addresses claims at three levels: substantive ("differences in the subject matter and characteristics of indigenous and Western knowledge"); methodological and epistemological ("the two forms of knowledge employ different methods to investigate reality"); and contextual ("traditional/indigenous knowledge is more deeply rooted in its environment") (Agrawal 1995a: 3).

Agrawal considers the distinctions untenable at all three levels. At the substantive level, he contests the claim that "indigenous knowledge is concerned with the immediate and concrete necessities of people's daily livelihoods, while Western knowledge attempts to construct general explanations and is one step removed from the daily lives of people" (Agrawal 1995a: 4). He finds that this dichotomy is too rigid and ignores both internal differences within each of the two categories and similarities between the two and also that indigenous and Western scientific knowledge systems have long been in contact with one another and open to mutual influences and exchange. At the methodological and epistemological level, he rejects the contention that science "is open, systematic, objective and analytical . . . [and] advances by building rigorously on prior achievements," whereas indigenous knowledge "is closed, non-systematic, holistic rather than analytical, without an overall conceptual framework, and advances on the basis of new experiences, not on the basis of a deductive logic" (p. 4). He argues that philosophers of science long since abandoned the attempt to tightly demarcate science from nonscience and also that it is because of anthropology's traditionally ahistorical approach that indigenous knowledge systems were portrayed as closed and static prior to Western contact. At the contextual level, Agrawal finds that indigenous knowledge is not alone in being embedded in a specific cultural context.

Although this critique raised some eyebrows, it may be fair to say that these are not earth-shattering objections (and several of the commentators on Agrawal's article [IKDM 1996a, 1996b, 1996c] concur). Many would agree that conceptual categories are not absolute and impermeable; that differences (and similarities) are often more a matter of degree than of quality and need to be understood in a historical perspective; that indigenous knowledge systems have always been open and dynamic; that Western science, like other knowledge systems, bears the marks of a specific epistemological tradition with identifiable roots in space and time and reflects the concerns and areas of expertise pertaining to that tradition; and that this is not in contradiction with the pursuit of reliable knowledge of the world. Witness Gould: "The true, insightful, and fundamental statement that science, as a quintessentially human activity, must reflect a surrounding social context does not imply either that no accessible external reality exists, or that science, as a socially embedded and constructed institution, cannot achieve progressively more adequate understandings of nature's facts and mechanisms" (2000: 261).

If anything, Agrawal's remarks should point to good prospects for integration of indigenous and Western scientific (and other) knowledge systems.[16] Perhaps what is more problematic in the relationship between indigenous knowledge and Western scientific knowledge relates to a point implied in Agrawal (1995a) and made explicit in his response to some of his commentators (1996; see also 1999). Agrawal argues that indigenous peoples and other marginalized poor need to acquire the power and right to assert, control, and protect their knowledge systems, threatened by the "pressures of modernization and cultural homogenization" (1995a: 5). Implicit in this assertion is that such pressures precipitate the replacement of local knowledge systems with others brought in from the outside, among which Western science is, of course, strongly represented in all quarters: from formal education to the media, applications in technology, and so forth. In other words, the power imbalance between local communities and outsiders—be they proponents of conservation and/or development projects or resource extraction—may be the greatest threat, not only to the integration of knowledge systems but also and in the first place to the very maintenance and continued development of indigenous systems, and thus to a deeper understanding of their nature and appreciation of their value.

Ironically, even a discipline like ethnobiology, which has contributed substantially, over the past several decades, to this understanding and appreciation, has not been immune to some of the implications of this power imbalance. The advent of bioprospecting and the direct or indirect use of traditional knowledge in the development of new drugs, crops, cosmetics, or dietary supplements have led to charges of biopiracy (for a review, see, e.g., Brush 2001), which have in some cases turned virulent and fueled distrust and conflict between indigenous communities and researchers, and even within and among indigenous communities. This and other aspects of the process of globalization raise new challenges for the practice of ethnobiology (Alexiades 2004) and pose complex problems for protection of and access to traditional knowledge (Maffi 2001c; Laird 2002; Bannister & Barrett 2004). Equally ironically, many of the (real or assumed) ultimate beneficiaries of the commercialization of traditional knowledge, such as major drug companies, have begun (at least officially) to pull out of the business of using this knowledge for product development, citing lack of profitability, excessive contentiousness and legal problems, and development of new screening technologies making the recourse to traditional knowledge obsolete. In what had all the features of a self-fulfilling prophecy, a 1999 editorial in *The Economist* commenting on this trend forecast "the end of ethnobotany."

A Way Forward: Ethnobiology's Role in the Conservation of Biocultural Diversity

Despite its involvement in this and the other controversies discussed above, ethnobiology is alive and well and, along with related fields such as ethnoecology and ecological anthropology, continues to present unique opportunities as a discipline because it is intrinsically positioned at the interface between nature and culture (Alexiades 2004). This makes ethnobiology (and related specialties) enduringly well suited to the investigation of the biocultural dynamics that shape and change the world's environments. The chapters in this volume are clear evidence of the breath and sophistication of current ethnobiological research. Careful long-term fieldwork is adding to the picture of humans' role in shaping ecosystems by acting as "agents of creative disturbance" through their subsistence behavior (E. L. Zent & S. Zent 2004) by refining our understanding of the nature and distribution of ethnobiological knowledge, its correlation with use of flora and fauna, and the sociocultural factors that affect knowledge change in the process of acculturation (Lizarralde 2004; S. Zent & E. L. Zent 2004); by testing the value of ethnobotanical knowledge as a "reality check" for remote sensing through the ground-truthing of satellite imagery (Shepard et al. 2004); by examining the ecological and social changes affecting and affected by traditional plant uses (Cunningham & Choge 2004; Sowerwine 2004; Sunderland 2004); and by highlighting the rural public health value of medical ethnobotanical systems (Carlson 2004).

This and other recent work in ethnobiology and ethnoecology (as well as in other areas of environmental social science such as the study of human ecosystems, e.g., Machlis et al. 1997; Stepp 2003), is bringing about an increasingly rigorous understanding of human–environment relationships. These advances should significantly help address many of the controversies discussed above, bridge the gap between the natural and social sciences, and contribute to the conservation of biocultural diversity. And indeed, although the relationship with conservation science and application may remain uneasy, there is reason to believe that a convergence with the natural sciences in theory and practice is beginning to take shape elsewhere. Other areas of the natural sciences, such as the field of ecosystem health, squarely embrace a humans-in-ecosystems perspective and appear to be promisingly open to interdisciplinary (and transdisciplinary) integration (see, e.g., Somerville & Rapport 2000; Wilcox 2001; Rapport 2003). The emerging field of sustainability science (Kates et al. 2001) professes as its goal that of understanding the fundamental character of nature–society interactions and of the nature-society system. One of its core questions is, "What determines the vulnerability or resilience of the nature-society system in particular kinds of places and for particular types of ecosystems and human livelihoods?"—a question that ethnobiology and related disciplines should be eminently qualified to tackle. Projects are beginning to be developed that seek to cross the natural–social science divide by adopting a biocultural perspective in assessing the state and promoting the restoration of given ecoregions (Center for Sustainable Environments . . . 2002).

Undeniably, to foster this integration, much needs to be done in terms of training for research. It should become customary for interdisciplinary (or transdisciplinary) training to be imparted across the natural and social sciences, so that, whatever any given student's main propensities, each will be reasonably conversant with issues in the various relevant fields. And it should become more and more common for interdisciplinary research teams to go to the field to study the relationships between biological and cultural diversity, having learned how to work together and create the necessary synergy. No less important, it should become more and more common for such teams to include indigenous and other local peoples as actual research partners, not just as guides, interpreters, and "informants" but as local scientists whose knowledge is sought, valued, and appropriately recognized for its genuine contributions, while fully respecting local ownership of this knowledge.

In this, as Alexiades (2004) further notes, ethnobiology also has a role, in that it operates at the intersection between science and society and between local and international communities. This gives ethnobiologists (and colleagues in related fields) a special position and special responsibilities in this arena. An instance of this, as discussed above, is the very active role taken by the International Society of Ethnobiology in promoting genuine and ethical partnerships with indigenous and traditional peoples in the conduct of research projects and in acknowledging and upholding the human rights dimension of the relationship between local communities, outside researchers, and national and global forces. Ethnobiologists also have played and can play a significant role in supporting indigenous and traditional peoples' efforts to protect and perpetuate their natural and cultural heritages (Bannister & Barrett 2004). Training in professional ethics and human rights should therefore also become a standard part of an integrated curriculum.

In other words, these new perspectives bear, on one hand, on the conduct of basic science. Basic science will greatly benefit from it—not only ethnobiology and related fields but also other natural and social sciences, from biotaxonomy to biogeography, ecology, anthropology, and linguistics. On the other hand, these new perspectives also bear on the relationship between basic and applied science—in this instance, between knowledge and conservation of biocultural diversity. The insights that have emerged so far from the understanding of the links between biological and cultural diversity have already begun to illuminate the work of conservation, and we can expect this to increase in the near future. In this process, an ethical and spiritual dimension is likely to come increasingly to the fore, as propounded in a recent international environmental document, the Earth Charter, <www.earthcharter.org>, and as poignantly stated by the ecologist Frank Golley:

> It is not clear to me where ecology ends and the study of the ethics of nature begins, nor is it clear to me where biological ecology ends and human ecology begins. These divisions become less and less useful. Clearly, the ecosystem, for some time at least, has provided the basis for moving beyond strictly scientific questions to deeper questions of how humans should live with each other and the environment. In that sense, the ecosystem concept continues to grow and develop as it serves a larger purpose. (Golley 1993: 205, quoted in Machlis et al. 1997: 363)

Notes

1. But see Naess (1990) for a different take from a deep ecology perspective.

2. The issue of what is meant by "indigenous" has been a vexing one for both some anthropologists and some of those to whom the term might be applied. It has also been the object of extensive debate in international processes and of definition in international instruments. As used in this chapter, "indigenous" is shorthand for "indigenous and tribal," following the definition in Article 1 of the International Labour Organization's Convention 169 on Indigenous and Tribal Peoples in Independent Countries (ILO 169), which states that the Convention applies to:

> (a) Tribal peoples in independent countries whose social, cultural and economic conditions distinguish them from other sections of the national community, and whose status is regulated wholly or partially by their own customs or traditions or by special laws or regulations;
>
> (b) Peoples in independent countries who are regarded as indigenous on account of their descent from the populations which inhabited the country, or a geographical region to which the country belongs, at the time of conquest or colonisation or the establishment of present State boundaries and who, irrespective of their legal status, retain some or all of their own social, economic, cultural and political institutions.

Article 1 of ILO 169 also states: "Self-identification as indigenous or tribal shall be regarded as a fundamental criterion for determining the groups to which the provisions of this Convention apply." These criteria are followed in various other international instruments and by many indigenous and tribal peoples themselves (for other useful criteria, see Toledo 2001). Official as well as self-appellation preferences for the use of "indigenous" versus "tribal" (as well as others such as "native," "first nation," "aboriginal," and "ethnic minority") vary from one region of the world to the other. In highly simplified terms, there is a general tendency toward the use of "indigenous" (or variants thereof) to refer in particular to the original inhabitants of the Americas, Australia, and the Pacific, whereas the terms "tribal" and "ethnic minority" are more common in Africa and Asia. Some of the official uses are beginning to be contested by some of the groups to which they are applied (e.g., indigenous groups in the United States rejecting "Native American" in favor of "Indian"). The expression "traditional communities," also used in this chapter, should be understood as referring to "local communities embodying traditional lifestyles," as per Article 8(j) of the CBD. See note 3 for a definition of "traditional."

3. According to a recent definition, TEK is a "cumulative body of knowledge, practice, and belief, evolving by adaptive processes and handed down through generations by cultural transmission, about the relationships of living beings (including humans) with one another and with their environment" (Berkes 1999: 8). The term "traditional," as used in this context (and as understood in this chapter), should not be taken to refer to something static and homogeneous. Rather, "tradition" should be understood as "a filter through which *innovation* occurs" (Posey 2001), a "tradition of invention and innovation" (Pereira & Gupta 1993). As explained in a report to the CBD Secretariat prepared by Canadian First Nations:

> What is "traditional" about traditional knowledge is not its antiquity, but *the way it is acquired and used*. In other words, the social process of learning and sharing knowledge, which is unique to each Indigenous culture, lies at the very heart of its "traditionality." Much of this knowledge is actually quite new, but it has a social meaning, and legal character, entirely unlike the knowledge indigenous people acquire from settlers and industrialised societies. (Four Directions Council 1996)

Traditional knowledge also varies according to age, gender, and other variables that make it nonhomogeneous and dynamic even within the same group of people.

4. The existence of an "inextricable link" between cultural and biological diversity—a link residing in indigenous peoples' knowledge of, and stewardship over, a large part of the world's most diverse ecosystems—was first affirmed in the Declaration of Belém, put forth in 1988 by the International Society of Ethnobiology (ISE), and is reflected in ISE's Code of Ethics. The texts of these two documents are found at <http://guallart.dac.uga.edu/ISE> (see also Posey 2001).

5. Cultural diversity can be understood as "the variety of human expression and organization, including that of interactions among cultural groups and between these groups and the environment" (Harmon 1998a: 353). Languages are the carriers of many cultural differences, and in their variety constitute "the building blocks of cultural diversity, arguably the fundamental 'raw material' of human thought and creativity" (Harmon 1996: 95).

6. An important caveat in these studies was that, as a proxy for cultural diversity, language richness is an imperfect tool. In many cases, distinctiveness of language does not correspond to distinctiveness of culture, or sameness of language to sameness of culture. It is in the nature of proxies to be imperfect tools, however, and the point of using them is to identify general trends. Similar considerations apply to the use of species richness as a proxy for biological diversity.

7. It is of interest to note that many of the factors commonly mentioned as threats to biodiversity conservation (see, e.g., WWF 1999) present close parallels with several factors considered to be threatening to cultural and linguistic diversity. Some examples: 1) Island ecosystems are fragile due to the sensitivity and endemicity of island species and to the severe threats native island biota face from introduced species and habitat loss. Islands' highly endemic cultures and languages are similarly fragile due to the influx of non-indigenous populations and loss of control over land by the local populations (Hawai'i being a characteristic example). 2) Habitat fragmentation is considered a prime indicator of an extinction-prone environment, due to species' inability to move in response to climate change or other disturbances. The fragmentation of the social "habitat" of human populations is correspondingly a significant factor in cultural and linguistic endangerment. 3) A region characterized by presence of many species with highly restricted distributions is at high risk for biodiversity loss under adverse conditions. Likewise, a large set of small, culturally distinct human communities living in a given region may be less buffered from outside human interference than may a single larger population.

8. This operational directive is currently under revision in view of turning it into an operational policy, now in draft form as Draft Policy on Indigenous Peoples, OP/BP 4.10.

9. One might, for instance, point to the contradiction of stating that "people—even indigenous people—will inevitably . . . become Westernized, adopting powerful, Western technologies for the exploitation and commercialization of plants and animals for the global market" (Soulé 2000: 61) and at the same time that "the major enemies in the campaign to save nature are negative, defeatist patterns of thought that poison optimism: despair about the fate of nature; fatalism about the potential for individuals to alter the heading of civilization" (p. 63).

10. Soulé 2000 seems to at least implicitly acknowledge this much.

11. This irony is fully brought out by comparing the following two quotations: "Many conservation biologists have proposed guidelines that minimize the loss of species diversity in [habitat] remnants. The newest of these guidelines calls for the protection or reintroduction of large keystone species—the major element of 'rewilding'" (Soulé 2000: 62); and "Native Americans were the ultimate keystone predator and the ultimate keystone species. . . . Until the importance of aboriginal land management is recognized and modern management practices change accordingly, our ecosystems will continue to lose the biological diver-

sity and ecological integrity they once had, even in national parks and other protected areas" (Kay 1998: 484, quoted in Shepard 2002: 126).

12. Although to a certain degree this debate may hinge specifically on the extent of anthropogenic intervention, the very anthropogenic nature of such "wildernesses" may be contentious in and of itself. See Dann (2000) for an analysis of how a "naturalist myth," developed in North America and pervasive in natural science circles, has lent support to a belief in, and quest for, such "pristine wildernesses" unspoiled by human presence and intervention. This belief may be the source of the "misanthropic anthropocentrism" detectable in some of the more radical biocentric positions in conservation, as discussed at the beginning of this chapter.

13. A discussion of arguments for and against the use of "traditional" versus "indigenous" knowledge would be lengthy and is beyond the scope of this chapter. Here they are used more or less interchangeably, in accordance with the definitions given in notes 2 and 3.

14. IKDM was published by the Netherlands-based Centre for International Research and Advisory Networks (CIRAN), a coordinating center for a worldwide network of indigenous knowledge research centers. This publication has now been discontinued and replaced by a bulletin.

15. Only a brief summary of this controversy can be provided here. See Maffi (forthcoming) for a more extensive discussion.

16. See Maffi (forthcoming) for an analysis of other epistemological reasons for not considering indigenous and Western scientific knowledge incommensurable and for trying to identify in these (and other) knowledge systems a mode of knowledge that might be recognized as "scientific," while not being insulated from underlying belief and value systems. One key aspect of this analysis points to the recognition that hypothetic-deductive logic cannot be taken to be the overarching epistemological model of "Western scientific knowledge" and that many domains of knowledge commonly recognized as scientific do not necessarily lend themselves to rigorous quantification and replicability. This would apply in the natural sciences as well, including Darwinian evolution. Maffi (forthcoming) suggests that the Darwinian method, with its unique combination of the inductive and deductive, applied repeatedly in a spiraling process, may approximate not only the way in which much of "Western science" proceeds but also the modus operandi of indigenous or traditional farmers, hunters, herders, fishers, gatherers, or artisans aiming to produce the reliable knowledge they need for the resolution of their daily necessities—and possibly for addressing some of their less immediate and more abstract queries about how the world works.

Literature Cited

Agrawal, A. 1995a. Indigenous and scientific knowledge: Some critical comments. Indigenous Knowledge and Development Monitor 3(3): 3–6.

———. 1995b. Dismantling the divide between indigenous and scientific knowledge. Development and Change 26: 413–439.

———. 1996. A response to certain comments. Indigenous Knowledge and Development Monitor 4(2): 17–18.

———. 1999. On power and indigenous knowledge. Pp. 177–180 *in* D. A. Posey (ed.), Cultural and Spiritual Values of Biodiversity. London and Nairobi: Intermediate Technology Publications and UNEP.

Alcorn, J. B. 1984. Huastec Mayan Ethnobotany. Austin: University of Texas Press.

———. 1993. Indigenous peoples and conservation. Conservation Biology 7(2): 424–426.

———. 1996. Is biodiversity conserved by indigenous peoples? Pp. 234–238 *in* S. K. Jain (ed.), Ethnobiology in Human Welfare. New Delhi: Deep Publications.

———. 1997. Indigenous peoples and protected areas. Pp. 44–49 *in* G. Borrini-Feyerabend (ed.), Beyond Fences: Seeking Sustainability in Conservation; vol. 2, A Resource Book. Gland, Switzerland: IUCN.

Alexiades, M. N. 2004. Ethnobiology and globalization: Science and ethics at the turn of the century. Pp. 281–303 *in* T. J. S. Carlson & L. Maffi (eds.), Ethnobotany and Conservation of Biocultural Diversity. Advances in Economic Botany, 15. Bronx: New York Botanical Garden Press.

Atran, S. 1990. Cognitive Foundations of Natural History. Cambridge: Cambridge University Press.

———. 1993. Itza Maya tropical agro-forestry. Current Anthropology 34: 633–700.

——— & D. Medin. 1997. Knowledge and action: Cultural models of nature and resource management in Mesoamerica. Pp. 171–208 *in* M. Bazerman, D. Messick, A. Tinbrunsel & K. Wayde-Benzoni (eds.), Environment, Ethics, and Behavior. San Francisco: New Lexington Press.

Bahn, P. & J. R. Flenley. 1992. Easter Island, Earth Island. London: Thames and Hudson.

Balée, W. 1994. Footprints of the Forest: Ka'apor Ethnobotany—The Historical Ecology of Plant Utilization by an Amazonian People. New York: Columbia University Press.

———. 2001. Environment, culture, and Sirionó plant names. Pp. 298–310 *in* L. Maffi (ed.), On Biocultural Diversity: Linking Language, Knowledge, and the Environment. Washington, D.C.: Smithsonian Institution Press.

Bannister, K. & K. Barrett. 2004. Weighing the proverbial "ounce of prevention" against the "pound of cure" in a biocultural context: A role for the precautionary principle in ethnobiological research. Pp. 304–336 *in* T. J. S. Carlson & L. Maffi (eds.), Ethnobotany and Conservation of Biocultural Diversity. Advances in Economic Botany, 15. Bronx: New York Botanical Garden Press.

Beltran, J. 2000. Indigenous and Traditional Peoples and Protected Areas: Principles, Guidelines, and Case Studies. Gland, Switzerland: IUCN and WWF International.

Berkes, F. 1999. Sacred Ecology: Traditional Ecological Knowledge and Resource Management Systems. Philadelphia: Taylor and Francis.

Berlin, B. 1992. Ethnobiological Classification: Principles of Categorization of Plants and Animals in Traditional Societies. Princeton, N.J.: Princeton University Press.

———, D. E. Breedlove & P. H. Raven. 1974. Principles of Tzeltal Plant Classification: An Introduction to the Botanical Ethnography of a Mayan Speaking Community in Highland Chiapas. New York: Academic Press.

Blackburn, T. C. & K. Anderson (eds.). 1993. Before the Wilderness: Environmental Management by Native Californians. Menlo Park, Calif.: Ballena Press.

Brokensha, D., D. M. Warren & O. Werner (eds.). 1980. Indigenous Knowledge Systems and Development. Washington, D.C.: University Press of America.

Brosius, J. P. 1997. Endangered forest, endangered people: Environmentalist representations of indigenous knowledge. Human Ecology 25(1): 47–69.

Brush, S. B. 2001. Protectors, prospectors, and pirates of biological resources. Pp. 517–530 *in* L. Maffi (ed.), On Biocultural Diversity: Linking Language, Knowledge, and the Environment. Washington, D.C.: Smithsonian Institution Press.

——— & D. Stabinsky (eds.). 1996. Valuing Local Knowledge: Indigenous People and Intellectual Property Rights. Washington, D.C.: Island Press.

Bulmer, R. N. H. 1982. Traditional conservation practices in Papua New Guinea. Pp. 59–77 *in* L. Morauta, J. Pernetta & W. Heaney (eds.), Traditional Conservation in Papua New Guinea: Implications for Today. Bokoro, Papua New Guinea: Institute of Applied Social and Economic Research.

Burger, J. 1987. Report from the Frontier: The State of the World's Indigenous Peoples. Atlantic Highlands, N.J.: Zed Books.

Carlson, T. J. S. 2004. Medical ethnobotany and therapeutic uses of *Croton lechleri* Müell.-Arg. for human health. Pp. 260–277 *in* T. J. S. Carlson & L. Maffi (eds.), Ethnobotany and Conservation of Biocultural Diversity. Advances in Economic Botany, 15. Bronx: New York Botanical Garden Press.

Center for Sustainable Environments, Northern Arizona University and Terralingua and Grand Canyon Wildlands Council. 2002. Safeguarding the Uniqueness of the Colorado Plateau: An Ecoregional Assessment of Biocultural Diversity. Flagstaff, Ariz.: Center for Sustainable Environments, Northern Arizona University and Terralingua and Grand Canyon Wildlands Council.

Chapin, M. 1992. The co-existence of indigenous peoples and environments in Central America. Research and Exploration 8(2): inset map.

———. 1994. Recapturing the old ways: Traditional knowledge and Western science among the Kuna Indians of Panama. Pp. 83–101 *in* C. D. Kleymeyer (ed.), Cultural Expression and Grassroots Development: Cases from Latin America and the Caribbean. Boulder, Colo., and London: Lynne Rienner Publishers.

Chicchón, A. 2000. Conservation theory meets practice. Conservation Biology 14(5): 1368–1369.

Clay, J. W. 1993. Looking back to go forward: Predicting and preventing human rights violations. Pp. 64–71 *in* M. S. Miller (ed.), State of the Peoples: A Global Human Rights Report on Societies in Danger. Boston: Beacon Press.

Colchester, M. 1994. Salvaging Nature: Indigenous Peoples, Protected Areas, and Biodiversity Conservation. Geneva: U.N. Research Institute for Social Development.

Conservation and Community. 2001. Results of the workshop on conservation and community. An initiative of the Conservation and Community Working Group, Anthropology and Environment Section, American Anthropological Association. Unpublished manuscript.

Cronon, W. 1995. Uncommon Ground: Rethinking the Human Place in Nature. New York: W. W. Norton.

Cunningham, A. B. & S. K. Choge. 2004. Crafts and conservation: The ecological footprint of international markets on an African resource. Pp. 212–231 *in* T. J. S. Carlson & L. Maffi (eds.), Ethnobotany and Conservation of Biocultural Diversity. Advances in Economic Botany, 15. Bronx: New York Botanical Garden Press.

Daes, E.-I. 1997. Protection of the Heritage of Indigenous People. Study Series 10, Human Rights. New York and Geneva: United Nations Publishing.

Dann, K. 2000. Across the Great Border Fault: The Naturalist Myth in America. New Brunswick, N.J., and London: Rutgers University Press.

Denevan, W. M. 1992a. The pristine myth: The landscape of the Americas in 1492. Annals of the Association of American Geographers 82(3): 369–385.

——— (ed.). 1992b. The Native Population of the Americas in 1492. Ed. 2. Madison: University of Wisconsin Press.

Diamond, J. M. 1986. The environmentalist myth. Nature 344: 19–20.

———. 1987. The worst mistake in the history of the human race. Discover, May, 64–66.

———. 1991. The Rise and Fall of the Third Chimpanzee. New York: HarperCollins.

———. 1993. Speaking with a single tongue. Discover, February, 78–85.

Durning, A. T. 1992. Guardians of the Land: Indigenous Peoples and the Health of the Earth. Worldwatch Paper 112. Washington, D.C.: Worldwatch Institute.

———. 1993. Supporting indigenous peoples. Pp. 80–100 *in* State of the World 1993: A Worldwatch Institute Report on Progress Toward a Sustainable Society. New York: W. W. Norton.

The Economist. 1999. The end of ethnobotany? The Economist, February, 20–26.
Eldredge, N. 1995. Dominion. New York: H. Holt.
Ellen, R. 1994. Rhetoric, practice and incentive in the face of changing times: A study in Nuaulu attitudes to conservation and deforestation. Pp. 127–143 *in* K. Milton (ed.), Environmentalism: The View from Anthropology. London and New York: Routledge.
European Commission. 1998. Working document of the European Commission on support for indigenous peoples in the development co-operation of the Community and the Member States. Brussels: European Commission.
European Union. 1998. Council resolution on indigenous peoples within the framework of the development cooperation of the Community and the Member States. Brussels: European Union.
Fishman, J. A. 1996. What do you lose when you lose your language? Pp. 80–91 *in* G. Cantoni (ed.), Stabilizing Indigenous Languages. Flagstaff: Northern Arizona University, Center for Excellence in Education.
Flannery, T. 1995. The Future Eaters: An Ecological History of the Australasian Lands and People. New York: George Braziller.
Florey, M. 2001. Threats to indigenous knowledge: A case study from eastern Indonesia. Pp. 325–342 *in* L. Maffi (ed.), On Biocultural Diversity: Linking Language, Knowledge, and the Environment. Washington, D.C.: Smithsonian Institution Press.
Four Directions Council. 1996. Forests, Indigenous Peoples and Biodiversity: Contribution of the Four Directions Council to the Secretariat of the Convention on Biological Diversity. Lethbridge, Alta.: FDC.
Goehring, B. 1993. Indigenous Peoples of the World: An Introduction to Their Past, Present, and Future. Saskatoon, Sask.: Purich Publishing.
Golley, F. B. 1993. A History of the Ecosystem Concept in Ecology. New Haven, Conn.: Yale University Press.
Gould, S. J. 2000. Deconstructing the "science wars" by reconstructing an old mold. Science 287: 253–261.
Gragson, T. & B. Blount (eds.). 1999. Ethnoecology: Knowledge, Resources and Rights. Athens: University of Georgia Press.
Gray, A. 1991. Between the Spice of Life and the Melting Pot: Biodiversity Conservation and Its Impact on Indigenous Peoples. IWGIA Document 70. Copenhagen: IWGIA.
———. 1999. Introduction. Pp. 61–66 *in* D. A. Posey (ed.), Cultural and Spiritual Values of Biodiversity. London and Nairobi: Intermediate Technology Publications and UNEP.
Greaves, T. (ed.). 1994. Intellectual Property Rights for Indigenous Peoples: A Sourcebook. Oklahoma City, Okla.: Society for Applied Anthropology.
Grinde, D. A. & B. E. Johansen. 1995. Ecocide of Native America: Environmental Destruction of Indian Lands and Peoples. Santa Fe, N.Mex.: Clear Light Publishers.
Hale, K. (ed.). 1992. Endangered languages. Special topic; Language 68(1): 1–42.
Hames, R. B. 1991. Wildlife conservation in tribal societies. Pp. 172–199 *in* M. L. Oldfield & J. B. Alcorn (eds.), Biodiversity: Culture, Conservation, and Ecodevelopment. Boulder, Colo.: Westview Press.
——— & W. T. Vickers (eds.). 1983. Adaptive Responses of Native Amazonians. New York: Academic Press.
Harmon, D. 1992. Indicators of the world's cultural diversity. Paper presented at the Fourth World Congress on National Parks and Protected Areas, Caracas, Venezuela, February 1992.
———. 1995. The status of the world's languages as reported in *Ethnologue*. Southwest Journal of Linguistics 14: 1–33.

———. 1996. Losing species, losing languages: Connections between biological and linguistic diversity. Southwest Journal of Linguistics 15: 89–108.

———. 1998a. The other extinction crisis: Declining cultural diversity and its implications for protected area management. Pp. 352–359 *in* N. W. P. Munro & J. H. M. Willison (eds.), Linking Protected Areas with Working Landscapes, Conserving Biodiversity. Wolfville, N.S.: Science and Management of Protected Areas.

———. 1998b. Sameness and silence: Language extinctions and the dawning of a biocultural approach to diversity. Global Biodiversity 8(3): 2–10.

——— & J. Loh. 2002. Draft framework for an index of biocultural diversity. Washington, D.C.: Terralingua. Unpublished manuscript.

——— & L. Maffi. 2002. Are linguistic and biological diversity linked? Conservation Biology in Practice 3(1): 26–27.

Heywood, V. H. (ed.). 1995. Global Biodiversity Assessment. Cambridge and New York: Cambridge University Press and UNEP.

Hill, J. H. 1997. The meaning of linguistic diversity: Knowable or unknowable? Anthropology Newsletter 38(1): 9–10.

———. 2001. Dimensions of attrition in language death. Pp. 174–189 *in* L. Maffi (ed.), On Biocultural Diversity: Linking Language, Knowledge, and the Environment. Washington, D.C.: Smithsonian Institution Press.

Hunn, E. S. 1977. Tzeltal Folk Zoology: The Classification of Discontinuities in Nature. New York: Academic Press.

———. 2001. Prospects for the persistence of "endemic" cultural systems of traditional environmental knowledge: A Zapotec example. Pp. 118–132 *in* L. Maffi (ed.), On Biocultural Diversity: Linking Language, Knowledge, and the Environment. Washington, D.C.: Smithsonian Institution Press.

IKDM. 1996a. Comments on article by Arun Agrawal. Indigenous Knowledge and Development Monitor 4(1): 12–19.

———. 1996b. A sequel to the debate. Indigenous Knowledge and Development Monitor 4(2): 17–33.

———. 1996c. Closing the debate. Indigenous Knowledge and Development Monitor 4(3): 13–20.

IUCN. 1997. Resolutions and Recommendations. World Conservation Congress, Montreal, Canada, 13–23 October 1996. Gland, Switzerland: IUCN.

——— & WWF. 1998. Implementation of Article 8(j) and Related Provisions. Policy recommendations on COP4 Agenda item 10. 4th Meeting of the COP to the Convention on Biological Diversity, Bratislava, Slovakia, 4–15 May 1998.

———, UNEP & WWF. 1991. Caring for the Earth: A Strategy for Sustainable Living. Gland, Switzerland: IUCN, UNEP, and WWF.

Johnson, A. 1989. How the Machiguenga manage resources: Conservation or exploitation of nature? Pp. 213–222 *in* D. A. Posey & W. Balée (eds.), Resource Management in Amazonia: Indigenous and Folk Strategies. Advances in Economic Botany, 7. New York Botanical Garden.

Kates, R. W., W. C. Clark, R. Corell, J. M. Hall, C. C. Jaeger, I. Lowe, J. J. McCarthy, H. J. Schellnhuber, B. Bolin, N. M. Dickson, S. Faucheaux, G. C. Gallopin, A. Grüber, B. Huntley, J. Jäger, N. S. Jodha, R. E. Kasperson, A. Mabogunje, P. Matson, H. Mooney, B. Moore III, T. O'Riordan & U.Svedin. 2001. Sustainability science. Science 292: 641–642.

Kay, C. E. 1998. Are ecosystems structured from the top-down or bottom-up? A new look at an old debate. Wildlife Society Bulletin 26(3): 484–493.

Kirch, P. V. 1997. Microcosmic histories: Island perspectives on "global" change. American Anthropologist 99(1): 30–42.

——— & T. L. Hunt (eds.). 1996. Historical Ecology in the Pacific Islands: Prehistoric Environmental and Landscape Change. New Haven, Conn.: Yale University Press.

Krauss, M. 1992. The world's languages in crisis. Language 68(1): 4–10.

———. 1996. Linguistics and biology: Threatened linguistic and biological diversity compared. Pp. 69–75 *in* CLS 32, Papers from the Parasession on Theory and Data in Linguistics. Chicago: Chicago Linguistic Society.

Krech, S., III. 1999. The Ecological Indian: Myth and History. New York: W. W. Norton.

Laird, S. A. (ed.). 2002. Biodiversity and Traditional Knowledge: Equitable Partnerships in Practice. London: Earthscan.

Lentz, D. (ed.). 2000. Imperfect Balance: Landscape Transformations in the Pre-Columbian Americas. New York: Columbia University Press.

Lévi-Bruhl, L. 1966. Primitive Mentality. Boston: Beacon Press.

Lévi-Strauss, C. 1966. The Savage Mind. Chicago: University of Chicago Press.

Lizarralde, M. 2001. Biodiversity and loss of indigenous languages and knowledge in South America. Pp. 265–281 *in* L. Maffi (ed.), On Biocultural Diversity: Linking Language, Knowledge, and the Environment. Washington, D.C.: Smithsonian Institution Press.

———. 2004. Indigenous knowledge and conservation of the rain forest: Ethnobotany of the Barí of Venezuela. Pp. 112–130 *in* T. J. S. Carlson & L. Maffi (eds.), Ethnobotany and Conservation of Biocultural Diversity. Advances in Economic Botany, 15. Bronx: New York Botanical Garden Press.

Machlis, G. E., J. E. Force & W. R. Burch. 1997. The human ecosystem, Part I: The human ecosystem as an organizing concept in ecosystem management. Society and Natural Resources 10: 347–367.

Maffi, L. 1998. Language: A resource for nature. Nature and Resources 34(4): 12–21.

——— (ed.). 2001a. On Biocultural Diversity: Linking Language, Knowledge, and the Environment. Washington, D.C.: Smithsonian Institution Press.

———. 2001b. Linking language and environment: A co-evolutionary perspective. Pp. 24–48 *in* C. L. Crumley (ed.), New Directions in Anthropology and Environment: Intersections. Walnut Creek, Calif.: AltaMira Press.

———. 2001c. Language, knowledge, and indigenous heritage rights. Pp. 412–432 *in* L. Maffi (ed.), On Biocultural Diversity: Linking Language, Knowledge, and the Environment. Washington, D.C.: Smithsonian Institution Press.

———. Forthcoming. On bringing modern scientific knowledge and traditional knowledge closer together. *In* G. Martin (ed.), Proceedings of the Seminar "Innovative Wisdom: The Impact of Traditional Knowledge on Science," White Oak, Florida, October 19–22, 2000. London: Earthscan.

———, T. Skutnabb-Kangas & J. Andrianarivo. 1999. Linguistic diversity. Pp. 21–57 *in* D. A. Posey (ed.), Cultural and Spiritual Values of Biodiversity. London and Nairobi: Intermediate Technology Publications and UNEP.

Mann, C. C. 2002. 1491. Atlantic Monthly 289(3): 41–53.

Marsh, D. M., P. C. Trenham & M. Colchester. 2000. Self-determination or environmental determinism for indigenous peoples in tropical forest conservation. Conservation Biology 14(5): 1365–1367.

Martin, C. 1993. Introduction. Pp. xv–xix *in* E. Kemf, The Law of the Mother: Protecting Indigenous Peoples in Protected Areas. San Francisco: Sierra Club Books.

McNeely, J. A. 1997. Interaction between biological and cultural diversity. Pp. 173–196 *in* S. Büchi, Ch. Erni, L. Jurt & Ch. Rüegg (eds.), Indigenous Peoples, Environment, and Development. IWGIA Document 85. Copenhagen: IWGIA; Zurich: University of Zurich, Department of Social Anthropology.

——— & K. R. Miller (eds.). 1984. National Parks, Conservation, and Development: The Role of Protected Areas in Sustaining Society. Washington, D.C.: Smithsonian Institution Press.
Medin, D. L. & S. Atran (eds.). 1999. Folkbiology. Cambridge and London: MIT Press.
Meffe, G. K. & C. R. Carroll. 1997. Principles of Conservation Biology. Ed. 2. Sunderland, Mass.: Sinauer Associates.
Miller, M. S. (ed.). 1993. State of the Peoples: A Global Human Rights Report on Societies in Danger. Boston: Beacon Press.
Minnis, P. E. & W. J. Elisens (eds.). 2000. Biodiversity and Native America. Norman: University of Oklahoma Press.
Mühlhäusler, P. 1995. The interdependence of linguistic and biological diversity. Pp. 154–161 *in* D. Myers (ed.), The Politics of Multiculturalism in the Asia/Pacific. Darwin, Australia: Northern Territory University Press.
———. 1996. Linguistic Ecology: Language Change and Linguistic Imperialism in the Pacific Rim. London: Routledge.
Nabhan, G. P. 1989. Enduring Seeds: Native American Agriculture and Wild Plant Conservation. San Francisco: North Point Press.
———. 1997. Cultures of Habitat: On Culture, Nature, and Story. Washington, D.C.: Counterpoint.
———. 2001. Cultural perceptions of ecological interactions: An "endangered people's" contribution to the conservation of biological and linguistic diversity. Pp. 145–156 *in* L. Maffi (ed.), On Biocultural Diversity: Linking Language, Knowledge, and the Environment. Washington, D.C.: Smithsonian Institution Press.
——— & S. St. Antoine. 1993. The loss of floral and faunal story: The extinction of experience. Pp. 229–250 *in* S. R. Kellert & E. O. Wilson (eds.), The Biophilia Hypothesis. Washington, D.C.: Island Press.
Naess, A. 1990. Sustainable development and deep ecology. Pp. 87–96 *in* J. R. Engel & J. G. Engel (eds.), Ethics of Environment and Development. Tucson: University of Arizona Press.
The Nation. 1999. Environmental destruction a threat to languages: UN Environment Programme. 7 Sep 1999. Distributed via Africa News Online by Africa News Service.
Nations, J. D. 1980. The evolutionary potential of Lacandon Maya sustained-yield tropical forest agriculture. Journal of Anthropological Research 36(1): 1–30.
Nettle, D. 1996. Language diversity in West Africa: An ecological approach. Journal of Anthropological Archaeology 15: 403–438.
Nevah, Z. & A. S. Lieberman. 1984. Landscape Ecology: Theory and Application. New York: Springer-Verlag.
Niec, H. (ed.). 1998. Cultural Rights and Wrongs. Paris: UNESCO; Leicester, U.K.: Institute of Art and Law.
Nietschmann, B. Q. 1992. The Interdependence of Biological and Cultural Diversity. Occasional Paper 21. Olympia, Wash.: Center for World Indigenous Studies.
Norgaard, R. B. 1981. Sociosystem and ecosystem coevolution in the Amazon. Journal of Environmental Economics and Management 8: 238–254.
OAS. 1997. Proposed American Declaration on the Rights of Indigenous Peoples, approved by the Inter-American Commission on Human Rights on February 26, 1997. Washington, D.C.: Organization of American States.
Odum, E. 1953. Fundamentals of Ecology. Philadelphia: W. B. Saunders.
Oldfield, M. L. & J. B. Alcorn (eds.). 1991. Biodiversity: Culture, Conservation, and Ecodevelopment. Boulder, Colo.: Westview Press.

Oviedo, G., L. Maffi & P. B. Larsen. 2000. Indigenous and Traditional Peoples of the World and Ecoregion-Based Conservation: An Integrated Approach to Conserving the World's Biological and Cultural Diversity. Gland, Switzerland: WWF-International and Terralingua.

Pereira, W. & A. K. Gupta. 1993. A Dialogue on indigenous knowledge. Honey Bee 4(4): 6–10.

Ponting, C. 1991. A Green History of the World. London: Sinclair-Stevenson.

Posey, D. A. (ed.). 1996. Traditional Resource Rights: International Instruments for Protection and Compensation for Indigenous Peoples and Local Communities. Gland, Switzerland, and Cambridge: IUCN.

——— (ed.). 1999. Cultural and Spiritual Values of Biodiversity. London and Nairobi: Intermediate Technology Publications and UNEP.

———. 2001. Biological and cultural diversity: The inextricable, linked by language and politics. Pp. 379–396 *in* L. Maffi (ed.), On Biocultural Diversity: Linking Language, Knowledge, and the Environment. Washington, D.C.: Smithsonian Institution Press.

——— & W. Balée (eds.). 1989. Resource Management in Amazonia: Indigenous and Folk Strategies. Advances in Economic Botany, 7. Bronx: New York Botanical Garden.

——— & G. Dutfield. 1996. Beyond Intellectual Property: Toward Traditional Resource Rights for Indigenous Peoples and Local Communities. Ottawa: International Development Research Centre.

Rapport, D. J. 2003. Regaining healthy ecosystems: The supreme challenge of our age. Pp. 5–10 *in* D. J. Rapport, W. L. Lasley, D. E. Rolston, N. O. Nielsen, C. O. Qualset & A. B. Damania (eds.), Managing for Healthy Ecosystems. Boca Raton, Fla., London, New York, Washington, D.C.: Lewis Publishers.

Redford, K. H. 1990. The ecologically noble savage. Orion 9(3): 24–29.

——— & S. E. Sanderson. 2000. Extracting humans from nature. Conservation Biology 14(5): 1362–1364.

——— & A. M. Stearman. 1993. Forest-dwelling native Amazonians and the conservation of biodiversity: Interests in common or in collision? Conservation Biology 7(2): 248–255.

Robins, R. H. & E. M. Uhlenbeck (eds.). 1991. Endangered Languages. Oxford: Berg.

Schwartzman, S., A. Moreira & D. Nepstad. 2000a. Rethinking tropical forest conservation: Perils in parks. Conservation Biology 14(5): 1351–1357.

———, D. Nepstad & A. Moreira. 2000b. Arguing tropical forest conservation: People versus parks. Conservation Biology 14(5): 1370–1374.

Shepard, G. H., Jr. 2002. Primates in Matsigenka subsistence and world view. Pp. 101–136 *in* A. Fuentes & L. Wolfe (eds.), Primates Face to Face: The Conservation Implications of Human and Nonhuman Primate Interactions. Cambridge: Cambridge University Press.

———, D. W. Yu & B. W. Nelson. 2004. Ethnobotanical ground-truthing and forest diversity in the western Amazon. Pp. 131–169 *in* T. J. S. Carlson & L. Maffi (eds.), Ethnobotany and Conservation of Biocultural Diversity. Advances in Economic Botany, 15. Bronx: New York Botanical Garden Press.

Shiva, V., H. Schücking, P. Anderson, A. Gray, L. Lohmann & D. Cooper. 1991. Biodiversity: Social and Ecological Perspectives. London: Zed Books.

Skutnabb-Kangas, T. 2000. Linguistic Genocide in Education—Or Worldwide Diversity and Human Rights? Mahwah, N.J.: Lawrence Erlbaum.

——— & R. Phillipson (eds.). 1994. Linguistic Human Rights: Overcoming Linguistic Discrimination. Berlin and New York: Mouton de Gruyter.

Somerville, M. A. & D. J. Rapport (eds.). 2000. Transdisciplinarity: ReCreating Integrated Knowledge. Oxford: EOLSS.

Soulé, M. E. 2000. Does sustainable development help nature? Wild Earth, Winter, 57–64.

Sowerwine, J. C. 2004. Effects of economic liberalization on Dao women's traditional knowledge, ecology, and trade of medicinal plants in northern Vietnam. Pp. 232–259 *in* T. J. S. Carlson & L. Maffi (eds.), Ethnobotany and Conservation of Biocultural Diversity. Advances in Economic Botany, 15. Bronx: New York Botanical Garden Press.

Steinmetz, R. 1998. The ecological science of the Karen in Thung Yai Naresuan Wildlife Sanctuary, western Thailand. WWF-Thailand Project Office.

Stepp, J. R., E. C. Jones, M. Pavao-Zuckerman, D. Casagrande, and R. K. Zarger. 2003. Remarkable properties of human ecosystems. Conservation Ecology 7(3). <http://www.consecol.org/vol7/iss3/art11>.

Stevens, S. (ed.). 1997. Conservation through Cultural Survival: Indigenous Peoples and Protected Areas. Washington, D.C.: Island Press.

Sunderland, T. C. H. 2004. Indigenous nomenclature, classification, and utilization of African rattans. Pp. 173–211 *in* T. J. S. Carlson & L. Maffi (eds.), Ethnobotany and Conservation of Biocultural Diversity. Advances in Economic Botany, 15. Bronx: New York Botanical Garden Press.

Taylor, P. M. 1990. The Folk Biology of the Tobelo People: A Study in Folk Classification. Smithsonian Contributions to Anthropology, 34. Washington, D.C.: Smithsonian Institution Press.

Ten Kate, K. & S. A. Laird. 1999. The Commercial Use of Biodiversity: Access to Genetic Resources and Benefit-Sharing. London: Earthscan.

Terborgh, J. 1999. Requiem for Nature. Washington, D.C.: Island Press.

———. 2000. The fate of tropical forests: A matter of stewardship. Conservation Biology 14(5): 1358–1361.

Toledo, V. M. 1994. Biodiversity and cultural diversity in Mexico. Different Drummer 1(3): 16–19.

———. 2001. Biodiversity and indigenous peoples. Pp. 3: 451–463 *in* S. A. Levin (ed.), Encyclopedia of Biodiversity. San Diego, Calif.: Academic Press.

Warren, D. M., L. J. Slikkerveer & D. Brokensha (eds.). 1995. The Cultural Dimension of Development: Indigenous Knowledge Systems. London: Intermediate Technology Publications.

WCPA, IUCN & WWF. 1999. Principles and Guidelines on Indigenous and Traditional Peoples and Protected Areas. Gland, Switzerland: WCPA, IUCN, and WWF.

Weber, R., J. Butler & P. Larson (eds.). 2000. Indigenous Peoples and Conservation Organizations: Experiences in Collaboration. Washington, D.C.: WWF–US.

Wilcox, B. A. 2001. Ecosystem health in practice: Emerging areas of application in environment and human health. Ecosystem Health 7(4): 317–325.

——— & K. N. Duin. 1995. Indigenous cultural and biological diversity: Overlapping values of Latin American ecoregions. Cultural Survival Quarterly 18(4): 49–53.

Wilkins, D. 1993. Linguistic evidence in support of a holistic approach to traditional ecological knowledge. Pp. 71–93 *in* N. M. Williams & G. Baines (eds.), Traditional Ecological Knowledge: Wisdom for Sustainable Development. Canberra: Australian National University, Centre for Resource and Environmental Studies.

Williams, N. M. & G. Baines (eds.). 1993. Traditional Ecological Knowledge: Wisdom for Sustainable Development. Canberra: Australian National University, Centre for Resource and Environmental Studies.

——— & E. S. Hunn (eds.). 1982. Resource Managers: North American and Australian Hunter-Gatherers. AAAS Selected Symposium 67. Boulder, Colo.: Westview Press.

Wilson, E. O. (ed.). 1988. Biodiversity. Washington, D.C.: National Academy Press.

Woodbury, A. C. 1993. A defense of the proposition, "When a language dies, a culture dies." Pp. 101–129 *in* R. Queen & R. Barrett (eds.), SALSA I: Proceedings of the 1st Annual Symposium about Language and Society. Austin, Tex.: TLF 33.

World Bank. 1991. Operational Directive OD 4.20 Indigenous Peoples. Washington, D.C.: World Bank.

WRI, IUCN & UNEP. 1992. Global Biodiversity Strategy: Guidelines for Action to Save, Study, and Use Earth's Biotic Wealth Sustainably and Equitably. Washington, D.C.: WRI, IUCN, and UNEP.

WWF. 1996. Indigenous Peoples and Conservation: WWF Statement of Principles. Gland, Switzerland: WWF-International.

———. 1998. WWF's Global Conservation Priorities. Gland, Switzerland: WWF-International.

———. 1999. Protect Forests for Future Generations. Gland, Switzerland: WWF-International.

Zent, E. L. & S. Zent. 2004. Amazonian Indians as ecological disturbance agents: The Hotï of the Sierra de Maigualida, Venezuelan Guayana. Pp. 78–111 *in* T. J. S. Carlson & L. Maffi (eds.), Ethnobotany and Conservation of Biocultural Diversity. Advances in Economic Botany, 15. Bronx: New York Botanical Garden Press.

Zent, S. 1999. The quandary of conserving ethnoecological knowledge: A Piaroa example. Pp. 90–124 *in* T. Gragson & B. Blount (eds.), Ethnoecology: Knowledge, Resources, and Rights. Athens: University of Georgia Press.

———. 2001. Acculturation and ethnobotanical knowledge loss among the Piaroa of Venezuela: Demonstration of a quantitative method for the empirical study of TEK change. Pp. 190–211 *in* L. Maffi (ed.), On Biocultural Diversity: Linking Language, Knowledge, and the Environment. Washington, D.C.: Smithsonian Institution Press.

——— & E. L. Zent. 2004. Ethnobotanical convergence, divergence, and change among the Hotï of the Venezuelan Guayana. Pp. 36–77 *in* T. J. S. Carlson & L. Maffi (eds.), Ethnobotany and Conservation of Biocultural Diversity. Advances in Economic Botany, 15. Bronx: New York Botanical Garden Press.

2

Ethnobotanical Convergence, Divergence, and Change among the Hotï of the Venezuelan Guayana

Stanford Zent and Egleé López-Zent

Introduction

Over the past 20 years numerous authors have pointed out the vital link between biological and cultural diversity, argued that the fast erosion of the earth's evolved endowment of biological species and of indigenous cultures are overlapping processes, and advocated the need for integrated conservation programs whose aim is to achieve simultaneously the preservation of distinct natural life-forms as well as unique human lifestyles, the so-called biocultural model of conservation (Bulmer 1982; Brownrigg 1985; McNeely & Pitt 1985; Oldfield & Alcorn 1987, 1991; Plotkin 1988; Posey & Overal 1990; Dasmann 1991; Nietschmann 1992; Meilleur 1994; Toledo 1994; Mühlhäusler 1995; Wilcox & Duin 1995; Alcorn 1996; Balick & Cox 1997; Nabhan 1997; Maffi 1998, 2001; Posey 1999; S. Zent 1999a). Ethnobiological research and researchers have been at the forefront of this nascent paradigm shift. Key contributions include: documenting the human utility of a rich variety of little-known biological species, demonstrating that indigenous peoples possess very detailed and scientifically accurate knowledge and are skillful manipulators of their local environments, explaining the importance of indigenous knowledge and resource management practices for biological conservation, and assuming the role of advocate/broker/intermediary between indigenous and world communities, in defense of the former's cultural, economic, and territorial rights (Schultes 1979, 1994; Nigh & Nations 1980; Morauta et al. 1982; Alcorn 1984, 1995; Posey et al. 1984; Brush 1986; Altieri & Merrick 1987; Posey & Balée 1989; Posey 1990a, 1990b; Schultes & Raffauf 1990; Plotkin & Famolare 1992; Redford & Padoch 1992; Toledo 1992; Cunningham 1993; Moran 1993; Martin 1995; Posey & Dutfield 1996).

A long list of other relevant topics or facets of ethnobotanical research could be mentioned in like regard, but we want to focus here on the specific conceptual and empirical contributions of quantitative ethnobotany to biocultural conservation. Quantitative ethnobotany refers to a relatively new research orientation that emphasizes quantitative data collection and statistical analysis, formal hypothesis formulation and testing, and methodological integration drawing from the fields of anthropology (ethnographic interviews and observations), systematic botany (specimen collections and taxonomic determinations), community forestry (botanical inventory, census, and measurement in sample plots), economics (market survey, cost-benefit analysis), cognitive science (projective, sorting, ranking, and reasoning tasks), geography (geographical information systems), and others (cf. Prance 1991; Phillips et al. 1994; Nazarea-Sandoval 1995; Phillips 1996; Zent 1996). Studies of this genre have advanced the notion of biocultural conservation in the following ways:

1. By documenting the impact of human activities on forest composition and structure as well as species richness and abundance, thus providing empirical support for the notion of anthropogenic landscapes (Anderson & Posey 1989; Irvine 1989; Balée 1993, 1994).
2. By measuring the extent to which indigenous peoples use the plant diversity of their local environment, thereby setting guidelines for land-reserve creation and protection (Prance et al. 1987; Bennett 1992; Kainer & Duryea 1992; Begossi 1996).
3. By estimating the cultural-use values of different botanical species and families, thus establishing culturally appropriate conservation targets (Stoffle et al. 1990; Phillips & Gentry 1993a, 1993b).
4. By identifying the habitat types or geographical areas of highest utility (and therefore highest conservation priority) for local populations (Stoffle et al. 1990; Bonet et al. 1992; Keel et al. 1993; Phillips et al. 1994; Zent 1995; Voeks 1996).
5. By assessing the (real or potential) economic value of local plant resources in order to show the comparative advantages of sustainable (i.e., nontimber) versus predatory uses (Padoch et al. 1985; Anderson 1988; Peters et al. 1989; Bennett 1991).

These examples of ethnobotanical research speaking to conservation issues may be categorized as either habitat directed or use directed; that is, plans for biocultural conservation are enhanced by defining what biological units (species, families, bioregions) are most worth protecting from cultural standpoints, or what use patterns optimize the balance between economic and ecological goals. Another key element in the equation is environmental knowledge, and that will be the focus of the present chapter. We argue that the landscape of knowledge needs to be surveyed in order to develop effective knowledge-based conservation strategies, such as cultural memory banks (cf. Nazarea 1998), environmental education programs (cf. Gough 1999), or community development projects (cf. Warren et al. 1995). This entails measuring and mapping the pattern of knowledge distribution and variation and the process of knowledge acquisition and evolution within a cultural community.

The quantitative study of ethnobotanical knowledge (EBK) variation and change, though still embryonic, is starting to yield insights into the significance of such knowledge for the conservation of cultural and biological diversity. Several authors (Adu-Tutu et al. 1979; Friedman et al. 1986; Trotter & Logan 1986; Johns et al. 1990; Berlin 1999) employ techniques of interinformant consensus analysis regarding medicinal plant use in order to identify culturally salient and pharmacologically effective remedies. Boster

(1980, 1984, 1986) analyzed patterns of agreement/disagreement in manioc identification among the Aguaruna ethnic group of Peru using a variety of statistical operations. His research found a single shared cultural model of Aguaruna manioc cultivar identification but also deviations from the model patterned by sexual division of labor, individual expertise (also related to age, social standing, and schooling), kinship and residential group membership, and participation in exchange networks. Phillips and Gentry (1993b) quantified the knowledge of plant-use values according to different age groups in the mestizo community of Tambopata, Peru. One important finding of their study is that "medicinal plant knowledge is largely confined to older people," which suggests, according to the authors, that medicinal plant lore is especially vulnerable to acculturation and therefore should be a research priority (pp. 41–43). S. Zent (1999a, 2001) examined the impact of the social variables of age, formal education, and bilingual speaking ability on knowledge of forest tree names and uses among Piaroa males in Venezuela. This study concluded that age, which correlates with intensity of acculturation pressure, as well as intrusive knowledge forms (formal education and bilingual ability), has a significant impact on individual knowledge levels. The broader theoretical significance of these examples of processually minded ethnobotanical research is to demonstrate that EBK is embedded in historically contingent social as well as ecological relationships. It follows from this that in order to understand how and why plant knowledge varies in time and space it is necessary to understand the complex and dynamic societal and environmental contexts in which it occurs. Accordingly, what emerges from the previous work cited here is a historical sociology and ecology of knowledge research perspective that focuses on the crucial evolving relationships among the adaptive ecological system, the sociocultural system, and the dynamic aspects (creation, transmission, distribution, and utilization) of EBK (cf. S. Zent 1999b).

The present chapter represents a continuation and, we hope, an advancement of this research direction by providing a systematically sampled and statistically supported analysis of EBK convergence, divergence, and change under controlled comparison conditions. We compare different aspects of plant lore (nomenclature, taxonomic classification, use-value categorizations) among a fairly large sample of people varying by age, sex, and community yet belonging to the same ethnic unit. The research was carried out among the Hotï, a relatively small (800+), culturally and linguistically related population living in the remote Sierra de Maigualida region of southern Venezuela. These people were recently contacted by westerners and are now undergoing dramatic cultural and ecological changes. Precisely because the Hotï have not been exposed to acculturating influences for very long, they offer an intriguing opportunity for assessing the sensitivity, adaptability, dynamic quality, intracultural variability, and contextual integrity of EBK in response to such changes.

Quantitative Ethnobotany of the Hotï

Inspired somewhat by the previous contributions of quantitative ethnobotany to culturally appropriate environmental conservation, in 1996 we began a quantitatively oriented field study of Hotï ethnobotany. Our original project had two general research objectives: botanical ethnography—to provide a comprehensive quantitative and qualitative description of Hotï EBK and behavior—and biocultural conservation —to inventory the range of plant resources upon which Hotï material and sociocultural reproduction depend and to contribute this information to the planning and creation of a protected area for the Maigualida region. The first objective responded to the fact that the Hotï culture and habitat were poorly known and described. In particular, their ethnobotany had not been studied, and the small population size, previous genetic and epidemiological isolation, and recent history of contact and acculturation suggested that they were highly vulnerable to biological and/or cultural extinction or modification. The second objective was motivated essentially by the fact that the Hotï homeland is not currently protected by any form of conservation law or restricted-use zonation, with the exception of areas above 800 meters above sea level (m asl) that are designated as natural monuments. However, most Hotï settlements and extensive parts of their home ranges are actually below this altitudinal level, and there has been a sharp influx of wildcat miners, ecotourists, and other Indian groups (not to mention the infectious diseases they carry) into the region in the past decade.

The Hotï were the last of Venezuela's 30 contemporary indigenous ethnic groups to be contacted by the Western world. The first sustained contact between the Hotï and non-native outsiders took place in 1969; and, due to the relative isolation of their homeland, the Hotï have persisted as one of the least acculturated or Westernized peoples in South America. Soon after the discovery of this "lost" tribe of "paleo-Indians" was announced, in the early 1970s, the Hotï were visited by several explorers, journalists, and anthropologists. Their visits to the region were very brief and their face-to-face interactions with the local people rather superficial and fleeting, but they did arrive on the scene fairly early on and produced the first ethnographic accounts of Hotï culture and environment (see Jangoux 1971; Corradini 1973; Eibl-Eibesfeldt 1973; Coppens & Mitrani 1974; Guarisma 1974; Coppens 1975, 1983).

The Hotï, circa 1970–1975, are typically described as displaying many of the classic traits associated with the South American Marginal culture type, within which they would seem to fit the profile of the forest foot nomad subtype (cf. Steward & Faron 1959: 424–437). The traditional Hotï cultural complex may be summarized as follows: interfluvial or upland adaptive orientation; primarily hunter-gatherers and secondarily incipient horticulturalists; dispersed and nomadic settlement pattern (including frequent trekking between different camp or house sites); very rudimentary native technology, including absence of watercraft; notable scarcity of Western trade goods; and small, atomistic, fluid band type of social organization.

The social, economic, and cultural panorama of the Hotï is very different in the early twenty-first century. Within a single generation, they have undergone considerable changes in habitat occupied, settlement pattern, subsistence focus, interethnic contact, social structure, accessible technology, material culture, socialization, and religion. However, not all groups have been equally affected by these sweeping changes. A few isolated settlement groups in the Upper Mosquito and Upper Cuchivero appear to be little altered, whereas other communities, especially the mission settlements, are obviously more acculturated. Our research project was specifically designed to capture the major gradients of this transition and indeed to assess what specific factors of the prevailing mode of cultural change have the greatest impact on their EBK and thus was carried out in four distinct Hotï communities: Caño Mosquito, Caño Majagua, Caño Iguana, and San José de Kayamá (Figure 2.1). The major social and ecological characteristics of these communities are described below.

Research Sites

Caño Mosquito

The Caño Mosquito region has a population of approximately 60 people distributed among five settlement groups. In the last decade or so, a few families have moved downriver and now occupy a lowland forest habitat (175 m asl), while the other groups inhabit the very hilly upriver zone (450–650 m asl). Our forest plot in this area was established near a small community of 12 people, situated in the mountainous upriver zone at an altitude of 450–500 m asl. The Hotï living in this area may be characterized as semi-isolated from interethnic contact, with only very sporadic contact with neighboring indigenous groups (mainly Yabarana-Piaroa and E'ñepa villages of the Parucito River). Our data on the residential histories of Hotï living in this and other areas seem to indicate that this region is a traditional habitation zone.

Although it is not uncommon to find the Mosquito Hotï wearing Western clothes (usually old and tattered), giving them a more acculturated appearance than the mission Hotï at first sight, they are undoubtedly the least acculturated of the four settlements described here. The cultural habits of the Mosquito groups bear a fairly close resemblance to the way Hotï culture was depicted during the early post-contact period (cf. Jangoux 1971; Guarisma 1974; Coppens 1983), notably the dispersed and seminomadic settlement pattern, hunting-gathering economic focus, and fluid band type of social organization. Also, during our time among them (a cumulative total of four months), we detected no hint of nonaboriginal religious or curing beliefs. Western metal and cloth goods are not uncommon, but quality is low (mostly old and worn items). The key trade products of the Mosquito people are the blowgun cane (*Arthrostylidium schomburgkii* Munro), a traditional trade item, and captive jungle animals (birds and monkeys).

The subsistence economy observed in the Upper Mosquito is heavily dependent on foraging, while the significance of horticulture is clearly secondary although variable according to settlement group. One group we encountered had no active gardens and reportedly were perpetually nomadic throughout the duration of our fieldwork. They did, however, harvest food from old fallows or from other people's gardens. By contrast, other groups kept elaborately polycultivated, occasionally weeded gardens. The main crops are plantain, sweet potato, yam, and maize, the importance of which varies by season. All groups are frequent trekkers/campers, during which time they rely mostly on wild resources, such as game (mostly arboreal animals), insects (honey and palm larvae), and forest fruits (see E. L. Zent & S. Zent 2004).

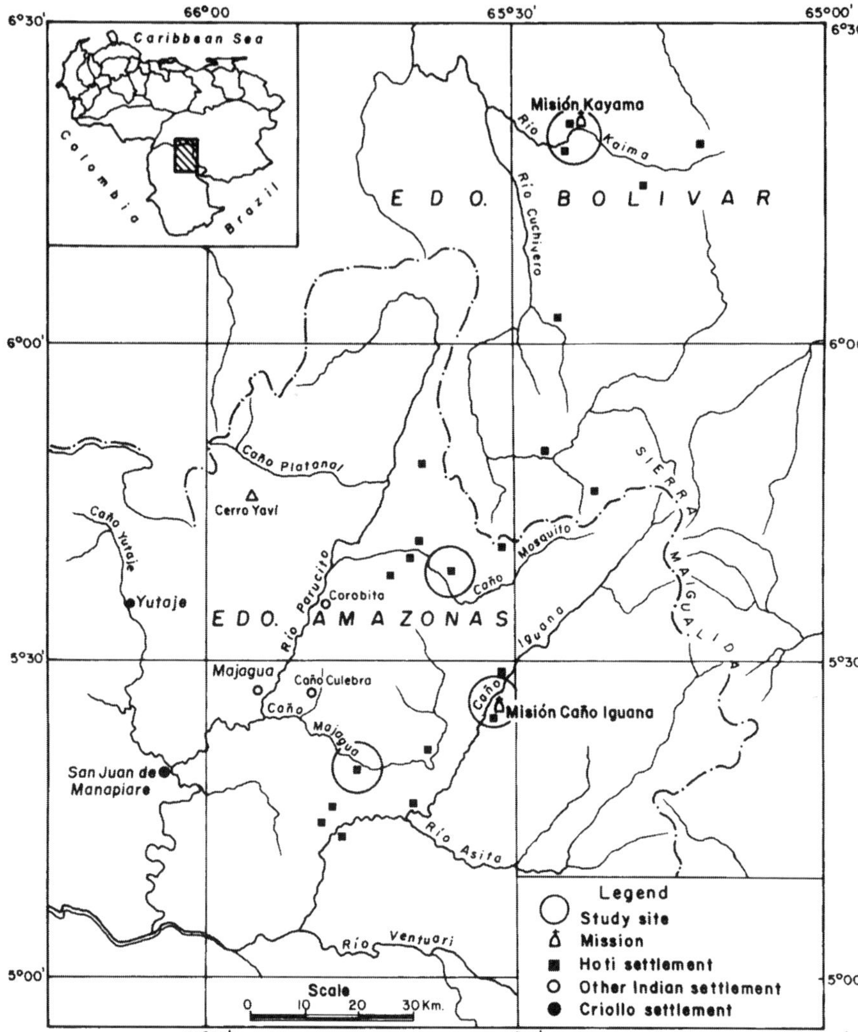

Figure 2.1. Hotï settlements and territory, southern Venezuela.

Caño Majagua

Scattered throughout the lowland forests of the middle-upper Caño Majagua and lower Asita River are a half-dozen small Hotï communities with an aggregate population of about 70 people. Most of them moved to their present location within the past 20 years, either from the upper reaches of Caño Majagua or Caño Iguana. The community where we based our forest plot study had been there for about five years, and the population varied from 5 to 25 people during the research period due to the frequent mobility of different family groups. Although ethnographic accounts place the

Hotï in the upper Majagua basin early on in post-contact period (Coppens & Mitrani 1974), such reports probably refer to locations farther upriver (J. Jangoux pers. comm.). Our data on residential histories indicate that the general vicinity of our study site (19 km from the mouth of Caño Majagua) really represents a zone of recent spontaneous colonization by members of this ethnic group. There is no missionary presence here, but some family groups formerly lived at the Caño Iguana mission settlement. A few individuals as children lived briefly and sporadically in Yekuana villages of the Upper Ventuari River or spent time in the nearby multiethnic town of San Juan de Manapiare, mainly for the purpose of attending school there. As a result of these experiences, they know some words and phrases of Spanish, the national language. The primary inter-ethnic contacts for the Majagua Hotï are with E'ñepa, Yabarana, and Piaroa Indians settled in the lower Parucito, itinerant (mostly indigenous) gold miners who occasionally enter the area, and the criollo and Indian residents of San Juan (Figure 2.1). Western goods such as steel tools, aluminum pots, and Western textiles (clothing, hammocks, and mosquito nets) are widespread among these groups. In spite of material acculturation, their settlement behavior, social structure, and religious practices appear to conform to traditional (i.e., early post-contact) customs.

The subsistence pattern of the Majagua Hotï reflects a relatively even and extensive mix of horticulture, collection, fishing, and hunting, the importance of each varying by season. Horticulture is the main provider of dietary energy and has obviously been elaborated through incorporation of several nontraditional plants, such as pineapple, watermelon, bitter manioc varieties, certain chili pepper varieties, peanut, mango, and avocado. Hunting and collecting are very productive in this lowland, seasonally flooded habitat where palms are the dominant botanical family. The preponderance of fish in the diet is especially notable during the dry season. Fish are caught using a variety of traditional and nontraditional technology: hand capture, *barbasco* (fish-poison) plants, lance, hook and line, scuba mask, and slingshot spear. Although settlements may be slightly more permanent than was previously the case, extended camping trips are still very common in both wet and dry seasons.

Caño Iguana

The New Tribes Mission station at Caño Iguana was founded in 1970. At first, the Hotï were only temporary and unpredictable visitors to the site, but over the years their presence gradually became more permanent and their numbers grew steadily. In 1996 there were approximately 165 resident Hotï, 10 E'ñepa, and 2–4 U.S. missionary families. Although a few small Hotï groups were apparently residing (at least on a seasonal basis) in this general area at the time of contact, many of the current residents moved there from the Iguana headwaters or upper Río Cuchivero during the 1980s and 1990s. There is no direct fluvial access to the community, so most foreign visitors and imported goods come in by mission-operated airplanes.

The Hotï at Caño Iguana have almost no contact with other ethnic people except for mission personnel and their family or friends. A few Hotï have visited Puerto Ayacucho, the capital city of Amazonas State, or Caracas, the national capital, in search of specialized medical treatment (usually escorted by a missionary). A greater number of people have also been to Kayamá, but visits are not very common due to the long distance (>100 km) and rugged, mountainous terrain. Despite their geographical isolation from surrounding indigenous and national populations, considerable material and ideological acculturation has nevertheless taken place due to their prolonged contact with the missionaries. One of the main impacts concerns improvement of the biological health of the population, which is growing rapidly (by 10% from 1996 to 1999). The Iguana community enjoys the best modern health-care services of any of the Hotï groups described here, including a well-trained native nurse, a well-stocked medical dispensary, emergency airplane flights in case of medical emergency, and effective access to specialized care in Caracas and elsewhere.

None of the Hotï at Caño Iguana speaks Spanish, nor does anyone even know more than a handful of words of that language. However, the missionaries have learned to speak Hotï and recently began to teach the Hotï about the Bible. In 1995 a school devoted to teaching Hotï adults and adolescents to read and write in their own language opened, as a result of which the vast majority of Iguana residents aged 10 and above are effective literates. In 1997 the missionaries initiated community-wide Bible-study classes. Printed handouts and/or cassette tape recordings of the class lessons are handed out to the learning members of the congregation so that they may study the lessons in their free time. Indeed, among the most common sedentary pastimes in this community are reading, writing, and listening to tapes.

In material matters, Iguana appears to be the most affluent community. Western trade goods are abundant, and the local people have learned to use Venezuelan money through their dealings with the missionaries.[1] The Hotï trade labor or food to the missionaries for cash and then give that money back to the missionaries in exchange for a wide range of imported goods—axes, machetes, knives, pots, kitchen utensils, fishhooks, fishing line, scuba masks, wristwatches, flashlights, batteries, clothes, beads, salt, soap, and so forth. Despite the relatively easy access to Western goods, the Iguana Hotï still fabricate and use various aboriginal craft items, such as cotton loincloths, hammocks, and beeswax torches. The main subsistence activity at Caño Iguana is agriculture; hunting, fishing, and collecting are important secondary pursuits. Most gardens are located within 2 km of the community, and secondary vegetation dominates in the surrounding valley bottomland. Consequently, it is now customary to cut new gardens in secondary forest and abandon the plots sooner than was previously the case. Short-term camping trips are very common (especially during weekends or other periods when school and mission activities are curtailed), motivated by the quest for wild food resources (fish, meat, honey, or fruit) that have become very scarce in the close vicinity of the settlement.

San José de Kayamá

In 1983 a mission under the auspices of the Missionary Order of Mother Laura (Misioneras de la María Inmaculada de la Madre Laura) of the Roman Catholic Church was founded among E'ñepa Indians in the lower Kayamá River in Bolívar State. The mission, later renamed San José de Kayamá, quickly became a magnet for scores of Hotï living on the northern edge of their territory, and immigration continued throughout the 1980s and 1990s. Kayamá is the largest Hotï community, with a population (in 1999) of approximately 300 Hotï, 200 E'ñepa, and 2–5 Laurita nuns who are of Venezuelan, Colombian, or Ecuadorian nationality. It is located in an ecologically heterogeneous area that ranges from savanna to mountain forest, clearly a nontraditional habitat for the Hotï. The infrastructure of the community includes an airstrip, a church, a chapel, missionary living quarters, a tourist lodge, a medical dispensary, several school buildings, and solar-powered electrical lighting.[2] The main form of communication with the outside is by airplane, originating from Ciudad Bolívar or Caracas. Occasional visitors include medical and other service personnel, Catholic clergy, the Venezuelan military, and foreign tourists.

The Kayamá Hotï may be regarded as the most acculturated Hotï community at the present time. In the first place, the Hotï living there have apparently adopted a number of cultural symbols, practices, and objects from the E'ñepa, although such influence may predate the founding of the mission. In any case, some degree of cultural assimilation is evident in mode of dress, household artifacts, architecture, agriculture, food preparation, vocabulary, and even ritual practices. The cultural influence of the missionaries has also been very significant. All of the Indian residents are nominally practicing Catholic Christians, and a regimen of frequent church services and ceremonies is observed. Virtually all children between the age of 6 and 14 attend the local school, which ranges from kindergarten to sixth grade. The school curriculum focuses on Spanish grammar and literacy skills, mathematics, and Venezuelan geography and history. Hotï schoolteachers teach most of the classes, so instruction is to some extent bilingual. The best students continue their education in boarding schools in distant national cities (Ciudad Bolívar, Caracas, or Barinas, a sizable city in western Venezuela). Due to this formal educational experience, a large number of Hotï at Kayamá are semifluent speakers of Spanish, and perhaps 15 or more people are fully fluent and literate in the national language. The missionaries provide Western medicines to the local populace, and a few Hotï have received paramedical training.

The food-production system at Kayamá is heavily dependent on horticulture, and collecting, hunting, and fishing are clearly secondary in food-energetic terms. A notable feature of the Kayamá agricultural landscape is the large, collectively managed and exploited *arboleda* (orchard garden), consisting of mostly non-native food crops (mango, orange, lemon, guava, guamo, and grapefruit), a clear indication of recent agricultural intensification as well as innovation.[3] Due to the localized dominance of savanna and extensive modification of gallery and foothill forests, where garden making is concentrated, one must travel a distance of 2–3 km from most houses before encountering primary forest. This means that most wild forest resources must be obtained by traveling considerable distances; hence camping away from the village is a common activity. A few members of the community, mostly the schoolteachers, now own shotguns. In addition to the schoolteachers, a number of people spend a good deal of time in nontraditional occupations required for the maintenance of the mission, such as nursing, carpentry, masonry, house construction, and housekeeping. These individuals are rewarded for their labor by Western material goods. However, due to the limited financial resources of the missionaries, local demand for such goods outstrips supply and, as a consequence, there are signs of incipient economic stratification that do not seem to exist in the other Hotï settlements.

Methods

The methods employed in this study in large part follow the strategy laid out in Zent (2001), consisting of the integrated use of ethnobotanical plot surveys, structured interviews, informant consensus analyses, and linear regression analyses. This strategy is specifically designed to elicit individual EBK in a culturally and naturally appropriate context, to facilitate systematic comparison of interinformant variation of such knowledge through a standardized interview instrument, to convert the corpus of verbal responses into numeric form, and to permit statistically meaningful analysis of the results.

One-hectare forest plots (20 by 500 m) were surveyed and marked in what was described by local inhabitants as primary forest in each of the four communities using a Suunto compass and 100-m measuring tape. Plot topography was measured using a Suunto clinometer, and length and width adjustments were made where slope changes were recorded, according to standard practice (Peters 1996). The distance from house settlement to plot varied from 100 m to 2.8 km. All large trees and lianas (>10 cm dbh) were measured, and a numbered aluminum tag was attached to the stem. Herbarium voucher specimens of all tagged plants were made over the course of five visits to each site spread over three years (May 1996–October 1999). The specimen collections were processed at the Herbario Manuel Ovalles, Universidad Central de Venezuela; sets of them were deposited in national herbaria in Caracas, Guanare, and Puerto Ayacucho and sent to the Missouri Botanical Garden. Angel Fernández, Gerardo Aymard, and other botanists based in Venezuela and Ronald Leisner at the Missouri Botanical Garden identified specimens.

Structured interviews were conducted with a sample of community members belonging to different age and sex groups. The interviews consisted of walking each respondent through the plot and asking him or her to name, classify, and state the uses of every tagged plant. This set of queries and responses was conducted entirely in the Hotï language. Due to the large number of plants in each plot, a single interview took between 5–15 hours to complete, spread out over one to three days. The entire sample analyzed up to this point consists of 104 interviews distributed as follows: Caño Mosquito, 12 males and 8 females; Caño Majagua, 11 males and 9 females; Caño Iguana, 16 males and 16 females; and Kayamá, 16 males and 16 females. Due to the small size of the Mosquito and Majagua communities, people from neighboring communities were also interviewed to make up the sample size stated above. Thus it is important to note that both the Mosquito and Majagua samples are actually made up of respondents who reside in three separate settlements within the same general area (no more than 8 km from the plot location), whereas the Iguana and Kayamá samples are made up of people who are more or less permanent residents of these large, nucleated communities.

The data were codified and analyzed using cultural consensus analysis as developed by Romney et al. (1986, 1987). Cultural consensus analysis is best described as a mathematical technique for measuring patterns of interinformant agreement/disagreement about selected culturally shared domains. It may be used to determine the correct (i.e., consensual) answers (when such answers are unknown beforehand) and to rate the individual knowledge levels, expressed in terms of competence scores. The reader is referred to the publications of Romney et al. (1986, 1987) and Weller and Romney (1988) for a complete description of the theoretical and methodological postulates of this technique and to Zent (2001) for an example of its specific application to the analysis of EBK variation. It should be emphasized that this method is specifically designed to quantify the extent and distribution of shared knowledge but is not appropriate for the precise measurement or validation of specialist knowledge.[4]

The results of the consensus analysis were then subjected to several descriptive and inferential statistical analyses for interpretive purposes, including calculations of sample population means and standard deviations, analyses of variance, t-tests, linear regressions, and coefficients of correlation and determination.[5] Community sample competence means and standard deviations were computed for the folk-generic and folk-specific ranks of categorization, as well as for the nomenclatural level of classification.[6] Those figures were then used to compare degrees of shared EBK across the different communities. Next, analysis of variance was performed on the results in order to test hypotheses about the environmental factors (habitat and settlement type) that explain intercommunity variation. Competence means and standard deviations for different taxonomic and use categories were also calculated according to gender, and these results were probed for significant differences using t-tests. Linear regression analysis was employed to determine the relationship between age and EBK in regard to folk-generic classification, food use, human medicine, and hunting medicine. Regression of the number of plant taxa identified as human medicines on individual use-value competence scores was also performed. The age of the sample respondents was estimated by using standard demographic methods developed for small-scale, illiterate societies (Howell 1979; Early & Peters 1990): community census, genealogies, kinship analysis, interviews about marriage and reproductive history, informant-supplied age ranking of coresidents, and inspection of missionary-kept birth records (only at Kayamá).

Results

Floristic Variation

A detailed description of the floristic composition and structure of the four forest plots will be presented in a future publication (Zent & Zent in press.), so only a minimal sketch of those parameters most relevant for the subsequent analysis of knowledge variation will be mentioned at this time. A quantitative description of the major phytosociological parameters of the forest plots is provided in Table 2.1. The floristic and

Table 2.1
Quantitative phytosociological parameters of the four forest plots

Plot Site	Number of families	Number of genera	Number of species	Absolute stem density	Total basal area (m^2)	Average basal area (cm^2)
Caño Mosquito	45	110	187	556	40.83	734.41
Caño Majagua	42	102	182	563	31.22	554.57
Caño Iguana	38	76	133	355	33.65	947.94
San José de Kayamá	51	120	191	538	20.56	382.1

Table 2.2
Sørenson coefficients of similarity for plot pairs

Plot Pairs	Number of species in common	Mean number of species per plot	Sørenson coefficient of similarity (%)
Caño Mosquito & Caño Majagua	38	165.5	23.0 %
Caño Mosquito & Caño Iguana	43	138.5	31.0 %
Caño Mosquito & San José de Kayamá	37	170.5	21.7 %
Caño Majagua & Caño Iguana	23	146	15.8 %
Caño Majagua & San José de Kayamá	38	167.5	22.7 %
Caño Iguana & San José de Kayamá	37	143.5	25.8 %

ecological analysis of the plot data, based on the plant-specimen collections, inventories, censuses, size measurements, and consultation of the relevant existing literature (cf. Huber & Riina 1997), reveals considerable variation of surrounding forest types among the communities (E. L. Zent 1999: 72–151). At Mosquito we find ombrophilous, evergreen, submontane forest (although some species are semideciduous), characterized by a medium-to-high (15–35 m) canopy. The plot at Majagua displays a mixed composition, encompassing ombrophilous, semideciduous, lowland forest and seasonally flooded, evergreen, riverine forest. Iguana is characterized by ombrophilous, evergreen basimontane forest (although some species are semideciduous), with a medium-to-high (15–30 m) canopy. The plot at Kayamá also contains a mixed association corresponding to savanna-forest ecotone, ombrophilous, subevergreen, basimontane forest, and ombrophilous, evergreen, submontane forest.

The important point to emphasize is the considerable floristic divergence of the four forest plots. The extent to which the plots are similar or different from each other in terms of number of shared species can measured using the Sørenson coefficient of similarity. This coefficient is defined as the number of species common to two areas expressed as a percentage of the mean number of species in the two areas (Greig-Smith 1983: 151). The results show that the number of shared species among plot pairs varies in absolute terms from 23 to 43 and in relative terms from 15.8% to 31.0% (Table 2.2). The two plots showing the highest degree of similarity are Iguana and Mosquito;

significantly enough, we identified these two areas as traditional habitation zones of this ethnic group (see the next section). The two plots most different from each other are Iguana and Majagua. Of the four forest types, Majagua is the most distant outlier (on average 80% different from any other), a not unexpected result given that it corresponds to a nontraditional, recently colonized, lowland habitat. The average percentage of shared species for the entire sample of six plot pairs is 23.3%. Expressed another way, the species composition of any plot is, on average, slightly more than three-quarters different from the species composition of any other. In short, the different settlement groups studied here are exposed to very different forest types.[7]

Knowledge Variation

The mean competence scores and corresponding standard deviations for folk-generic, folk-specific, and nomenclatural classification among the four communities are presented in Figure 2.2. The first important result that should be highlighted is that all settlement groups exhibited high enough consensus (no mean is less than 40%) at all levels of classification to suggest that the consensus model is indeed appropriate for the present analysis (i.e., plant classification does constitute a monocultural domain). A common pattern observed in all locations is that group consensus descends progressively as one moves from generic to specific to nomenclatural level of classification. The average difference between generic and specific classification means is .123 (with a range of .098 to .166) and between specific and nomenclatural classification means is .088 (with a range of .067 to .121). In other words, group consensus at the generic level is about 12% higher than at the specific level, and specific level consensus is 9% higher than at the nomenclatural level. Standard deviations reflect this general trend—higher to lower values are obtained as one moves from higher (i.e., generic) to lower (i.e., nomenclatural) levels of classification, with one exception: Caño Iguana. Here we observe that the group variance in competence is almost identical for generic and specific classification and somewhat greater at the nomenclatural level.

The observed competence means are highest for the Iguana group and lowest for the Majagua group. In Iguana we find a nearly 80% agreement on plant classification at the generic level, 67% agreement at the specific level, and 55% agreement at the nomenclatural level. In Majagua consensus is 61% at the generic level, 50% at the specific level, and 42% at the nomenclatural level. Mosquito and Kayamá occupy the middle range here, displaying nearly identical means at the generic level (.7), but specific and nomenclatural level means were higher among the latter group (.6 and .535, respectively). One obvious reason for the higher group consensus achieved in Caño Iguana is the significantly lower number of taxa in this plot compared with the others.[8] Thus collective agreement decreases as a partial function of the diversity of the query set (in this case the number of plant taxa) that makes up the test. Beyond this difference, we interpret the spread of consensus across our multicommunity sample in terms of two key environmental variables: habitat familiarity and settlement nucleation. Specifically, we propose that consensus in EBK is positively related to the historical depth of settlement in a given floristic region and to the degree of settlement size and nucleation. In the first place, we reason that groups long resident in a phytogeographical region should

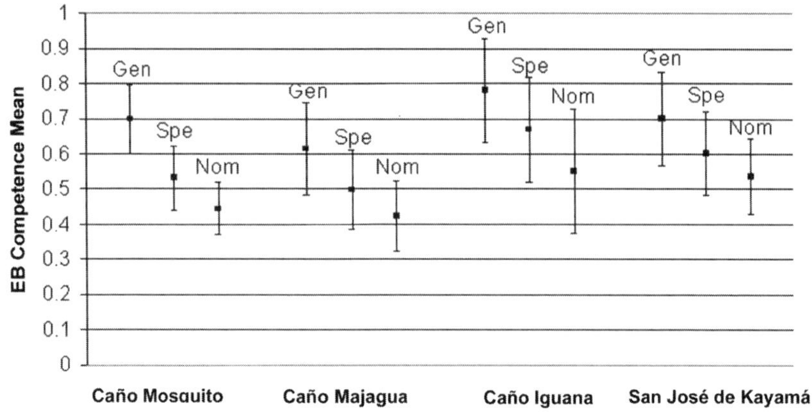

Figure 2.2. Comparison of ethnobotanical competence means, by Hotï community. *Key:* Gen = folk-generic; Spe = folk-specific; Nom = folk nomenclature.

be more familiar with the local flora and therefore display higher consensus than groups who inhabit areas where the flora is less well known. Second, we believe that consensus should be higher in nucleated settlements due to closer proximity, more frequent contact and interaction, and more intensive communication about plants among coresident members. Accordingly, we note that Iguana (where consensus is highest) corresponds to a nucleated settlement situated within the traditional biogeographical range. By contrast, the Majagua population (where consensus is lowest) has recently colonized a lowland zone floristically divergent from the traditional upland habitat, in which the pattern of small, dispersed settlements is maintained. Mosquito is characterized by traditional upland forest habitat and dispersed settlement; Kayamá, by nontraditional mixed forest-savanna habitat and nucleated settlement.

The hypothesis that habitat and settlement types affect group consensus regarding plant classification was tested using a 2-by-2 between-subjects analysis of variance. The test design consists of two factors (habitat and settlement), each of which has two levels (traditional/nontraditional and dispersed/nucleated, respectively), thus creating four cells that correspond to the four community samples mentioned above. Only generic classification competence was tested here, because it is considered to be the core level of the ethnobiological classification system (cf. Berlin 1992) and because the competence scores observed at the other two levels are consistently different and therefore predictable from it. The results show a significant main effect of habitat ($F(1, 100) = 9.8$, $MS_e = .17$, $p < .01$) and a significant main effect of settlement ($F(1, 100) = 10.3$, $MS_e = .18$, $p < .01$). However, the interaction of these two factors was not significant ($F(1, 100) = .003$, $MS_e = .00$, $p > .05$). Hence, the hypothesis is statistically confirmed, so we can conclude that living in traditional environments or in nucleated settlements is independently associated with higher group consensus for plant taxonomic knowledge.

A comparison of ethnobotanical competence scores by gender is presented in Figure 2.3. This comparison focused on five significant ethnobotanical categories: generic

a. Caño Mosquito

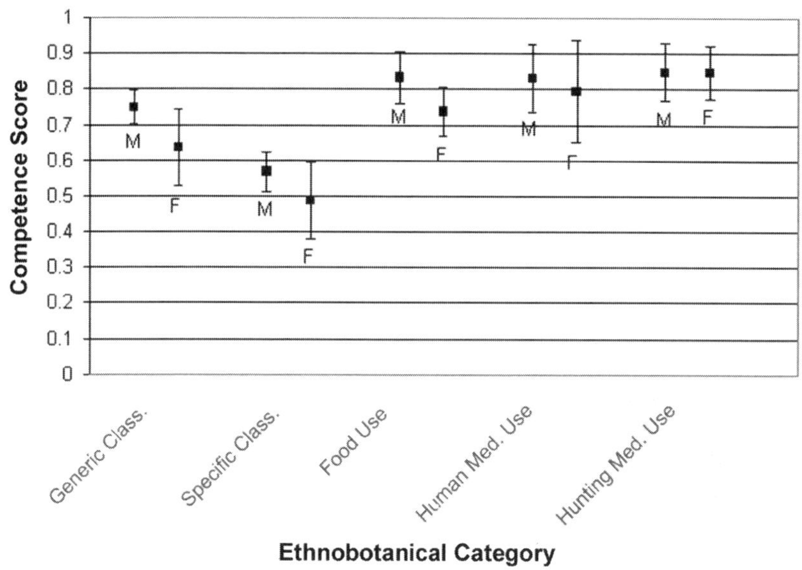

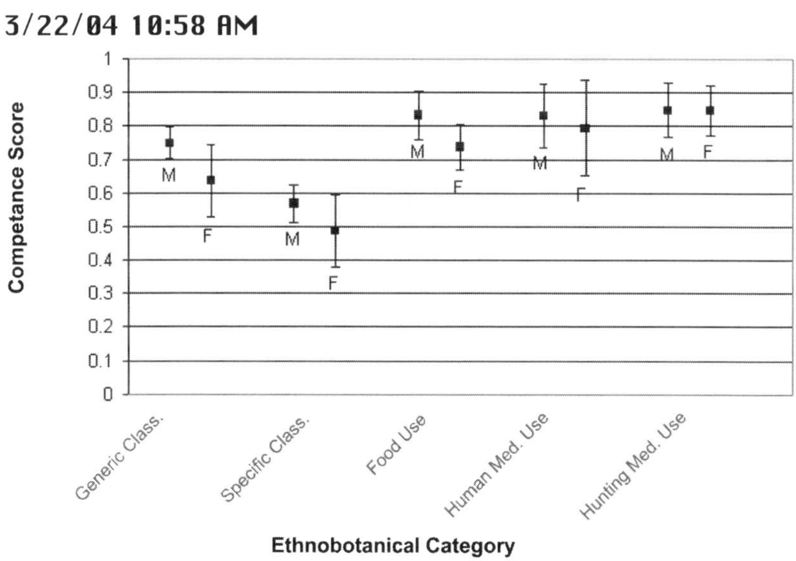

Figure 2.3. Average ethnobotanical competence, by gender. **a.** Caño Mosquito. **b.** Caño Majagua.

(Figure 2.3 continued on next page)

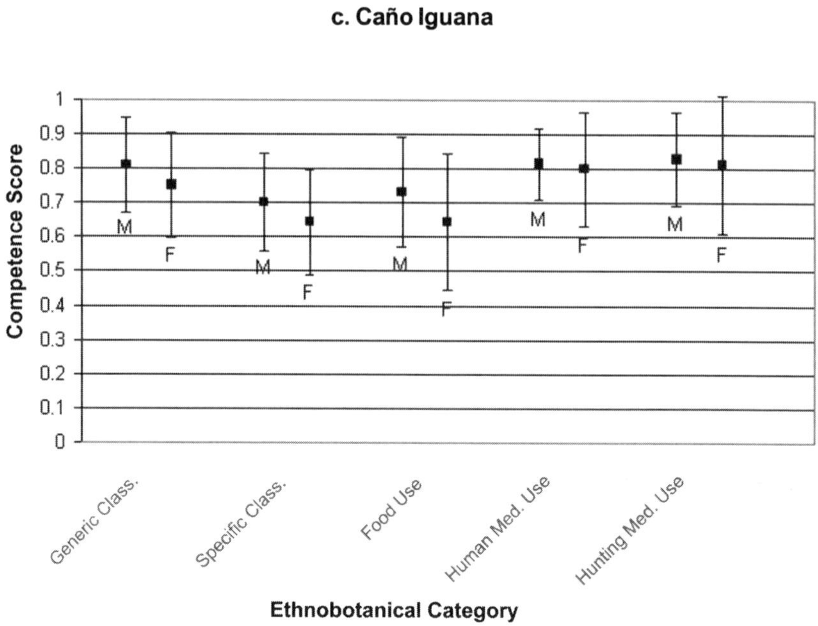

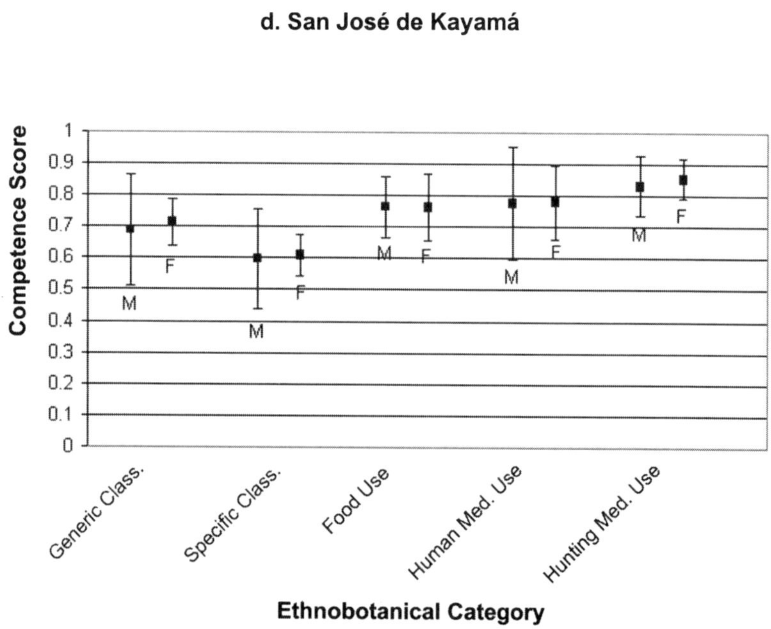

Figure 2.3. **(continued)** **c.** Caño Iguana. **d.** San José de Kayamá.

classification, specific classification, food use, human medicinal use, and hunting medicinal use. The first discernible pattern we noticed was that males have a modest but distinct advantage over females with regard to generic classification, specific classification, and food-use categorization among the Mosquito, Majagua, and Iguana groups. The observed differentials ranged from .06 (Iguana) to .16 (Majagua), .057 (Iguana) to .136 (Majagua), and .088 (Iguana) to .117 (Majagua) for generic, specific, and food-use classification, respectively. Furthermore, the males in these communities also display lower variation in competence score (i.e., standard deviation) for these types of classification. In order to test the hypothesis that males are more competent ethnobotanists than females, we performed independent samples t-tests on male versus female competence scores. First, we tested for homogeneity of variance and were unable to find significant differences in the variances in any of our samples. Therefore, the differences in standard deviation by gender reflected in Figure 2.2 cannot at this time be interpreted as factor or effect related. Second, the results of the t-tests do in fact show that males have significantly higher competencies than females for generic, specific, and food classification, but only at Mosquito ($p < .01$, $p < .05$, and $p < .01$, respectively) and Majagua ($p < .01$, $p < .01$, and $p < .05$, respectively). Why should women and girls living in the dispersed, nomadic, nonmission settlements have inferior consensual plant knowledge than their male counterparts? Conversely, why should greater cross-gender parity be the norm at the more sedentary, mission communities? We are unable to think of a convincing explanation at this time. Experimental error may have influenced the results, however. Females at the nonmission communities are less used to being and speaking with outsiders, and our impression is that in general they were less communicative and hence less informative interviewees than their missionized counterparts.[9] Another possible reason is that the economic role of men in the traditional cultural setting is that of being hunter-gatherers of forest resources, whereas at least some women skew their activities more toward gardening, domestic chores, and childcare. In this situation, men as a group could be expected to acquire a higher level of knowledge of wild plant resources. By contrast, male activities have become more varied in the mission communities, such that some of them are no longer devoted full-time to foraging for wild resources out in the forest, and one result is that their EBK is diminished accordingly. This would lower the overall mean competence scores for all males and thus account for greater parity among the sexes. This trend seems most evident at Kayamá, where several men are moving into specialized Westernized (i.e., nonsubsistence) economic roles, such as schoolteacher, paramedic, carpenter, or builder.[10] Obviously the whole issue of gendered knowledge differentials in this ethnic context deserves a more in-depth examination, but at least these hypotheses point the way to future research designs.

The next phase of the analysis utilizes linear regression to explore the predictive relationship between age and EBK, which is especially critical for the light it may shed on variation and change in the ethnobotanical learning process. Figures 2.4–2.6 and 2.8 depict this relationship in regard to different types of knowledge: folk-generic categorization, food-use categorization, human medicinal use categorization, and hunting medicinal use categorization, respectively. In order to reveal the significant learning trends across the sample of respondents, we divided each sample into two groups according to the variable of age (younger group / older group). Thus, assuming that age is related to differences in EBK, which in turn is a function of the normal learning process, the purpose of this binary analytical procedure is to identify at what age the active learning process reaches its culmination or at least enters a new phase (i.e., less active learning or nonlearning), which may be signified graphically by a noticeable turn in the slope of the regression line. The point at which this division was made was determined by calculating the two best fitting lines to the spread of data points (i.e., lowest combined standard error of estimate), which allows for the fact that the active learning phase may vary according to community and category of EBK. We also built a two-year overlap into our definitions of the two sample groups, such that the last two years in the younger group are included as the first two years in the older group, in order to provide greater continuity between the lines corresponding to each group and therefore account for the fact that the transition from active to nonactive (or less active) learning phases is not always abrupt. This procedure compels us to define the critical turning point between the two learning phases as an age range of at least two years rather than one precise age, which we believe is a more accurate depiction of the empirical reality of this process.

The resulting regressions of generic classification competence on age observed in Majagua, Iguana, and Kayamá (Figure 2.4) show a strongly convergent pattern that, in turn, is indicative of a similar learning trend and rate. We can characterize this trend generally as consisting of a very active learning primary phase during childhood and early adulthood that reaches its culmination by the late teen (18–19) years and then is followed by an essentially nonlearning secondary phase through the rest of adulthood. Regarding the primary phase, we predict a rapid increase in the ability of the average Hotï to discriminate and label correctly the surrounding inventory of forest plant specimens at the generic level of taxonomic classification, from a base level of 40% (Majagua) to 55% (Iguana) correct identifications at 8–9 years of age to a peak level of 70% (Majagua) to 85% (Iguana) correct identifications by 18–19 years of age. In all three community samples mentioned above, the explanitory relationship between age and generic classification knowledge is strong, positive, and statistically significant among the younger cohort (Majagua: $r^2 = .7, p = .003$; Iguana: $r^2 = .622, p = .000$; Kayamá: $r = r^2 = .496, p = .007$), whereas there is no statistically significant relationship between these variables among the older cohort (Majagua: $r^2 = .099, p = .32$; Iguana: $r^2 = .051, p = .354$; Kayamá: $r^2 = .004, p = .782$). Although these results focus on the generic level of classification, parallel trends were observed for specific and nomenclatural levels, albeit at lower levels of agreement. To summarize, the data from Majagua, Iguana, and Kayamá predict that before a person reaches the age of 20, he or she will have acquired essentially all of his or her knowledge about plant classification and nomenclature that is widely shared among the group.

The observed relationship between age and generic classification competence in Mosquito (Figure 2.4a) exhibits a strongly divergent pattern, and it is precisely this difference that provides the counterpoint and context needed for interpreting how social and environmental factors influence and structure the previous pattern. In Mosquito there is no marked difference between younger and older cohorts in regard to the above-mentioned statistical relationship. The computed regression lines show no clear distinction in the learning trends between the

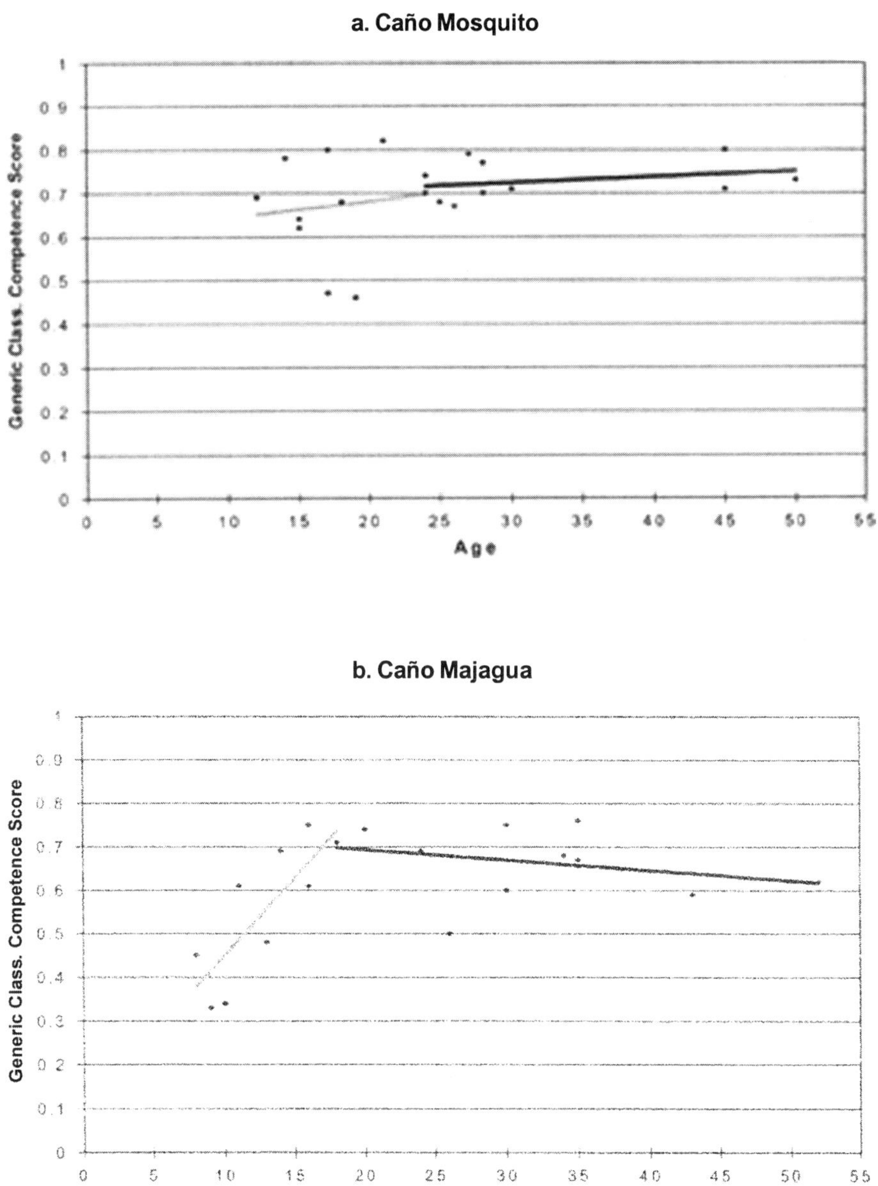

Figure 2.4. Folk-generic classification competence, by age. **a.** Caño Mosquito. Gray tendency line, representing the subgroup <25 years old: $n=11$, $r=.126$, $r^2=.016$, $p=.712$, $Y'=.604+.004X$. Black tendency line, representing the subgroup ≥ 23 years old: $n=11$, $r=.291$, $r^2=.085$, $p=.386$, $Y'=.686+.001X$. **b.** Caño Majagua. Gray tendency, line representing the subgroup <20 years old: $n=10$, $r=.836$, $r^2=.7$, $p=.003$, $Y'=.094+.036X$. Black tendency line, representing the subgroup 18 years old: $n=12$, $r=-.314$, $r^2=.099$, $p=.32$, $Y'=.74+.002X$.

(Figure 2.4 continued on next page)

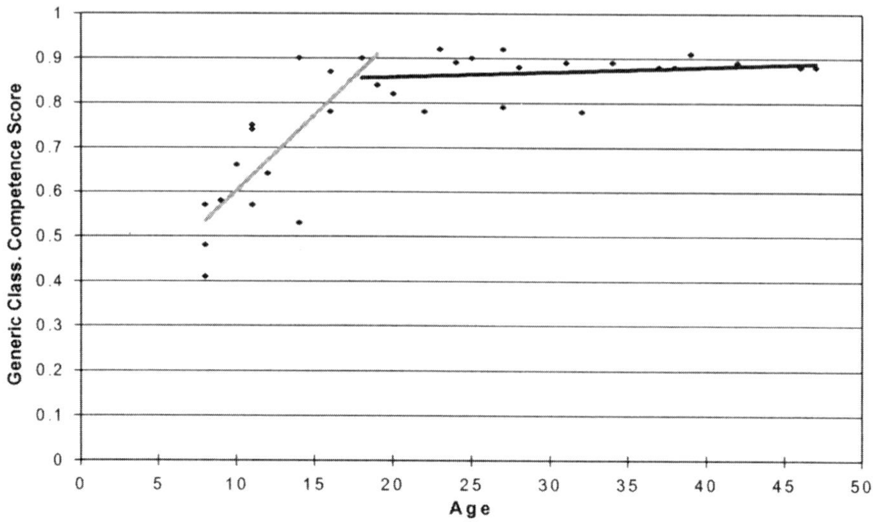

d. San José de Kayamá

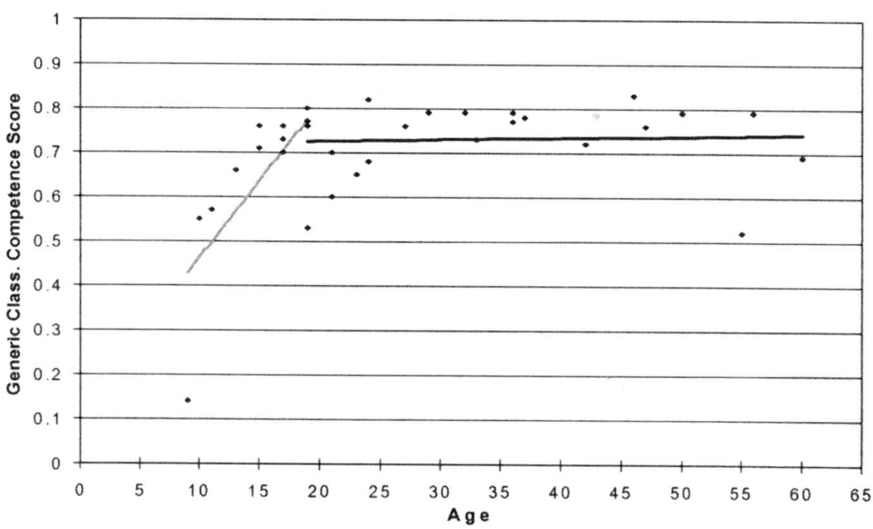

c. Caño Iguana. Gray tendency line, representing the subgroup <20 years old: $n = 15$, $r = .789$, $r^2 = .622$, $p = .000$, $Y' = .262 + .034X$. Black tendency line, representing the subgroup 18 years old: $n = 19$, $r = .225$, $r^2 = .051$, $p = .354$, $Y' = .835 + .001X$. **d.** San José de Kayamá. Gray tendency line, representing the subgroup <20 years old: $n = 13$, $r = .704$, $r^2 = .496$, $p = .007$, $Y' = .115 + .035X$. Black tendency line, representing the subgroup 18 years old: $n = 23$, $r = .061$, $r^2 = .004$, $p = .782$, $Y' = .718 + .0004X$.

two groups; instead, there is only a very gradual increase of such knowledge across the entire age spectrum, from a 65% competence level at the age of 12 to a 75% competence level at the age of 50. However, in no case is the linear relationship statistically significant. This rather slight, almost flat, trend of increasing competence with age contrasts markedly with the pattern described above for the other communities. What can explain such divergence? We propose that the active learning phase in Mosquito probably occurs at a younger age interval because of more intensive direct contact and interaction with forest biota at an earlier age. Unfortunately, due to the small size and dispersed distribution of settlements in this region, we were unable to find children below the age of 12 to interview in our forest plot. Therefore, we are unable to demonstrate this thesis with conclusive empirical data. Nevertheless, the relatively high competence displayed by the youngest individuals in our sample does point to this interpretation. It is significant to note that Mosquito is the least acculturated Hotï community in our sample, characterized by greater dominance of the foraging component of the subsistence economy, a more dispersed and seminomadic residential pattern, more restricted access to Western technology, the absence of nontraditional activities, and a traditional upland forest habitat. Meanwhile, the relatively lower competences recorded for younger individuals in Majagua, Iguana, and Kayamá suggest that the active learning phase is comparatively delayed. Here the settlements are larger and more sedentary, more dependent on Western goods, more engaged in nontraditional activities (e.g., school, church), and/or located in nontraditional habitats. The apparent ethnobotanical impact of these cultural changes is that young people have less direct contact with culturally encoded forest plants and that, therefore, the active learning process is extended through the teen years.

The predictive relationship between age and competence concerning food-use value, arguably the most important utilitarian category, is represented in Figure 2.5. The most salient result in this series is that the same general intersettlement pattern observed for generic classification prevails for food-use categorization, but not without some interesting variations. That is, Majagua, Iguana, and Kayamá display a distinctly dualistic trend of knowledge variation according to age, consisting of a statistically significant positive relationship in the younger cohort of respondents followed by the absence of such relationship in the older cohort. By contrast, Mosquito once again lacks any such dichotomous distinction or change in trend of food-use knowledge across the entire sample of respondents. Once more we assume that the divergent pattern observed at Mosquito points to aquisition of most knowledge about wild food plants in earlier childhood, conditioned by the more strongly nomadic hunting-gathering lifestyle actualized in a historically familiar biogeographical range. And while it may be said that Majagua, Iguana, and Kayamá share the same general pattern of age-dependent knowledge accumulation in the younger group and no further significant accumulation in the older group, it is also important to consider whether their respective hypothesized learning trends are effectively convergent or not. To this end, a test of equivalence of the regression models produced for the younger groups at Majagua, Iguana, and Kayamá was performed using indicator variables (see Chatterjee et al. 2000:132–140) and it was found that the learning trend observed at Kayamá is significantly different from the other two ($F = 4.68, p < .01$) while a paired comparison of Majagua and Iguana indicated that they come close to being (but are not quite) significantly different ($F = 3.22, .05 < p < .1$). Probing further this result, we are able to note several possibly significant intersettlement differences in terms of the following parameters: the amount of knowledge

a. Caño Mosquito

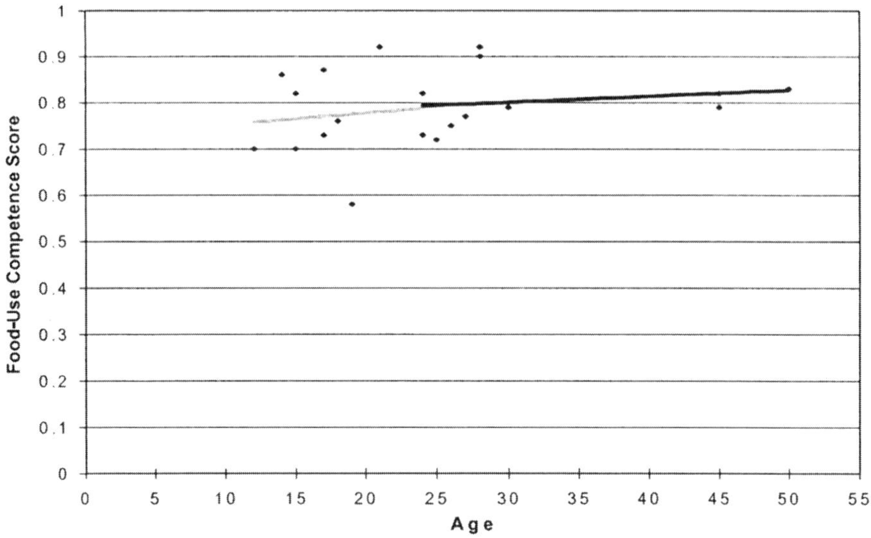

c. Caño Iguana

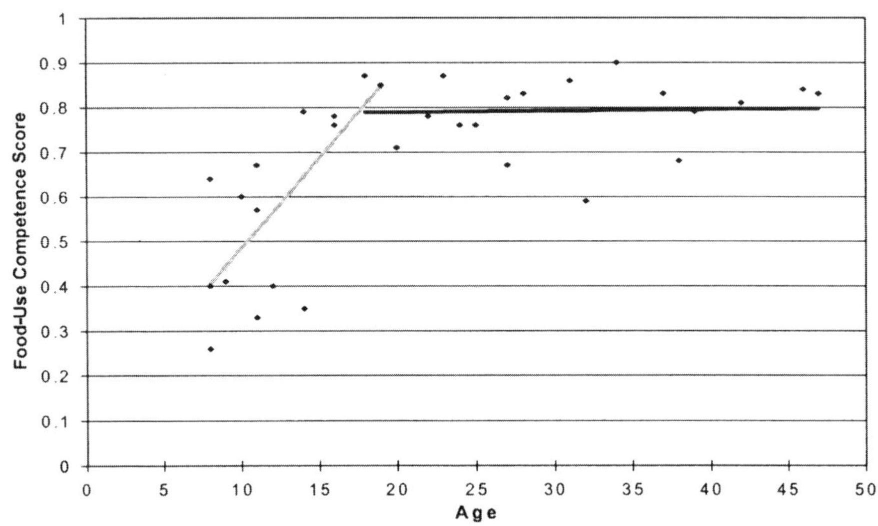

Figure 2.5. Food-use classification competence, by age. **a.** Caño Mosquito. Gray tendency line, representing the subgroup <25 years old: $n = 11, r = .101, r^2 = .01, p = .769, Y' = .727 + .0025X$. Black tendency line, representing the subgroup $\geq$ 23 years old: $n = 11, r = .197, r^2 = .039, p = .561, Y' = .762 + .001X$. **b.** Caño Majagua. Gray tendency, line representing the subgroup <21 years old: $n = 11, r = .729, r^2 = .531, p = .011, Y' = .392 + .026X$. Black tendency line, representing the subgroup 19 years old: $n = 10, r = -.423, r^2 = .179, p = .223, Y' = .922 - .003X$. **c.** Caño

b. Caño Majagua

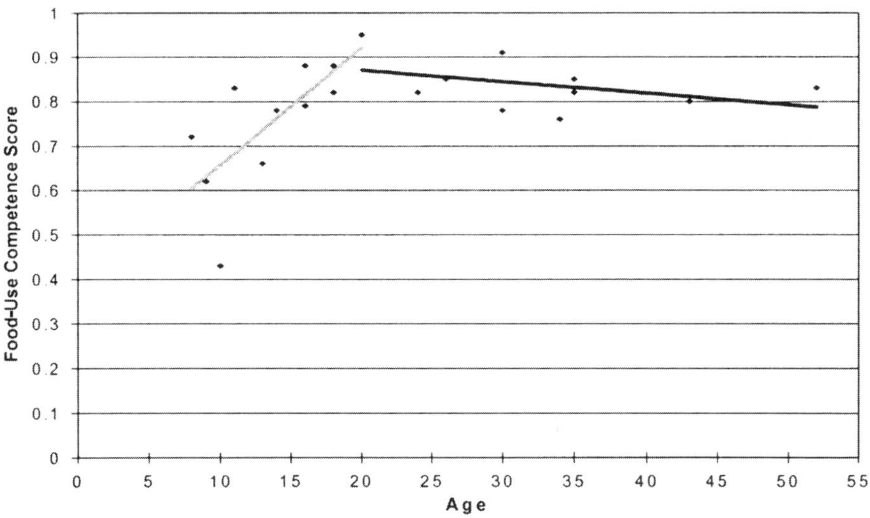

d. San José de Kayamá

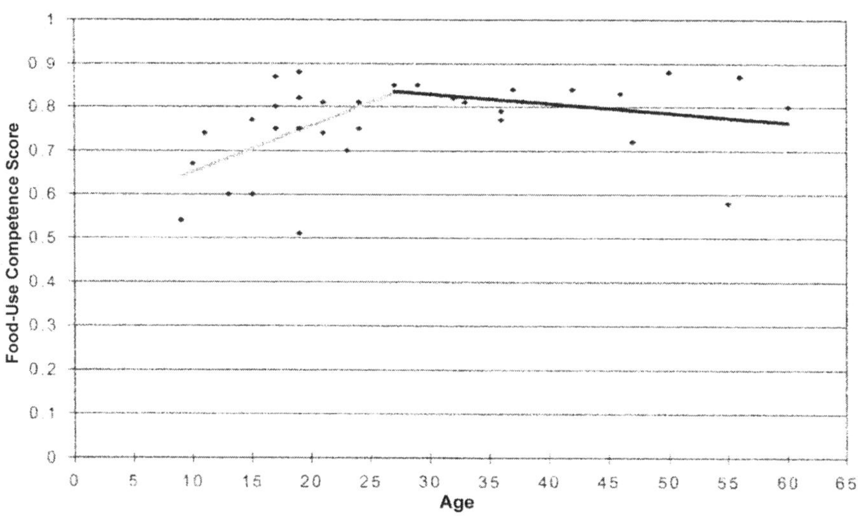

Iguana. Gray tendency line, representing the subgroup <20 years old: $n = 15$, $r = .711$, $r^2 = .505$, $p = .003$, $Y' = .084 + .04X$. Black tendency line, representing the subgroup 1&years old: $n = 19$, $r = .033$, $r^2 = .001$, $p = .893$, $Y' = .783 + .0003X$. **d.** San José de Kayamá. Gray tendency line, representing the subgroup <28 years old: $n = 19$, $r = .488$, $r^2 = .238$, $p = .034$, $Y' = .545 + .011X$. Black tendency line, representing the subgroup 26 years old: $n \geqq 14$, $r = -.299$, $r^2 = .089$, $p = .3$, $Y' = .894 - .002X$.

variation determined by age among the active learning group, the rate increase or slope of the learning curve, the age at which knowledge peaks, the peak competence level, and the knowledge trend in the older cohort. The explanatory significance of age for knowledge estimation was somewhat stronger in Iguana ($r^2 = .505, p = .003$) and Majagua ($r^2 = .531, p = .011$) than in Kayamá ($r^2 = .238, p = .034$). Here we remind the reader that the Iguana settlement corresponds to the traditional natural environment and Majagua, to the traditional social (i.e., nonmission) environment. The weaker relationship observed for Kayamá may be explained conversely by considering the effects of its being a nontraditional settlement situated within a nontraditional habitat. On one hand, the greater variance of individual competence scores from the statistically averaged regression line reflects the fact that knowledge of food plants in some individuals lags far behind that of others, as may be expected in a large community where various nontraditional community-based activities (e.g., school, church, organized games, social visits) now compete with traditional forest-based activities. On the other hand, a few members of the younger cohort appear to have equal or greater knowledge of food plants than do some members of the senior age group. Here we note that edibility is one of the first distinctive semantic features the child learns about a plant, sometimes even before names are reliably memorized, and, given the experimental nature of children, it should be expected that individuals motivated by hunger and amply exposed to this environment from an early age quickly learn what is edible. At the same time, most of the adults interviewed at this locality immigrated there as adults, which means after the age interval when most ethnobotanical learning occurs, leading us to assume that their learning capacity is not as great as it would have been had they arrived at an earlier age. We also observed that certain nonmajor food plants are commonly regarded as eaten by juveniles but not by mature adults, so the latter may not admit that some plants are theoretically edible due to their current age and social status.

The learning curve of edible plants among the younger cohort is most dramatic (i.e., steeper and faster) in Iguana, where relative competence rises from the low 40s at 8–9 years of age to the mid-80s at 18–19 years of age. A relatively high level of consensus is achieved fairly rapidly in this nucleated community located within the traditional upland forest environment. Younger children at Iguana display lower competence than do comparably aged individuals at Majagua and Kayamá, possibly because of the degraded vegetation in much of the river valley surrounding the settlement and the tendency for children under 10 years of age to wander within the relatively close quarters of the house compound and garden lands. However, by the age of 11–13 they have crossed a key threshold of social and spatial independence, are marriageable, and begin to go out on more distant camping trips on weekends and other holidays from the mission-organized activities. This period is marked by longer and farther ventures into the surrounding forest and, coincidentally, by a remarkable surge in their knowledge of edible forest trees.

The age at which knowledge of food plants peaks is lowest in Iguana (18–19), higher in Majagua (20–21), and highest in Kayamá (26–27). The prolonged learning phase implied by the latter may reflect diminished and delayed foraging opportunities as well as more dissonant patterns of knowledge transmission, given the more shallow historical occupation of this botanical environment. The relatively early peak registered in Majagua, despite its being a recently colonized habitat, may be explained by the maintenance of the traditional social and economic lifestyle, in which foraging trips in the forest constitute the primary pastimes of young and old alike. The trend line plotted for the Majagua youngsters also attains a higher peak (.92) than that modeled for the other two communities (.84). We speculate that this result may be conditioned on one hand by the lower relative number (25%) of taxa deemed edible in the Majagua plot that the corresponding amount observed in Iguana (44%), thus there are fewer edible taxa to learn, and on the other hand by the closer proximity of primary forest and greater participation in foraging activities by the young people at Majagua in comparison with their Kayamá counterparts.

The knowledge trend observed among the older cohort also varies by community. The regression line computed for Iguana is basically flat, whereas it is slightly (though not significantly) declining in the Majagua and Kayamá samples. We attribute this result to the fact that the latter are recently colonized, hence less familiar, habitats. We have already noted that older adults who immigrated there at more advanced ages are probably less inclined to expand their knowledge of newly encountered edible plants, which may explain why food-use competencies actually decline with age among the older cohort.

The distribution of individual knowledge of human medicinal plants according to age is portrayed in Figure 2.6. Human medicinal plants are defined here as plants that are utilized to cure specific human ailments (e.g., headache, fever) or to improve the general health, strength, and well-being of people. Here we observe a marked division in prevailing knowledge patterns, in which those of Majagua and Mosquito are contrasted with those of Iguana and Kayamá. In the former there is no statistically significant relationship to be found among any age grade, whereas in the latter age was found to determine competence score in both younger and older sample groups. The significant trends observed in Iguana and Kayamá are positive in the case of the younger subgroup and negative in the case of the older subgroup. Thus the results show that competence in medicinal plant knowledge definitely increases with age until the age of 18–19 in Iguana and 26–27 in Kayamá, after which there is a moderate but nevertheless significant decline in competence as people age. The quicker assimilation of this kind of EBK at Iguana is probably conditioned by the more traditional habitat type and by the longer term of continuous residence at this site in comparison with Kayamá. Although it may appear, upon superficial inspection, that younger people learn more about medicinal plants than do older people, we believe that this pattern really expresses a dual learning path characterized by the generalized learning of widely recognized and used therapeutic plants up through early adulthood and by the specialized learning of less commonly known plants during later adulthood. This hypothesis is sustained by the finding that the relationship between competence score and the number of plant taxa deemed to have medicinal value is strongly negative (Figure 2.7). In other words, the more medicinal plants a person learns the lower his or her competence score will be. Therefore, we suggest that this area of EBK is better described as specialist rather than group knowledge, or at least as a combination of both aspects. As mentioned before, the consensus analysis employed here is not designed to perform analyses of specialist-type knowledge. However, the model is capable of determining whether a particular domain of knowledge has a specialistic orientation (i.e., concentrated among a minority of individuals), precisely when the domain is defined by mostly negative criteria (i.e., most plants are judged as having no medicinal use value). Accordingly, it is important to point out that the average consensus scores in the category of human medicinal plants are higher than those registered for generic classification and food-use categories (Figure 2.3), but most of the "right" answers under this category chosen by the consensus analytical program that we employed (Anthropac) are in fact negative ones (i.e., the plant has no medicinal value). Even though the results from Mosquito and Majagua manifest no patterned relationship between age and knowledge of medicinal plants, there we also find that average consensus levels are high and that most program-determined "correct" answers are negative ones. We can only conclude that medicinal plant knowledge among the Hotï is at least partially the intellectual province of specialists. Such a conclusion is also consistent with our unsystematic observations, interactions, and dialogues with different Hotï while living among them: Certain individuals are indeed recognized by their peers to be more knowledgeable healers.

Turning now to the question of why knowledge of human medicinal plants should be significantly determined by age in Iguana and Kayamá and not in Majagua and Mosquito, we believe that the key factor is the availability of modern medicines and health-care services. The former pair of communities count on in situ modern medical services provided by or through the missionaries, whereas the latter pair are isolated from such services. The dual learning path of medicinal plants described above is more pronounced in Iguana and Kayamá due to the ascendant reliance on Western medicines and consequently diminished use of traditional herbal medicines. It is likely that such impact is more strongly felt by younger people who were brought up in the mission settlements and more exposed to Western medical technology during the period of their primary intellectual development.[11] Thus we interpret the results as indicating that most young people gradually acquire a basic knowledge of a few widely known medicinal plants but are less motivated to continue learning about more exotic ones as they grow older, thus converting the few older adults who do possess more knowledge of this kind into de facto specialists (even though the latter register lower competence scores). The pattern is muted in Majagua and Mosquito, probably because there is a more generalized need to learn about medicinal plants for lack of therapeutic alternatives. Consequently, such knowledge is more randomly or evenly distributed throughout the population, and no significant differences within or between age groups can be found.

The knowledge of hunting medicinal plants by age is depicted in Figure 2.8. This ethnobotanical category refers to plants that are ingested orally or nasally or bathed with in order to improve one's hunting success.[12] In distinct contrast to the previous ethnobotanical categories, no statistically significant relationships were observed in any age group in any community. We consider this lack of systematic relationship anywhere an indication of a high level of individuality in the knowledge and use of this category of plant. Although some plants are widely recognized to be effective hunting medicines, it is also normal to find that different hunters depend on highly idiosyncratic suites of plants for this purpose; indeed, this is an area of very active ethnobotanical experimentation. Somewhat surprisingly, there are no significant differences between male and female knowledge of hunting medicines that could be detected through consensus analysis.

Text continues on p. 69

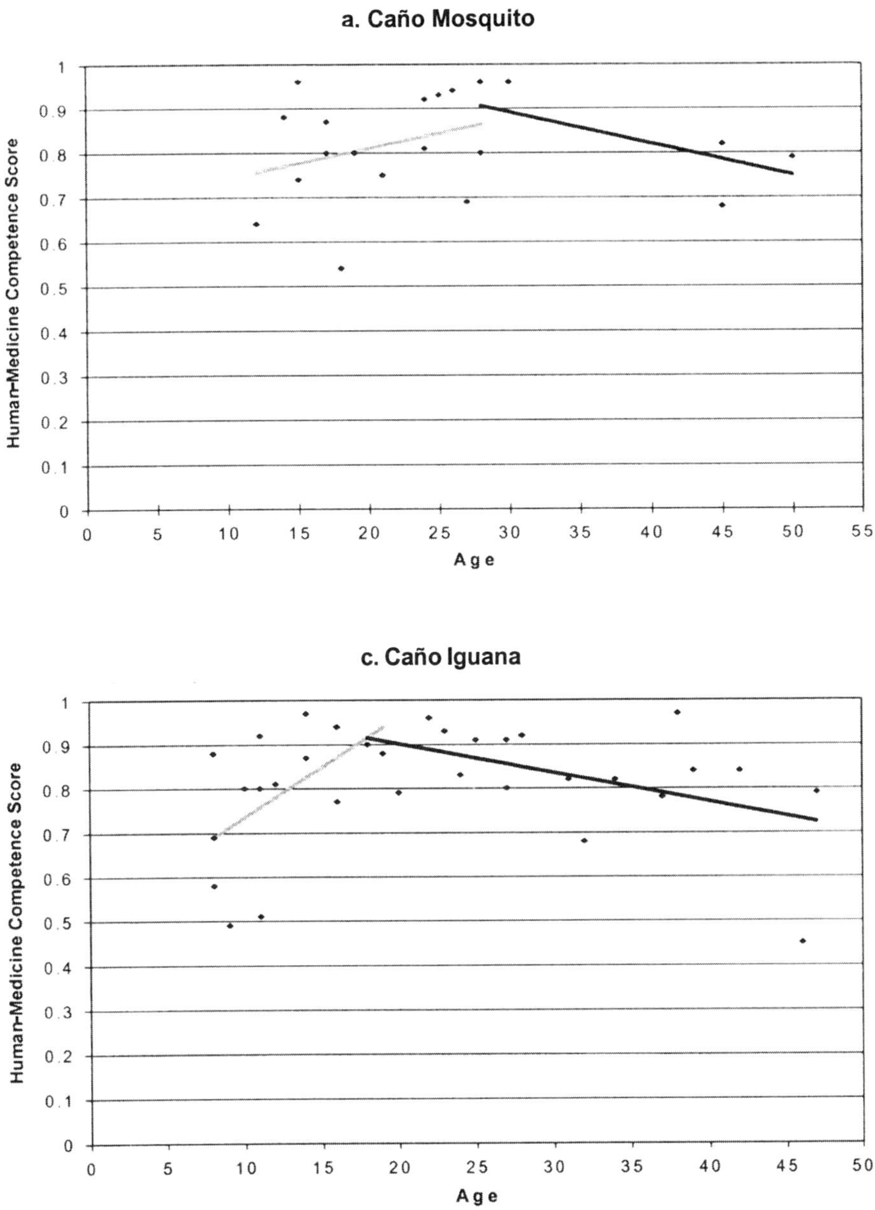

Figure 2.6. Human medicinal use classification competence, by age. **a.** Caño Mosquito. Gray tendency line, representing the subgroup <29 years old: $n = 16$, $r = .289$, $r^2 = .089$, $p = .262$, $Y' = .676 + .007X$. Black tendency line, representing the subgroup ≥27 years old: $n = 6$, $r = -.638$, $r^2 = .407$, $p = .173$, $Y' = 1.104 - .007X$. **b.** Caño Majagua. Gray tendency, line representing the subgroup <21 years old: $n = 11$, $r = .348$, $r^2 = .121$, $p = .295$, $Y' = .739 + .008X$. Black tendency line, representing the subgroup ≥19 years old: $n = 10$, $r = -.119$, $r^2 = .014$, $p = .744$, $Y' = .894 - .002X$.

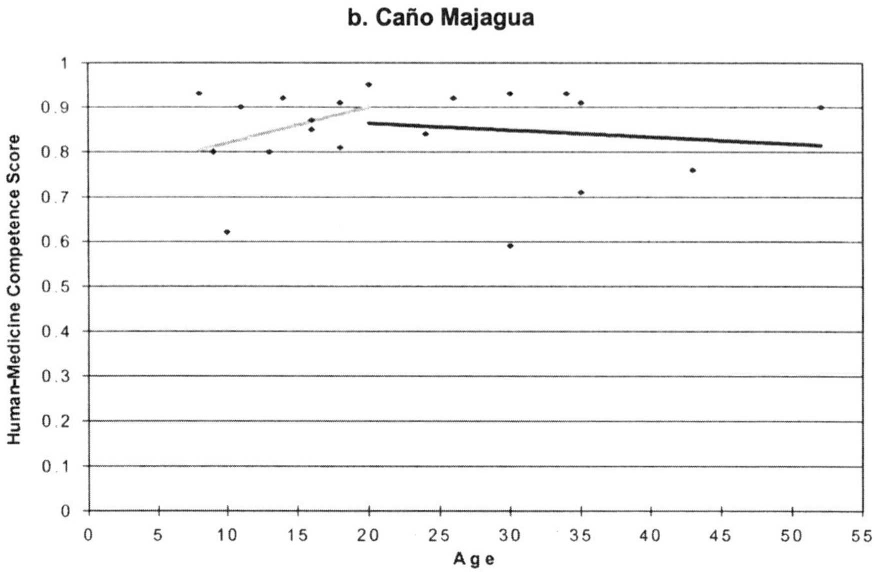

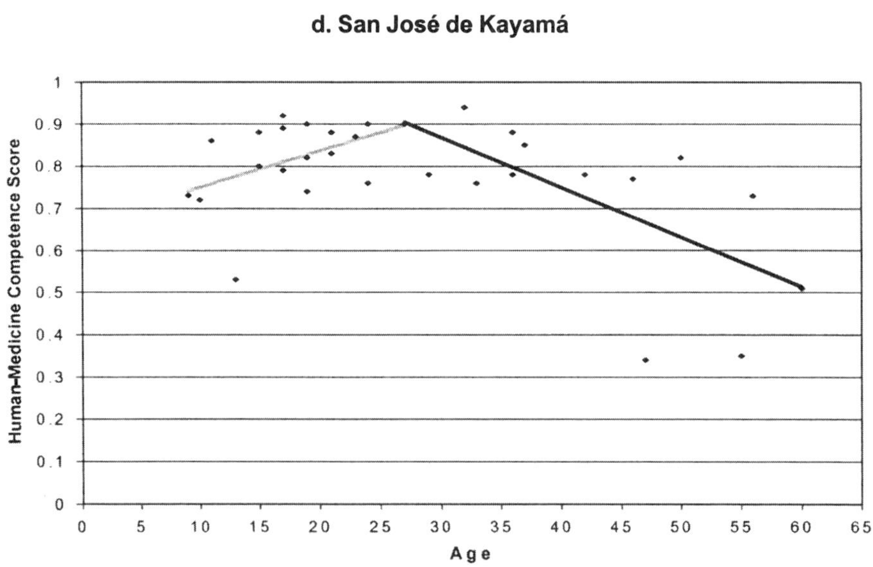

c. Caño Iguana. Gray tendency line, representing the subgroup <20 years old: $n = 15$, $r = .54$, $r^2 = .291$, $p = .038$, $Y' = .508 + .023X$. Black tendency line, representing the subgroup ≥18 years old: $n = 19$, $r = -.505$, $r^2 = .255$, $p = .028$, $Y' = 1.033 - .007X$. **d.** San José de Kayamá. Gray tendency line, representing the subgroup <28 years old: $n = 19$, $r = .456$, $r^2 = .208$, $p = .049$, $Y' = .663 + .009X$. Black tendency line, representing the subgroup ≥26 years old: $n = 14$, $r = -.627$, $r^2 = .393$, $p = .022$, $Y' = 1.223 - .012X$.

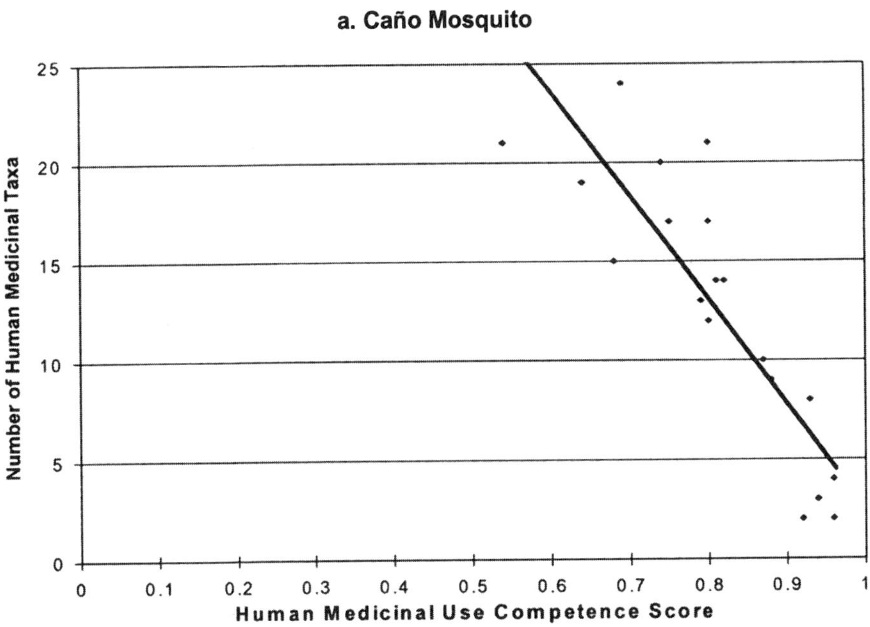

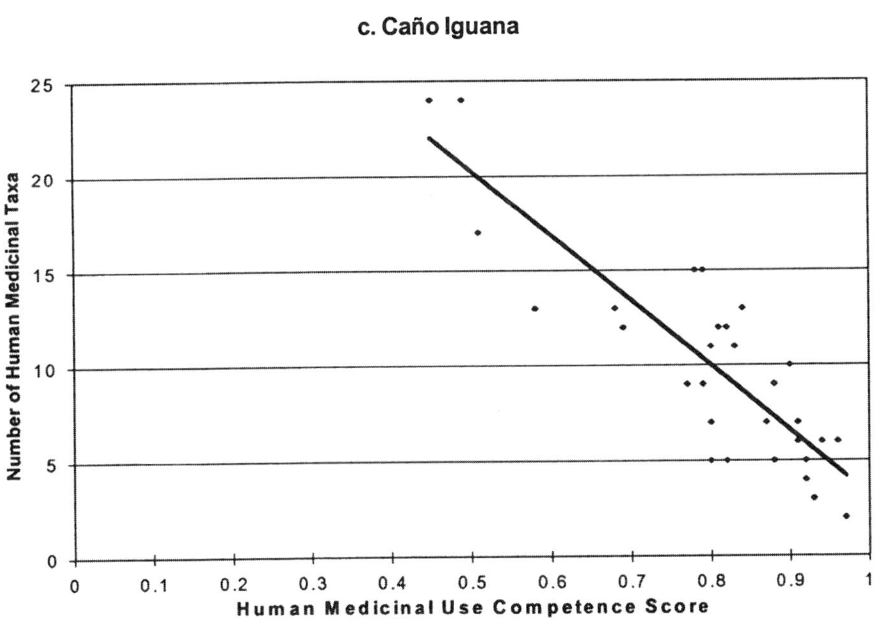

Figure 2.7. Regression of human medicinal use competence and number of human medicinal taxa named by each informant. **a.** Caño Mosquito: $n = 20$, $r = -.869$, $r^2 = .755$, $p = .000$,

b. Caño Majagua

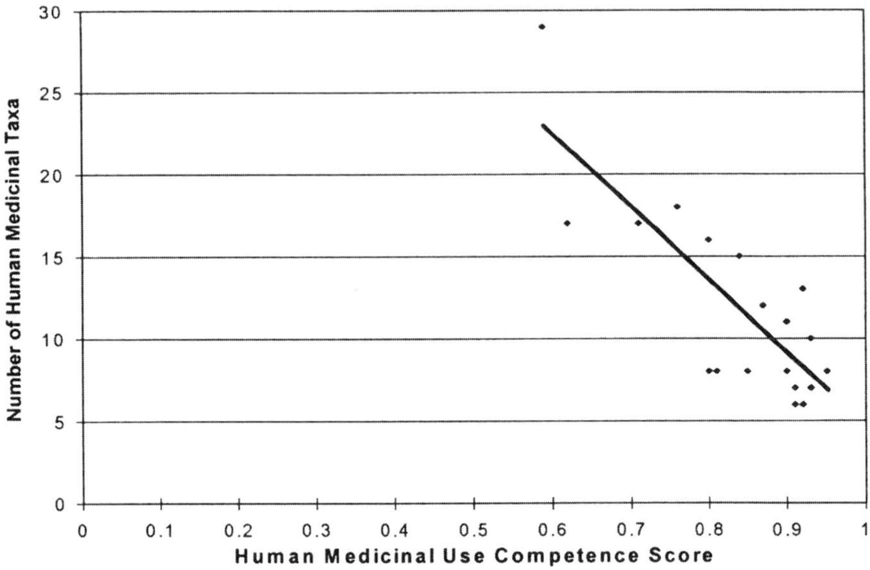

d. San José de Kayamá

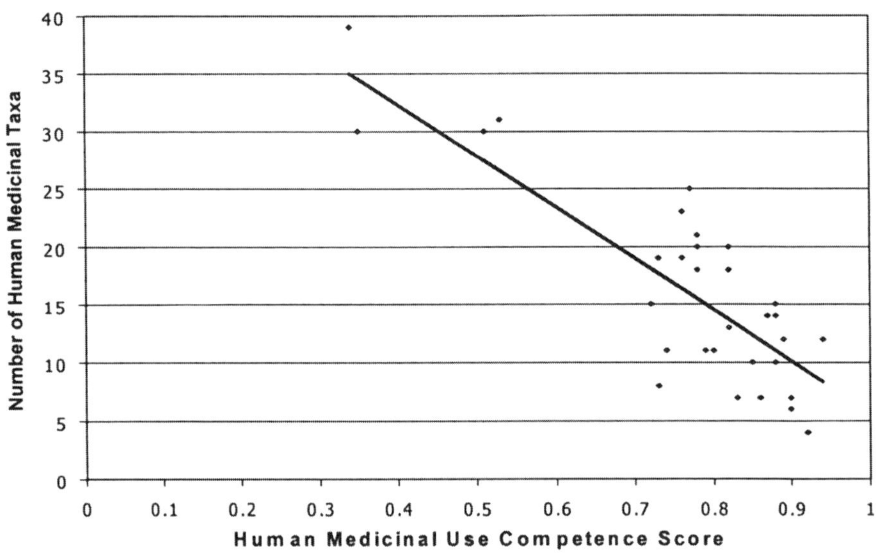

$Y' = 55.2 - 52.6X$. **b.** Caño Majagua: $n = 20$, $r = -.819$, $r^2 = .67$, $p = .000$, $Y' = 49.3 - 44.6X$. **c.** Caño Iguana: $n = 32$, $r = .54$, $r^2 = -.856$, $p = .000$, $Y' = 37.4 - 34.2X$. **d.** San José de Kayamá: $n = 32$, $r = -.802$, $r^2 = .644$, $p = .000$, $Y' = 50.1 - 44.4X$.

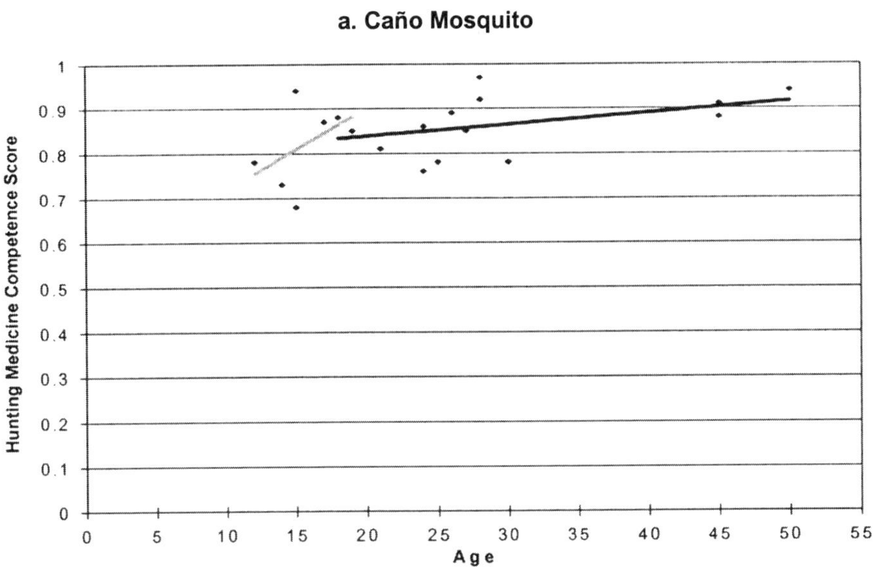

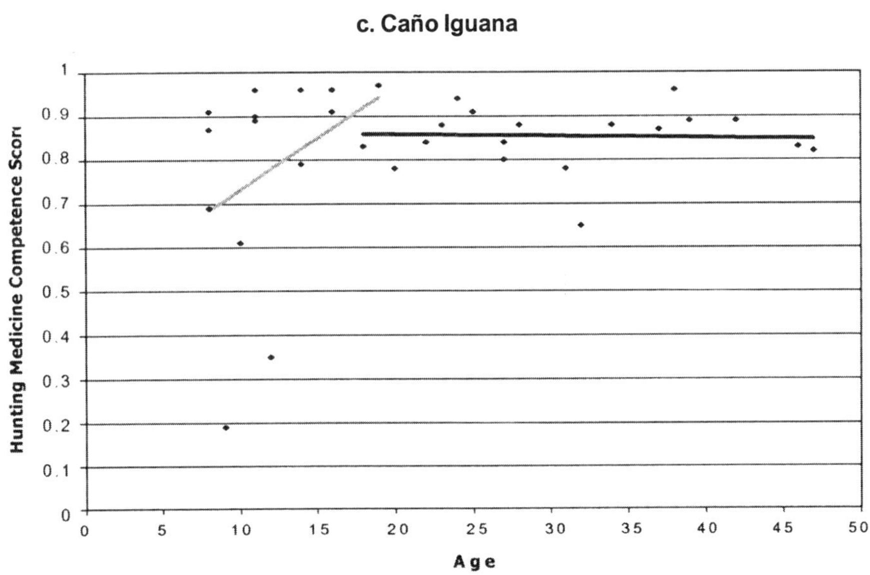

Figure 2.8. Hunting medicinal use classification competence, by age. **a.** Caño Mosquito. Gray tendency line, representing the subgroup <20 years old: $n = 8$, $r = .483$, $r^2 = .233$, $p = .226$, $Y' = .534 + .018X$. Black tendency line, representing the subgroup ≥18 years old: $n = 14$, $r = .411$, $r^2 = .169$, $p = .144$, $Y' = .787 + .003X$. **b.** Caño Majagua. Gray tendency, line representing the subgroup <20 years old: $n = 10$, $r = -.425$, $r^2 = .181$, $p = .221$, $Y' = .996 - .005X$. Black tendency line, representing the subgroup≥ 18 years old: $n = 12$, $r = -.137$, $r^2 = .019$, $p = .671$, $Y' = .894 - .002X$.

b. Caño Majagua

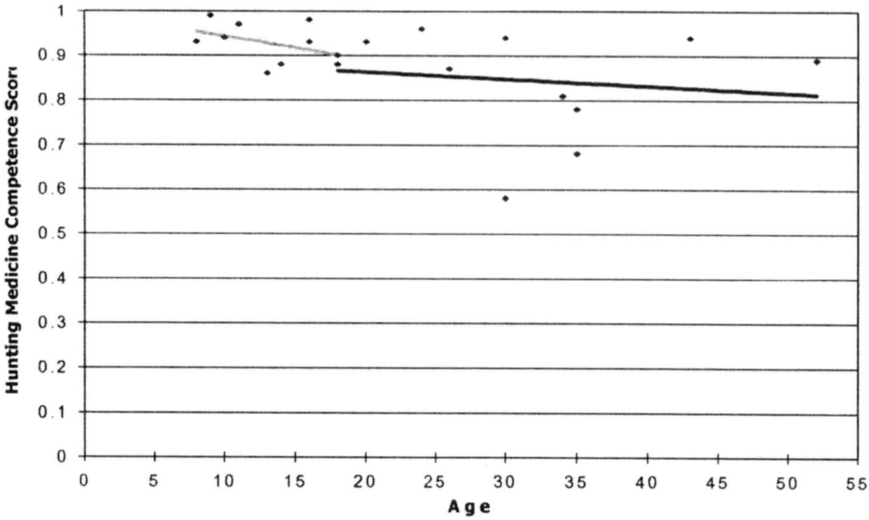

d. San José de Kayamá

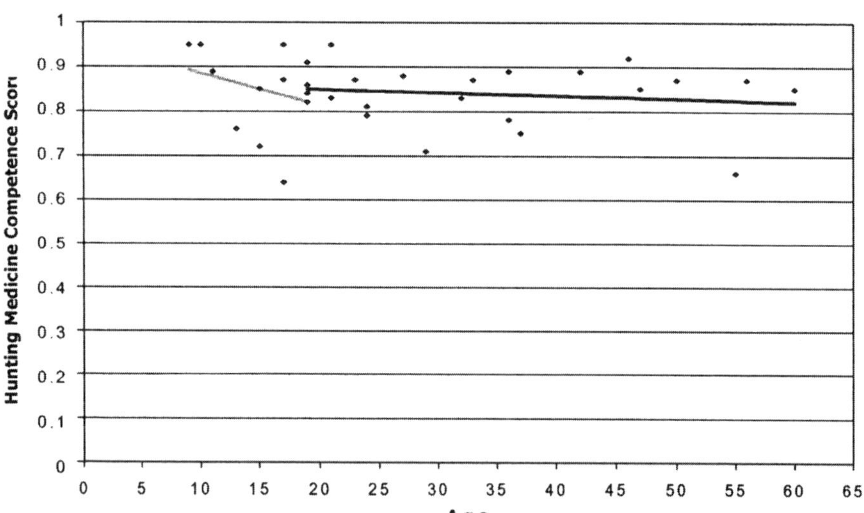

c. Caño Iguana. Gray tendency line, representing the subgroup <20 years old: $n = 15$, $r = .365$, $r^2 = .134$, $p = .18$, $Y' = .496 + .023X$. Black tendency line, representing the subgroup≥18 years old: $n = 19$, $r = -.049$, $r^2 = .002$, $p = .841$, $Y' = .867 - .0004X$. **d.** San José de Kayamá. Gray tendency line, representing the subgroup <20 years old: $n = 13$, $r = -.278$, $r^2 = .077$, $p = .358$, $Y' = .958 - .007X$. Black tendency line, representing the subgroup≥ 18 years old: $n = 23$, $r = -.138$, $r^2 = .019$, $p = .531$, $Y' = .863 - .0007X$.

Summary and Conclusions

The present study was intended to provide a quantitative description of the systematic variation of EBK within and between four Hotï communities. The empirical results were analyzed and interpreted by means of the cultural consensus analytical model and some standard descriptive and inferential statistical procedures. One of our major findings was that the semantic domain of Hotï EBK is anchored by a consensual core of widely shared categorical and attributive definitions, expressed through common local names, taxonomic classes, and use values attached to certain plant species. In this sense, the consensus model does seem to be an appropriate analytical framework to use in the present case. Nevertheless, we also found that the EBK system pertaining to this rather small ethnic population (<1000 people) scarcely beyond the first generation of contact with the Western world exhibits considerable variation across communities and is in fact very dynamic and responsive to the social and environmental changes encompassing them. Under the influence of external agents of change, the Hotï are undergoing major modifications of settlement pattern, ecological zone inhabited, subsistence focus, and typical activity routines, which in turn have important consequences for EBK change. In particular, it appears that settlement nucleation is leading to greater EBK convergence, whereas territorial migration and occupation of a new habitat promotes greater EBK divergence.

What are the diachronic implications of these countervailing trends? First, subcultural diversity of knowledge should be diminished and, hence, the capacity for self-generated evolutionary change reduced in areas where settlement nucleation is greater or more permanent. On the other hand, the mission-centered, nucleated settlements that face greater acculturation pressures are precisely those that, alone or in combination, constitute potential sources of new EBK variation and change that may be manifested at the local level. Acculturating forces of significance in this regard include personal contacts with outsiders, exposure to foreign cultural ideas and practices, increasing dependence on domesticated resource production, access to Western technological options, anthropogenic impact on the surrounding vegetation, and some degree of occupational role specialization that permits some individuals to concentrate on nontraditional activities. Second, local-level divergence of knowledge conducive to a more rapid rate of change will prevail to the extent that the regions inhabited are floristically more distant from the traditional home range. Colonization of a new territory may in fact enhance the overall fund of EBK due to contact with a greater range of diverse floristic communities, but there is also the danger that the traditional knowledge base (which we presume to be more empirically accurate, detailed, and useful than recently acquired knowledge because it was accumulated and refined over a longer time) will be seriously eroded, especially if these demographic shifts lead to serious depopulation of the traditional habitats. In the case of the Hotï, more than half of the population now lives in nontraditional ecological zones, and out-migration and settlement nucleation are continuing processes.

Another important result of our study was that subsistence and lifestyle changes appear to cause EBK changes by altering the time span and age range when most knowledge about plants is acquired by the individual. The active learning process is retarded in the more settled, more acculturated, more horticulture-dependent communities where children have less contact and exposure to forest plants than is the case in the less acculturated regions. A similar delayed learning curve is also observed in the communities that are newly settled in nontraditional ecological zones, presumably due to greater cognitive dissonance (i.e., lower consensus) about plant names, classifications, and uses.

Culturally and biologically appropriate conservation programs should take into account the prevailing trends of EBK change, such as those mentioned above, by anticipating what effect they may have on program goals and by incorporating an understanding of them into program design. For example, the case described here should make it clear that simple ethnic territorial demarcation and protection is not enough to ensure preservation of traditional plant knowledge and use habits. Hotï cultural (and by extension ethnobotanical) change is the product of the interaction of external influences (the arrival of outsiders and their material and symbolic accessories) and internal mechanisms (the Hotï desire to gain access to foreign knowledge, trade goods, and medical technology), which for the most part are beyond the scope of intervention capabilities and indeed may be irreversible. So intervention actions must work through the existing structures of change, such as incorporating ethnobotanical study components into school curricula or integrating herbal medicinal gardens into community-organized medical dispensaries. Furthermore, the creation of an adequate land reserve for the Hotï must allow for the fact that the population is currently undergoing a significant ecogeographical transition.

A proper understanding of the social organization and distribution of EBK within the target population is equally important for conservation initiatives in the sense of shedding light on several key issues: the cultural validity or acceptability of knowledge, the contextual constraints on knowledge use and communication, the social networks that constrain the flow of information, the appropriate social segments or actors to whom intervention measures should be addressed, the viable points of integration of local with scientific knowledge, and the fair sharing and distribution of resources that should devolve back to knowledge holders (cf. Bell 1979; Borofsky 1987; Benfer & Furbee 1996; Atran 1999). Thus, from the standpoint of salvage recovery and revitalization programs, it is crucial to establish: Who are the experts? Who are the novices? Who are the central social actors? Who are the peripheral social actors? What sectors of society are most affected by culture changes? When are the crucial learning phases? What factors are altering the processes of acquisition, assimilation, storage, and transmission? What types of knowledge are most endangered and therefore need to be included in memory banking or special educational programs? Which individuals should be targeted for remedial EBK training? In this vein the results of the present study lead us to make the following specific recommendations:

1. The learning process is slower and more drawn out in acculturated or unfamiliar settings but eventually reaches a comparatively high level in regard to shared knowledge. The implications for education programs are that the ethnobotanical component of a school curriculum should be offered as early as possible (in elementary school). In the nucleated settlements, camping behavior is crucial for eventual EBK acquisition, and, therefore, community activities (church, school) should be flexible enough to allow for frequent camping by family groups.

2. Schoolteachers and others who spend a good deal of time in nonaboriginal occupations exhibit lower EBK competence than do people who are more committed to traditional activities. Therefore, ethnobotanical education programs should not depend on those people who have been trained to teach Western types of knowledge (and who have embraced Western ways to a greater extent) but should, instead, specifically attempt to enlist the participation of people who depend more on traditional occupations.
3. Adults who have spent more time in a locality may be the best sources of information and instruction about plants, especially in regard to food plants, since they are more likely to learn more about them if they grew up in the area. Knowledge of food plants appears to be common knowledge and may therefore be an appropriate topic for inclusion in the general school curriculum through the elaboration of bilingual manuals (cf. Wilbert 1998).
4. Knowledge of human medicinal plants appears to be, at least in part, a specialist type of knowledge. Therefore, appropriate training programs should probably be tailored to individuals, as with the shaman's apprentice programs (cf. Plotkin 1993). At the present time, there is a greater need for this type of knowledge augmentation in Kayamá because the resident population has already become dependent on imported Western medicines; yet in recent years the supply and availability of such medicines have deteriorated (due mainly to government spending reductions). At the same time, there is an urgent need for modern medical services and training capabilities in the communities of Majagua and Mosquito in order to ensure their long-term viability.
5. Frequent experimentation and a high level of individuality characterize the knowledge and use of hunting medicinal plants. Thus the number of plant types exploited for this purpose by the community is a factor several times greater than the local population of hunters. The implication of this finding for culturally appropriate conservation planning is that communities require relatively large resource exploitation areas in order for all hunters to have access to the full suite of hunting medicinal plants that they know and use.

Acknowledgments

Special thanks to Robert Storrie for sharing his knowledge of the Hotï language with us and for other forms of assistance during our first field season. Luisa Maffi and Tom Carlson, Charles Peters, and Sally Kerr provided useful commentary on the first draft of this text, and Carenne Ludeña gave advice on statistical analysis. Carlos Quintero skillfully drew the map. The fieldwork was made possible by financial support from the Consejo Nacional de Investigaciones Científicas y Tecnológicas, the National Science Foundation, the Wenner-Gren Foundation for Anthropological Research, and the Instituto Venezolano de Investigaciones Científicas. Herbarium support was provided by the Herbario Victor Manuel Ovalles, the Herbario Nacional de Venezuela, and the Herbario Julian Steyermark del Estado Amazonas.

Notes

1. This not to say that they have become experts at counting money or understanding the exact monetary value of different denominations, but nevertheless they have become accustomed to using it as an exchange medium in their dealings with the missionaries.

2. All of the mission buildings at Kayamá are made from mostly local materials and based loosely on aboriginal architectural styles. This contrasts markedly with the appearance of Caño Iguana, where the missionaries' houses and outbuildings are made of imported materials and reflect Western design.

3. The creation of the arboleda was instigated by the missionaries, but the garden is presently managed and exploited by the local people.

4. This limitation of cultural consensus analysis should be kept in mind when considering the particular meaning of the key concepts associated with this method. Thus the concept of "correct answer" usually refers simply to the most popular answer, although weighted significance is given to the answer(s) provided by more knowledgeable respondents (i.e., those with higher demonstrated competence). In a similar vein, the notion of "competence" really refers to the extent to which one's answers form part of the majority opinion. We can, of course, imagine other criteria by which correctness and competence can be defined and still be valid from a cultural standpoint; that is, when the majority of people defer to and accept the knowledge of the few or the individual.

5. The statistical calculations and significance tests were performed using the SPSS 8.0 for Windows program.

6. Following Berlin's principles of ethnobiological classification (1992: 21–31), we make distinctions between folk-generic and folk-specific ranks of ethnobiological categorization and between ethnobiological categorization and ethnobiological nomenclature. Generic taxa are the most numerous categories of the system, the most basic level of categorization in the sense of being the first learned by children, taxonomically included in taxa of life-form rank or unaffiliated, inductively perceived (as a gestalt image), and labeled by primary (i.e., monomial) names. Specific taxa partition generic taxa into two or more members, are deductively perceived (i.e., distinguished on the basis of one to a few particular characters), and usually labeled by secondary (i.e., polynomial) names. Nomenclature refers to the linguistic expressions used to label plant categories. The Hotï themselves distinguish ethnobiological categorization from nomenclature by virtue of recognizing that the same category can be named by more than one linguistic form. Thus various consultants mentioned that they preferred a given name while someone else, sometimes even a close relative, preferred another name but also insisted that "it is the same (thing)." We were able to determine taxonomic ranks and nomenclatural equivalences mostly through structured elicitations and free listing exercises that were performed with a sample of approximately 20 Hotï informants in the early stages of the fieldwork. The exercises consisted of asking the informant to name serially and subordinately all of the plant taxa he or she could think of, starting with the Unique Beginner taxon and ending with lower ranking terminal taxa. For each name so elicited, the informant was asked to provide alternative names (if any) for the same category. Based on our analysis of Hotï biotaxonomic organization and synonymy, we segregated the plot interview response data according to whether the answers were the same or different in terms of folk-generic category (e.g., *hãlĩ wai*^{h}*ka* = *balo wai*^{h}*ka* = *waiyo wai*^{h}*ka*), folk-specific category (e.g., *hãlĩ wai*^{h}*ka* = *balo wai*^{h}*ka* ⊭ *waiyo wai*^{h}*ka*), or strict nomenclature (e.g., *hãlĩ wai*^{h}*ka* ≠ *balo wai*^{h}*ka* ≠ *waiyo wai*^{h}*ka*).

7. The significance of these results is reinforced by comparing them with similar results obtained elsewhere in Amazonia. Balée (1993) used the Jaccard coefficient of similarity (number

of shared species / sum of the total number of species in each plot minus the shared species) to evaluate the degree of botanical divergence among different plots in primary and fallow forest in Maranhão, Brazil. He found that the average "similarity for pairs of hectares of high forest/ fallow is only 10.9%" (p. 239). Using the Jaccard measure, the range of similiarity for our plots is 7.9–15.5%, for an average similarity of approximately 11.7%. In other words, the botanical divergence among the four forest plots described in our study is less than one percentage point greater than the divergence between primary and secondary forests found in Maranhão.

8. In order to mitigate the effect of interplot variability in plant density on the consensus test results, and thus enhance the intersettlement comparability of the results, we based the analysis on folk taxa and not on stems. Nevertheless, there resulted a considerable spread in the number of taxa per site, as follows: Mosquito, 141; Majagua, 132; Kayamá, 120; and Iguana, 88.

9. In order to limit the experimental error that may be caused by interviewer sex bias or distrust, the first author (who is male) interviewed most of the males and the second author (who is female), most of the females.

10. Perhaps of significance as well, more boys than girls were sent to boarding schools outside the community in earlier years of the mission.

11. One of our consultants in Caño Iguana explained that some of the plants formerly thought to have human medicinal use value are no longer considered as such because effective modern medicines are now available through the community dispensary, so there is no good reason to teach children about them.

12. Like Shepard (1998), we prefer to refer to the functionality of this class of plants as medicinal rather than magical because we cannot preclude the possibility that the plant substance may have a real psychoactive or physiological effect on the "patient" that enhances his or her ability to hear or track game, shoot straighter or farther, reduce fatigue, and so forth. By the same token, we do not assume a priori that the curative effect of human medicinal plants is physiological or psychological or some combination of both.

Literature Cited

Adu-Tutu, M., Y. Afful, K. Asante-Appiah, D. Liberman, J. B. Hall & M. Elvin-Lewis. 1979. Chewing stick usage in southern Ghana. Economic Botany 33: 320–328.

Alcorn, J. B. 1984. Huastec Mayan Ethnobotany. Austin: University of Texas Press.

———. 1995. Economic botany, conservation, and development: What is the connection? Annals of the Missouri Botanical Garden 82: 34–46.

———. 1996. Is biodiversity conserved by indigenous peoples? Pp. 233–238 *in* S. K. Jain (ed.), Ethnobiology in Human Welfare. New Delhi: Deep Publications.

Altieri, M. A. & L. C. Merrick. 1987. In situ conservation of crop genetic resource through maintenance of traditional farming systems. Economic Botany 41(1): 86–96.

Anderson, A. B. 1988. Use and management of native forests dominated by agai palm (*Euterpe oleracea* Mart.) in the Amazon estuary. Pp. 144–154 *in* M. J. Balick (ed.), The Palm—Tree of Life: Biology, Utilization, and Conservation. Advances in Economic Botany, 6. Bronx: New York Botanical Garden.

——— & D. A. Posey. 1989. Management of a tropical scrub savanna by the Gorotire Kayapó of Brazil. Pp. 159–173 *in* D. A. Posey & W. Balée (eds.), Resource Management in Amazonia: Indigenous and Folk Strategies. Advances in Economic Botany, 7. Bronx: New York Botanical Garden.

Atran, S. 1999. Managing the Maya Commons: The Value of Local Knowledge. Pp. 190–214 *in* V. D. Nazarea (ed.), Ethnoecology: Situated Knowledge / Located Lives. Tucson: University of Arizona Press.

Balée, W. 1993. Indigenous transformation of Amazonian forests. L'Homme 126–128: 231–254.
———. 1994. Footprints of the Forest: Ka'apor Ethnobotany—The Historical Ecology of Plant Utilization by an Amazonian People. New York: Columbia University Press.
Balick, M. J. & P. A. Cox. 1997. Plants, People, and Culture: The Science of Ethnobotany. New York: Scientific American Library.
Begossi, A. 1996. Use of ecological methods in ethnobotany: Diversity indices. Economic Botany 50(3): 280–289.
Bell, M. 1979. The exploitation of indigenous knowledge or the indigenous exploitation of knowledge: Whose use of what for what? IDS Bulletin 10(2): 44–50.
Benfer, R., Jr. & L. Furbee. 1996. Can indigenous knowledge be brokered without scientific understanding of the community structure and distribution of that knowledge? Indigenous Knowledge Monitor 4(2). <http://www.nuffic.nl/ciran/ikdm/4-2/articles/benfer.html>.
Bennett, B. C. 1991. Aspectos económicos y sociológicos de la etnobotánica. Pp. 359–365 *in* M. Ríos & H. B. Pederson (eds.), Las plantas y el hombre. Quito: Abya-Yala.
———. 1992. Plants and people of the Amazonian rainforests. BioScience 42: 599–607.
Berlin, B. 1992. Ethnobiological Classification: Principles of Categorization of Plants and Animals in Traditional Societies. Princeton, N.J.: Princeton University Press.
———. 1999. Lexical Reflections on the Cultural Importance of Medicinal Plants among Tzotzil and Tzeltal Maya. Pp. 12–23 *in* T. Gragson & B. Blount (eds.), Ethnoecology: Knowledge, Resources, and Rights. University of Georgia Press.
Bonet, M. A., C. Blanché & J. Vallés X. 1992. Ethnobotanical study in River Tenes valley (Catalonia, Iberian Peninsula). Journal of Ethnopharmacology 37: 205–212.
Borofsky, R. 1987. Making History: Pukapukan and Anthropological Constructions of Knowledge. Cambridge: Cambridge University Press.
Boster, J. 1980. How the Exceptions Prove the Rule: An Analysis of Informant Disagreement in Aguaruna Manioc Identification. Ann Arbor, Mich.: University Microfilms International.
———. 1984. Classification, cultivation, and selection of Aguaruna cultivars of *Manihot esculenta* (Euphorbiaceae). Pp. 34–47 *in* G. T. Prance & J. A. Kallunki (eds.), Ethnobotany in the Neotropics. Advances in Economic Botany, 1. Bronx: New York Botanical Garden.
———. 1986. Exchange of varieties and information between Aguaruna manioc cultivators. American Anthropologist 88(2): 428–436.
Brownrigg, L. A. 1985. Native cultures and protected areas: Management options. Pp. 33–44 *in* J. McNeely & D. Pitt (eds.), Culture and Conservation: The Human Dimension in Environmental Planning. London: Croom Helm.
Brush, S. B. 1986. Genetic diversity and conservation in traditional farming systems. Journal of Ethnobiology 6(1): 151–167.
Bulmer, R. N. H. 1982. Traditional conservation practices in Papua New Guinea. Pp. 59–77 *in* L. Morauta, J. Pernetta & W. Heaney (eds.), Traditional Conservation in Papua New Guinea: Implications for Today. Boroko, Papua New Guinea: Institute of Applied Social and Economic Research.
Chatterjee, S., A. S. Hadi & B. Price. 2000 Regression Analysis by Example. Third Edition. New York: John Wiley& Sons.
Coppens, W. 1975. Contribución al estudio de las actividades de subsistencia de los Hotis del río Kaima. Boletín Indigenista Venezolano, n.s. 16(12): 65–78.
———. 1983. Los Hoti. Pp. 2: 243–302 *in* W. Coppens (ed.), Los aborígenes de Venezuela. Caracas: Fundación La Salle / Monte Avila Editores.
——— & P. Mitrani. 1974. Les Indiens Hoti: Compte rendu de missions. L'Homme 14(3–4): 131–142.
Corradini, H. 1973. Los indios chicanos. Venezuela Misionera 35(405): 6–9; 35(406): 42–44; 35(407): 88–91.

Cunningham, A. 1993. Ethics, Ethnobiological Research, and Biodiversity. Gland, Switzerland: WWF International.

Dasmann, R. F. 1991. The importance of cultural and biological diversity. Pp. 7–15 *in* M. L. Oldfield & J. B. Alcorn (eds.), Biodiversity: Culture, Conservation, and Ecodevelopment. Boulder, Colo.: Westview Press.

Early, J. D. & J. F. Peters. 1990. The Population Dynamics of the Mucajai Yanomami. New York: Academic Press.

Eibl-Eibesfeldt, I. 1973. Die Waruwádu (Yuwana), ein kurzlich entdeckter, noch unerforschter Indianerstamm Venezuelas. Anthropos 68(1–2): 137–144.

Friedman, J., Z. Yaniv, A. Dafni & D. Palewitch. 1986. A preliminary classification of the healing potential of medicinal plants, based on a rational analysis of an ethnopharmacological field survey among the Bedouins in the Negev Desert, Israel. Journal of Ethnopharmacology 16: 275–287.

Gough, A. 1999. Module 5: Indigenous knowledge for the environment. *In* J. Fien, D. Heck & J. Ferreira (eds.), Learning for a Sustainable Environment: A Project of the UNESCO–Asia Pacific Centre of Educational Innovation and Development and Griffith University. <http://www.ens.gu.edu.au/ciree/LSE/mod5.htm>.

Greig-Smith, P. 1983. Quantitative Plant Ecology. Ed. 3. Berkeley: University of California Press.

Guarisma P., V. 1974. Los Hoti: Introducción etno-lingüística. Thesis, Universidad Central de Venezuela.

Howell, N. 1979. Demography of the Dobe !Kung. New York: Academic Press.

Huber, O. & R. Riina. 1997. Glosario fitoecológico de las Américas, vol. 1: América del Sur: Paises hispanoparlantes. Caracas: Fundación Instituto Botánico de Venezuela / UNESCO.

Irvine, D. 1989. Succession management and resource distribution in an Amazonian rain forest. Pp. 223–237 *in* D. A. Posey & W. Balée (eds.), Resource Management in Amazonia: Indigenous and Folk Strategies. Advances in Economic Botany, 7. Bronx: New York Botanical Garden.

Jangoux, J. 1971. Observations on the Hoti Indians of Venezuela. Unpublished manuscript.

Johns, T., J. O. Kokwaro & E. K. Kimanani. 1990. Herbal remedies of the Luo of Siaya District, Kenya: Establishing quantitative criteria for consensus. Economic Botany 44(3): 369–381.

Kainer, K. A. & M. L. Duryea. 1992. Tapping women's knowledge: Plant resource use in extractive reserves, Acre, Brazil. Economic Botany 46(4): 408–425.

Keel, S., A. Gentry & L. Spinzi. 1993. Using vegetation analysis to facilitate the selection of conservation sites in eastern Paraguay. Conservation Biology 7(1): 66–75.

Maffi, L. 1998. Language: A resource for nature. Nature and Resources 34(4): 12–21.

——— (ed.). 2001. On Biocultural Diversity: Linking Language, Knowledge, and the Environment. Washington, D.C.: Smithsonian Institution Press.

Martin, G. J. 1995. Ethnobotany: A Methods Manual. London: Chapman & Hall.

McNeely, J. & D. Pitt. 1985. Culture: The missing element in conservation and development. Pp. 1–9 *in* J. McNeely & D. Pitt (eds.), Culture and Conservation: The Human Dimension in Environmental Planning. London: Croom Helm.

Meilleur, B. A. 1994. In search of "Keystone Societies." Pp. 259–279 *in* N. L. Etkin (ed.), Eating on the Wild Side: The Pharmacologic, Ecologic, and Social Implications of Using Noncultigens. Tucson: University of Arizona Press.

Moran, E. 1993. Managing Amazonian variability with indigenous knowledge. Pp. 753–766 *in* C. M. Hladik, A. Hladik, O. F. Linares, H. Pagezy, A. Semple & M. Hadley (eds.), Tropical Forests, People and Food: Biocultural Interactions and Applications to Development. Man and the Biosphere Series, 13. Paris: UNESCO and the Parthenon Publishing Group.

Morauta, L., J. Pernetta & W. Heaney (eds.). 1982. Traditional Conservation in Papua New Guinea: Implications for Today. Boroko, Papua New Guinea: Institute of Applied Social and Economic Research.

Mühlhäusler, P. 1995. The interdependence of linguistic and biological diversity. Pp. 154–161 *in* D. Myers (ed.), The Politics of Multiculturalism in the Asia/Pacific. Darwin, Australia: Northern Territory University Press.

Nabhan, G. 1997. Cultures of Habitat: On Culture, Nature, and Story. Washington, D.C.: Counterpoint.

Nazarea, V. D. 1998. Cultural Memory and Biodiversity. Tucson: University of Arizona Press.

Nazarea-Sandoval, V. D. 1995. Local Knowledge and Agricultural Decision Making in the Philippines. Ithaca, N.Y.: Cornell University Press.

Nietschmann, B. Q. 1992. The Interdependence of Biological and Cultural Diversity. Occasional Paper 21. Olympia, Wash.: Center for World Indigenous Studies.

Nigh, R. & J. Nations. 1980. Tropical rainforests. Bulletin of the Atomic Scientists 36(3): 12–19.

Oldfield, M. L. & J. Alcorn. 1987. Conservation of traditional agroecosystems. BioScience 37(3): 199–208.

——— & ——— (eds.). 1991. Biodiversity: Culture, Conservation, and Ecodevelopment. Boulder, Colo.: Westview Press.

Padoch, C., J. Chota I., W. DeJong & J. Unruh. 1985. Amazonian agroforestry: A market-oriented system in Peru. Agroforestry Systems 3: 47–58.

Peters, C. M. 1996. Beyond nomenclature and use: A review of ecological methods for ethnobotanists. Pp. 241–276 *in* M. N. Alexiades (ed.), Selected Guidelines for Ethnobotanical Research: A Field Manual. Advances in Economic Botany, 10. Bronx: New York Botanical Garden.

———, A. H. Gentry & R. Mendelsohn. 1989. Valuation of an Amazonian rainforest. Nature 339: 655–656.

Phillips, O. 1996. Some quantitative methods for analyzing ethnobotanical knowledge. Pp. 171–197 *in* M. N. Alexiades (ed.), Selected Guidelines for Ethnobotanical Research: A Field Manual. Advances in Economic Botany, 10. Bronx: New York Botanical Garden.

——— & A. H. Gentry. 1993a. The useful plants of Tambopata, Peru, I: Statistical hypotheses tests with a new quantitative technique. Economic Botany 47(1): 15–32.

——— & ———. 1993b. The useful plants of Tambopata, Peru, II: Additional hypothesis testing in quantitative ethnobotany. Economic Botany 47(1): 33–43.

———, ———, C. Reynel, P. Wilkin & C. Gálvez-Durand B. 1994. Quantitative ethnobotany and Amazonian conservation. Conservation Biology 8(1): 225–248.

Plotkin, M. J. 1988. Ethnobotany and conservation in the Guianas: The Indians of southern Suriname. Pp. 87–109 *in* F. Almeda & C. Pringle (eds.), Tropical Rainforests: Diversity and Conservation. San Francisco: California Academy of Sciences.

———. 1993. Tales of a Shaman's Apprentice: An Ethnobotanist Searches for New Medicines in the Amazon Rain Forest. New York: Viking.

——— & L. Famolare (eds.). 1992. Sustainable Harvest and Marketing of Rain Forest Products. Washington, D.C.: Island Press.

Posey, D. A. 1990a. Introduction to ethnobiology: Its implications and applications. Pp. 1: 1–8 *in* D. A. Posey & W. L. Overal (eds.), Ethnobiology: Implications and Applications. Belém, Brazil: Museu Paraense Emílio Goeldi.

———. 1990b. The application of ethnobiology in the conservation of dwindling natural resources: Lost knowledge or options for the survival of the planet. Pp. 1: 47–59 *in* D. A. Posey & W. L. Overal (eds.), Ethnobiology: Implications and Applications. Belém, Brazil: Museu Paraense Emílio Goeldi.

——— (ed.). 1999. Cultural and Spiritual Values of Biodiversity. London and Nairobi: Intermediate Technology Publications and UNEP.

——— & W. Balée (eds.). 1989. Resource Management in Amazonia: Indigenous and Folk Strategies. Advances in Economic Botany, 7. Bronx: New York Botanical Garden.

——— & G. Dutfield. 1996. Beyond Intellectual Property: Toward Traditional Resource Rights for Indigenous Peoples and Local Communities. Ottawa: International Development Research Centre.

——— & W. L. Overal (eds.). 1990. Ethnobiology: Implications and Applications. 2 vols. Belém, Brazil: Museu Paraense Emílio Goeldi.

———, J. Frechione, J. Eddins, L. F. da Silva, D. Myers, D. Case & P. MacBeath. 1984. Ethnoecology as applied anthropology in Amazonian development. Human Organization 43: 95–107.

Prance, G. T. 1991. What is ethnobotany today? Journal of Ethnopharmacology 32: 209–216.

———, W. Balée, B. M. Boom & R. L. Carneiro. 1987. Quantitative ethnobotany and the case for conservation in Amazonia. Conservation Biology 1(4): 296–310.

Redford, K. & C. Padoch (eds.). 1992. Conservation of Neotropical Forests. New York: Columbia University Press.

Romney, A. K., S. C. Weller & W. H. Batchelder. 1986. Culture as consensus: A theory of culture and informant accuracy. American Anthropologist 88(2): 313–318.

———, W. H. Batchelder & S. C. Weller. 1987. Recent applications of cultural consensus theory. American Behavioral Scientist 31(2): 163–177.

Schultes, R. E. 1979. The Amazon as a source of new economic plants. Economic Botany 33: 259–266.

———. 1994. The importance of ethnobotany in environmental conservation. American Journal of Economics and Sociology 53(2): 202–206.

——— & R. F. Raffauf. 1990. The Healing Forest: Medicinal and Toxic Plants of the Northwest Amazonia. Portland, Ore.: Dioscorides Press.

Shepard, G. 1998. Gift of the harpy eagle: Hunting medicines of the Machiguenga. South American Explorer 51(Spring): 9–11, 21.

Steward J. & L. Faron. 1959. Native Peoples of South America. New York: McGraw-Hill.

Stoffle, R. W., D. B. Halmo, M. J. Evans & J. E. Olmsted. 1990. Calculating the cultural significance of American Indian plants: Paaiute and Shoshone ethnobotany at Yucca Mountain, Nevada. American Anthropologist 92: 416–432.

Toledo, V. M. 1992. What is ethnoecology? Origins, scope and implications of a rising discipline. Etnoecológica 1(1): 5–21.

———. 1994. Biodiversity and cultural diversity in Mexico. Different Drummer, Summer: 16–19.

Trotter, R. T. & M. H. Logan. 1986. Informant consensus: A new approach for identifying potentially effective medicinal plants. Pp. 91–112 *in* N. Etkin (ed.), Plants in Indigenous Medicine and Diet: Biobehavioral Approaches. Bedford Hills, N.Y.: Redgrave.

Voeks, R. A. 1996. Tropical forest healers and habitat preference. Economic Botany 50(4): 381–400.

Warren, D. M., L. J. Slikkerveer & D. Brokensha (eds.). 1995. The Cultural Dimension of Development: Indigenous Knowledge Systems. London: Intermediate Technology Publications.

Weller, S. C. & A. K. Romney. 1988. Systematic Data Collection. Newbury Park, N.J.: Sage.

Wilbert, W. 1998. Odau Yarokota: Plantas medicinales Warao. Instituto Caribe de Antropología y Sociología, Monografía 45. Caracas: Fundación La Salle de Ciencias Naturales.

Wilcox, B. A. & K. N. Duin. 1995. Indigenous cultural and biological diversity: Overlapping values of Latin American ecoregions. Cultural Survival Quarterly 18(4): 49–53.

Zent, E. L. 1999. Etnobotánica Hotï: Explorando las interacciones entre la flora y el ser humano del Amazonas venezolano. Ph.D. diss., University of Georgia.

——— & S. Zent. 2004. Amazonian Indians as ecological disturbance agents: The Hotï of the Sierra de Maigualida, Venezuelan Guayana. Pp. 78–111 *in* T. J. S. Carlson & L. Maffi (eds.), Ethnobotany and Conservation of Biocultural Diversity. Advances in Economic Botany, 15. Bronx: New York Botanical Garden.

——— & ———. In press. Floristic composition, structure, and diversity of four forest plots in the Sierra Maigualida, Venezuelan Guayana. Biodiversity and Conservation.

Zent, S. 1995. Clasificación, explotación y composición de bosques secundarios en el Alto Río Cuao, Estado Amazonas, Venezuela. Pp. 79–113 *in* H. D. Heinen, J. J. San José & H. Caballero A. (eds.), Naturaleza y ecología humana en el neotrópico. Scientia Guaianae, 5. Caracas: REFOLIT.

———. 1996. Behavioral orientations toward ethnobotanical quantification. Pp. 199–239 *in* M. N. Alexiades (ed.), Selected Guidelines for Ethnobotanical Research: A Field Manual. Advances in Economic Botany, 10. Bronx: New York Botanical Garden.

———. 1999a. The quandary of conserving ethnoecological knowledge: A Piaroa example. Pp. 90–124 *in* T. Gragson & B. Blount (eds.), Ethnoecology: Knowledge, Resources, and Rights. Athens: University of Georgia Press.

———. 1999b. Los elementos paradigmáticos de la ecología histórica: Pautas para la renovación de la etnobotánica. Memorias del Instituto de Biología Experimental 2: 27–30.

———. 2001. Acculturation and ethnobotanical knowledge loss among the Piaroa of Venezuela: Demonstration of a quantitative method for the empirical study of TEK change. Pp. 190–211 *in* L. Maffi (ed.), On Biocultural Diversity: Linking Language, Knowledge, and the Environment. Washington, D.C.: Smithsonian Institution Press.

3

Amazonian Indians as Ecological Disturbance Agents: The Hotï of the Sierra de Maigualida, Venezuelan Guayana

Egleé López-Zent and Stanford Zent

Introduction

During the past two decades, several authors have advanced the notion that the Amazonian-Orinoco lowlands consist of eminently anthropogenic rather than pristine ecosystems (Smith 1980; Posey 1983a, 1984; Stocks 1983; Eden et al. 1984; Clark & Uhl 1987; Denevan & Padoch 1987; Balée 1988, 1989, 1993; Anderson & Posey 1989; Irvine 1989; Smole 1989; Denevan 1992a; Roosevelt 1992; Gragson 1995; Heinen et al. 1995; Junk 1995; Zent 1997). According to this viewpoint, the historical hand of human intervention is evident in various key aspects of contemporary landscapes, including vegetation, fauna, soils, topography, and hydrology. In a previous article, E. L. Zent (1998) reviewed several case studies in support of this argument and concluded that indigenous human populations that maintain a traditional (i.e., subsistence-dominated, low technology, low population density) lifestyle not only have a nondetrimental impact on their surrounding environment but also are agents of creative disturbance in dynamically configured Amazonian ecosystems. The theoretical perspective adopted in the previous study and continued in the present one conceives of the human-nature relationship as art (in its literal meaning from the Latin *ars,* "ability," "expertise," "skill"), which is creative in the sense of triggering ecosystemic outcomes beyond those aimed solely at satisfying needs.

The representation of Amazonia as an anthropogenic habitat remains controversial, mainly because of the relative paucity of historical and field studies of the ecological consequences of human-natural interactions in the area. The present study offers further empirical support for the working hypothesis that traditional native human populations are agents of creative beneficial modification of Amazonian ecosystems by considering data collected among the Hotï Indians of the Sierra de Maigualida region of Venezuela. More specifically, we describe several resource-exploitation practices of the Hotï that are not necessarily detrimental to the environment, that may even have the effect of enhancing local and regional biodiversity, and that, indeed, may be interpreted as vital inputs into the dynamic maintenance of the Maigualida ecosystem. Before we narrate the ethnographic facts, however, it will be useful to consider how the notion of anthropogenic ecosystems is linked to the concept of biocultural diversity.

Integrated Concepts: Anthropogenic Ecosystems, Creative Disturbance, Historical Ecology, and Biocultural Diversity

Acceptance of the notion of Amazonia as an anthropogenic ecosystem is closely tied to shifting theoretical perspectives on the stability and evolution of tropical forest ecosystems, the concept of ecological disturbance, and the nature of human-environment interactions in Amazonia and elsewhere. The previous view that tropical forest biomes are pristine relicts and evolve toward relatively stable climax states (Richards 1952) has been overturned in favor of a view that stresses considerable biological heterogeneity over space and time and the critical role of natural and human disturbances in proliferating and perpetuating that diversity (Sanford et al. 1985; Meggers 1987, 1994, 1995; Schubert 1988; Colinvaux 1989; Kauffmann & Uhl 1990; Sponsel 1992; Cavelier et al. 1995; Stahl 1996). Biologists reckon that the extraordinary biodiversity observed in present-day tropical forest biomes is largely due to the disturbance and discontinuity of ecological conditions at different spatial and temporal scales (Haffer 1969; Vanzolini 1973; Connell 1978; Prance 1978, 1982; Whitmore & Prance 1987; Goldammer 1992; Phillips et al. 1994). At both macro- and microanalytical scales, this kind of ecosystem appears to exist in a state of dynamic nonequilibrium (i.e., has no indefinite stable point), consisting of constantly shifting patches and mosaics of different habitat types. However, using a different scale of observation (e.g., middle range), it is apparent that community structure and interspecific interactions exhibit some degree of stability, in which case it is precisely the fluctuation and balance among disturbed, immature, and mature patches that define the ecosystem and allow for its continuity and prevalence. Conservation biology within this paradigm focuses on dynamic processes and the physical contexts in which they unfold (Meffe & Carroll 1994).

Disturbance is the key conceptual tool employed for understanding the dynamic or nonequilibrial character of ecosystems. Disturbance is defined as "any relatively discrete event in time that disrupts ecosystem, community, or population structure and changes resources, substrate availability, or the physical environment" (White & Pickett 1985: 7). This concept includes environmental fluctuations and destructive events related to the spatial-temporal dimensions of a particular ecosystem as well as processes of both autogenesis (change triggered by the inherent biological properties of the ecosystem) and allogenesis (change triggered by an outside agent) (White & Pickett 1985). The viewpoint that disturbance constitutes an integral component of many so-called natural ecological systems clashes with the previously held understanding, expressed in much of the earlier ecological literature, of ecosystems as tending toward stability or equilibrium (Botkin & Sobel 1975). The conceptualization of disturbance as a creative event is relatively new in ecological thinking and ethic (Sprugel 1991). The more recent connotation of disturbance—as the trigger of ecological processes enhancing and providing the medium for different biotic structures and composition of natural environments—invalidated previous assumptions of disturbances as events that impeded or changed the natural regeneration of ecological systems. This creative perspective remains controversial (Hoopes & Harrison 1998), even when there is no dispute that disturbances are autogenetic.

The notion of anthropogenic landscapes forms a centerpiece of recent interpretations of Amazonian cultural ecology under the research paradigm of "historical ecology" (Balée 1989, 1992, 1995; Zent 1992; Posey 1994, 1998; Rival 1998; Whitehead 1998), which emphasizes the dialectical interaction (i.e., mutual causation) of human cultural systems and natural ecosystems over time (Crumley 1994; Balée 1998b). Using this perspective, Balée (1989) advanced the argument that a minimum of 12% of Amazonia is covered by anthropogenic habitats, to which he refers as "cultural forests" in order to convey their simultaneous significance as cultural artifact and natural life form. The dual meaning follows from the theoretical postulate that there is an interpenetration of human culture and nonhuman nature, which in turn leads to the methodological principle of considering cultural practices and natural organisms as belonging to a single unit of analysis; that is, as a "total phenomenon" (Balée 1998a). In short, the ontological and epistemological separation of human culture from infrahuman nature is rejected.

Viewed as an example of the interpenetration and analytical inseparability of cultural and natural spheres, the conceptual construct of cultural forests is important for understanding what is meant by the concept of biocultural diversity. Some researchers identified scenarios (e.g., managed successions) in which the human-modified habitats exhibit greater specific and structural plant diversity than do natural ecosystems (Posey 1983a, 1984; Irvine 1989; Balée 1993, 1994); other studies established a link between human management and greater abundance of useful plant species in forest communities (Unruh & Alcorn 1987; Unruh & Flores 1988; Anderson & Ioris 1992; Salick 1992; Balée 1993). Aboriginal Amazonian human populations are also thought to be responsible for developing a good deal of crop genetic diversity (Clement 1989, 1999a, 1999b) and for dispersing useful plants over great distances (Stone 1984; Posey 1988; Sánchez & Jaffé 1992). Perhaps more significant, however, empirical descriptive as well as comparative studies are now bringing out the fact that native Amazonian resource-management systems display considerable diversity and distinctiveness in terms of the specific kinds and amounts of resources harvested and the particular combination of exploitative and manipulative behaviors of the biotic and abiotic environment, both between and within ethnic domains and among different groups who occupy similar ecological ranges (Salick 1989; Salick & Lundberg 1990; Milton 1991; Kaplan & Kopischke 1992; Padoch & De Jong 1992; Beckerman 1993, 1994; Descola 1996; S. Zent & E. L. Zent 2004). We venture the hypothesis that the diversity of disturbance behaviors among culturally distinct human groups in Amazonia will result in a diverse range of possible impacts on the environment, and in this sense we consider biological diversity to be dependent upon cultural diversity (and vice versa). But in order to prove this hypothesis, the ecological disturbance behavior of various ethnic Amazonian groups must be described in detail and then compared. We begin this enormous task by describing the disturbance behaviors of the Hotï and their possible ecological impacts.

Ethnobotanical and Human Ecological Research in the Sierra de Maigualida Region

The Sierra de Maigualida is a large, mountainous formation that provides a natural border between Amazonas and Bolívar States in southern Venezuela, between 5° and 6° N and between 65° and 66° W. The principal rivers that drain this area include the Cuchivero, Nichare, Erebato, and Caura on the Bolívar side, which flow north into the Middle Orinoco, and the Parucito and Asita-Iguana on the Amazonas side, which flow south toward the Ventuari River basin (Figure 3.1). The indigenous human inhabitants of the Sierra de Maigualida and surrounding lands refer to themselves as "Hotï." Contact between Hotï and the West was established relatively recently (1969), due principally to the inaccessibility of the Maigualida region, so neither the people nor the land are well known from scientific or lay standpoints. Previous studies and reports of them are based on brief fieldwork experiences (Cruxent 1961; Bodin 1969; Bou 1970; Dye 1970; Jangoux 1971; Corradini 1973; Eibl-Eibesfeldt 1973; Coppens & Mitrani 1974; Guarisma 1974; Coppens 1975, 1983; Guarisma & Coppens 1978; Vilera 1985; Keogh 1995), and, naturally, this anthropological and botanical research provided a major motivation for our research project.[1]

From May 1996 to October 1999 we conducted an intensive field study of Hotï ethnobotany and behavioral ecology. The information presented in this chapter is based on approximately 21 months of direct observation carried out in four Hotï communities: San José de Kayamá, Caño Iguana, Caño Majagua, and Caño Mosquito. The former two are larger, more permanent, mission-based settlements; the latter two, small, mobile, independent communities (for additional details, see S. Zent & E. L. Zent 2004). The methods used in this research included ethnobotanical plot studies (again, for more details, see S. Zent & E. L. Zent 2004), general plant specimen collections, formal and informal interviews, participant observation of a diverse range of activities, instantaneous scan sampling of activities, focal person observations of subsistence behaviors, resource accounting studies, geographical coordinate mapping of resource activities and sites, and daily conversational interactions.

Biophysical Parameters

The Sierra de Maigualida displays considerable geological, geomorphological, and ecological diversity. Geologically, the region is dominated by moderately tectonized granites, with lesser extensions of pyroclastic-acidic volcanic rock, and localized regions of amphibolitic gneisses and metasedimentary green schists (MARNR–ORSTOM 1988: 25, 95). Altitudes range from 150 to 2400 meters above sea level (m asl) and include

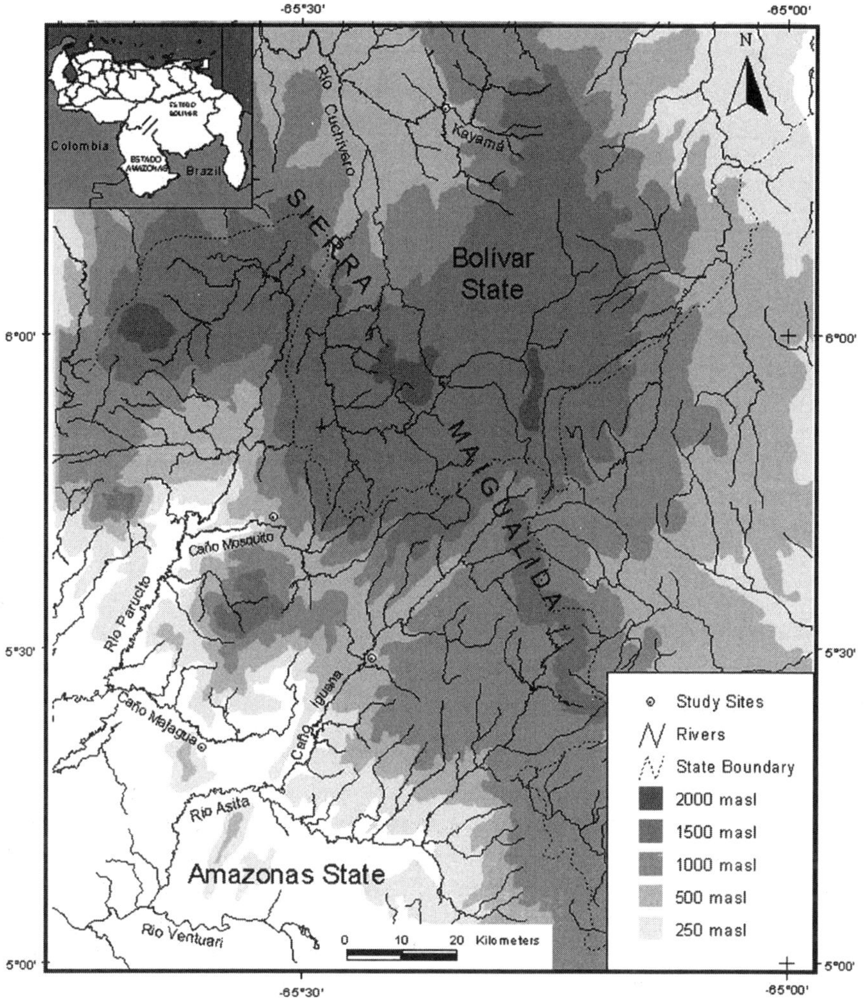

Figure 3.1. Topography of the Sierra Maigualida region of Venezuela.

various landforms: rocky and broken high plateaus (>1000 m asl), mountain slopes, cliffs, ridges and valleys (400–1000 m asl), convex and half-orange peneplains (300–600 m asl), and alteration plain surfaces (150–200 m asl) (pp. 27–29). Rainfall varies between 1900 and 2700 mm per year, and the climate for much of the region is classified under the Köppen system as Am, defined as wet tropical with a dry season of two months (pp. 17–19). The dominant vegetation is mostly semideciduous and evergreen forests, which are described as medium to high and dense, composed of two to four arboreal strata and a relatively open understory, and varied in structure and composition according to an altitudinal gradient ranging from basimontane to montane (Huber & Alarcón 1988). It is assumed that the Maigualida regional fauna corresponds in general with the characteristic fauna of the neotropical biome of northern South America (see Phelps & de Schauensee 1978; Gremone et al. 1986; Eisenberg 1989; Gorzula & Celsa 1998; Linares 1998), but field studies in this precise area are lacking.

Cultural History and Population

The language affiliation, as well as the cultural history, of the Hotï is largely a matter of debate due to the lack of definitive data. The Hotï language has been classified as an isolate (Durbin 1977: 34), and there have been divergent hypotheses that the Hotï may be affiliated with Yanomami (Coppens 1983: 253), Makú (Henley et al. 1994–1996), or Piaroa (Jangoux 1971; Eibl-Eibesfeldt 1973; Guarisma 1974; Vilera 1985; Zent et al. 2001). In terms of material culture, the Hotï bear some striking resemblances to the E'ñepa (Carib language family), their northern neighbors, which are most likely the result of borrowing and historical interaction between the two groups.

The Hotï population appears to be growing rapidly following contact with Western society. The first comprehensive Venezuelan Indian Census, realized in 1982, provides a figure of 398 Hotï (OCEI 1985: 475). The subsequent Indian Census of 1992 recorded 643 Hotï distributed in about 25 communities (OCEI 1993: 42). We estimate that the Hotï population easily exceeded 800 people by October 1999, when our fieldwork ended, based on our census of 14 communities that we visited personally, on information provided to us by missionaries, and on interviews with Hotï about the locations and sizes of other communities that we were not able to visit.

Settlement and Subsistence Orientation

At the time of contact the Hotï exhibited a strong interfluvial, upland, adaptive orientation, with most settlements at 500–1000 m asl (Corradini 1973: 42; Coppens 1983: 255). The early postcontact ethnographic accounts portray Hotï settlements as typically small (5–30 people), single dwellings, scattered in space, and frequently nomadic (including trekking; Jangoux 1971; Coppens & Mitrani 1974; Guarisma 1974). By contrast, the Hotï settlements we encountered during our fieldwork displayed considerable variation in size, number of houses, sedentism, distance from neighbors, interethnic contacts, internal composition, architectural style, and ecogeographical zone inhabited. Consequently, it may no longer be possible to speak of a "typical" Hotï settlement pattern by way of a simple checklist of traits. For example, there is a notable gradient in regard to settlement nucleation and sedentarization, from relatively permanent, densely populated communities to very nomadic, small, dispersed settlements made up of a single nuclear family. The two mission-based communities at Kayamá and Iguana are larger and more permanent, whereas the nonmission settlements, Majagua, Mosquito, and others that we visited, resemble the traditional pattern in terms of size and mobility. In spite of the more sedentary appearance of the mission settlements, camping and trekking are still very common practices in all communities, but in the case of the former two these activities usually involve very small groups and shorter stays away from the main settlement. In any case, two basic kinds of settlements can be distinguished: base settlements, where larger and more durable housing is built and the main swidden plots are located, with a relatively stable duration of five or more years, to which the coresidential unit or segment returns on a recurrent basis from trekking or camping trips throughout the annual cycle; and camp settlements, temporary shelters, most commonly lean-tos, where individuals, family groups, or occasionally the entire settlement unit camp out during foraging outings, trekking, trade expeditions, kin reunions, ritual observances, and other forays.

The Hotï subsistence system rests on an extensive mix of hunting, gathering, fishing, and horticultural activities, the relative importance of each varying according to the size, permanence, surrounding habitat, and economic acculturation of the community (for more details, see S. Zent & E. L. Zent 2004). Medium-sized and large terrestrial mammals (e.g., tapirs, peccaries, pacas) are hunted mainly with steel-tipped lances, and arboreal mammals (mostly monkeys) and birds are shot with blowguns and curare-tipped darts. The most prominent gathered food items include palm larvae, honey, and numerous species of tree fruit (see below). Fishing was not an important activity in the traditional upland habitat and today is significant only in settlements next to the larger waterways. The main cultigens found in Hotï gardens are plantain/banana (*Musa* x *paradisiaca*), yams (*Dioscorea alata* and *D. trifida*), maize (*Zea mays*), sweet potato (*Ipomoea batatas*), and sweet manioc (*Manihot esculenta*) (Zent et al. 2001).

The scheduling, duration, and location of different subsistence activities is directly dependent on the Hotï's knowledge of the phenology and geography of key botanical resources and the behavioral ecology and demography of certain animal species.[2] In a similar vein, the fluctuation from sedentary to nomadic settlement modes is determined primarily by the perceived abundance and availability of wild resources and secondarily by the labor demands of the horticultural cycle. Among the more nomadic settlement groups, swidden plots appear to be relatively neglected, and garden labor is often deferred or simply forgotten. For this reason, we consider the Hotï subsistence strategy to be characterized by a dominant foraging orientation, even though horticulture may be the principal provider of dietary energy in some communities. As may be expected, horticultural activity has been intensified in the more nucleated and sedentary mission communities. Nevertheless, a huge variety of wild plant and animal species are collected on an almost daily basis in all of the communities we studied, making a significant contribution to the diet and also furnishing raw materials for the manufacture of household utensils, tools, ornaments, soaps, medicines, hunting magic, fish bait, firewood, building materials, and so forth (Zent et al. 2001).

The heavy dependence of the Hotï on wild forest resources for their material and social reproduction is now well documented (E. L. Zent 1999; Zent et al. 2001). Still to be determined is whether this dependency has in turn conditioned or stimulated dependent responses in at least some of the botanical or zoological species that have been—and continue to be—subject to human exploitative manipulation for many years, decades, or even centuries. In short, to what extent have the Hotï domesticated their forest environment? It is significant to note that most research to date on the cultural forest question in Amazonia has emphasized the impact of slash-and-burn or shifting cultivation on forest composition and structure (Denevan & Padoch 1987; Balée 1989, 1993; Irvine 1989). However, without intending to downplay the importance of this form of human disturbance, we have chosen to focus our present description on the creative disturbances associated with the exploitation of so-called natural resources or natural spaces. Our main reason for taking this alternative approach is because, as we expressed earlier, the organizational focus of the Hotï subsistence system is on foraging, which seems to indicate the historical dominance of this economic sector.[3] Therefore, we assume that the cumulative ecological impact of nonswidden activities is greater in historical perspective. Another reason for focusing on the ecological disturbances resulting from foraging behavior is to call attention to the possible historical role of human foragers in shaping Amazonian landscapes. Five kinds of human disturbance of natural resources are explored here: harvest and dispersal of edible fruit trees, exploitation and manipulation of palms, management of palm grubs, gap cultivation, and honey extraction.

Harvest and Dispersal of Edible Fruit Trees

Environmental disturbances can be classified as events that cause the death of one or more individuals, dynamic events that change the structure and possibly the composition of a biotic community, or phenomena that liberate resources previously used and thereby increase the resource availability to other organisms (cf. Barbour et al. 1987; Petraitis et al. 1989; Pickett et al. 1992). Disturbances can take place at several levels of ecological organization: individual, population, community, ecosystem, region, and so on. The many types of environmental manipulation practiced by the Hotï correspond to the three classes of disturbance mentioned above, and their effects may be observed at different levels.

The most common disturbance activity observed among the Hotï involves the harvesting and dispersal of edible fruit trees. The collection and consumption of wild tree fruits are flexible events, in which persons of all ages (>5 years old) and both sexes participate actively. However, some fruit types are considered to be eaten only by children, which may explain why knowledge of edible plant species develops faster than do other areas of ethnobotanical knowledge (see S. Zent & E. L. Zent 2004). The frequency of such gathering bouts in a particular settlement depends on the abundance and distribution of edible tree species in the resource catchment area, season, settlement mobility patterns, and other activities being engaged in. Whereas some wild fruit foraging expeditions are consciously planned with the goal of harvesting a specific species, in which case the forager knows ahead of time exactly what plant individual(s) will be exploited and/or their general location, other harvesting events are best described as opportunistic, in that a person ingests edible fruits if he or she happens to come across them when walking through the forest, often while engaged in some other activity. The former type of harvesting strategy is usually reserved for certain species that are especially appreciated for their unique flavor or that provide a certain minimum yield (we estimate about 3 kg) of harvestable fruit during one harvesting event. Trekking or camping trips are sometimes motivated by the principal desire to consume the fruit of a very productive, seasonally available tree species, such as *Humiria balsamifera, Chrysophyllum sanguinolentum,* or *Dacryodes* spp. (see below). By contrast, the opportunistically exploited species are obviously accorded lower harvesting priority, but this does not mean that their habits and habitats are less well known by the Hotï. In fact, such knowledge allows the Hotï to modulate their subsistence practices as the need arises, such that these resources are more intensively exploited during periods of environmental or social fluctuation, when the more common food items may become scarcer. We recorded a total of 222 wild botanical species that are considered edible by the Hotï, the vast majority of which are rarely eaten when normal subsistence conditions prevail. The number of plant foods actually harvested during the research period was considerably fewer, according to our resource accounting records. The most important fruit tree species exploited by the Hotï are listed in Table 3.1.

The harvesting of tree fruits usually involves one or more of the following standard collection techniques: picking up the fallen mature fruits directly from the ground; climbing trees or vines with the aid of a climbing ring, *hudibu,* which is fashioned from the inner bark of several trees (*lɨ̃ ʰkutile ʰyaɨ, Guatteria foliosa* Benth, Annonaceae; *eloli ʰyaɨ, Cordia sericicalix* A.D.C., Boraginaceae; *baʰtu hõlõ ʰyaɨ, Lecythis corrugata* Poiteau, Lecythidaceae; *totohu ʰyaɨ, Brosimum utile* [Kunth] Pittier, Moraceae) or the outer stem of *ʰkãlãwa ʰwaʰwa* (*Phenakospermum guyanense* [L. C. Rich.] Endl. ex Miq., Strelitziaceae); building scaffolds, if the base of the trunk is too voluminous and difficult to climb; removing tree branches; and chopping down the tree. After the fruit is collected, it may be eaten immediately and the seeds discarded at the same spot. Any leftover fruit is carried back to the settlement or campsite. It is not uncommon for the collectors to continue to eat the fruit and throw away the seeds as they walk home. If the fruit is consumed at the settlement, inedible material, including the seeds, is either deposited in refuse piles along the margins of house yards or thrown into a nearby creek. Sometimes fruit that is brought down from the tree canopy is neither eaten nor hauled away by humans (for example, if not yet ripe), in which case it may be dispersed by animals. All of these practices related to the collection and dispersal of tree fruits entail rather low-level disturbance events, in the sense of affecting an individual organism, the immediate environs of the harvest site, the route from harvest to settlement sites, and/or the settlement site surroundings. However, if one considers the considerable frequency and density with which such events occur in time and space under normal subsistence practice, one can imagine the potential impact at higher levels of ecological organization.[4] This point is illustrated through description of the exploitation of the fruit tree, *walikye ʰyaɨ* (*Ecclinusa guianensis* Eyma).

During the first two months of the rainy season (April–May), the Hotï collect and eat large quantities of *walikye* fruit. This member of the Sapotaceae family is found mostly in montane and premontane forests. Its presence in lower-altitude forests is attributed by the Hotï to the dispersal behavior of human and animals. For example, it is a common trailside inhabitant around the community of Kayamá, although it is not considered by the people living there to be a native species of this savanna-forest ecotone

Table 3.1
Edible fruit trees most commonly harvested by the Hotï

Family	Scientific name	Hotï name	Season of harvest
Anacardiaceae	*Anacardium giganteum* Hanc.	*uli eloli ʰyaɨ*	July–September
Apocynaceae	*Couma macrocarpa* Barb. Rodr.	*ʰkalibɛ̃lẽ ʰyaɨ*	May–August
Burseraceae	*Dacryodes* sp.	*uli ʰlute ʰyaɨ*	August–November
Burseraceae	*Dacryodes chimantensis* (Steyerm.) Cuatrec.	*hãlĩ ʰlute ʰyaɨ*	July–November
Burseraceae	*Protium tenuifolium* (Engler) Engler	*oʰtewaka bãu ʰyaɨ*	May–July
Burseraceae	*Protium opacum* Swart	*uli bãu ʰyaɨ*	May–July
Caesalpinaceae	*Hymenaea courbaril* L.	*bũye ʰyaɨ*	November–January
Caryocaraceae	*Caryocar microcarpum* Ducke	*uli ʰkawale ʰyaɨ*	April–August
Cecropiaceae	*Pourouma melinonii* Benoist	*uli weʰkao ʰyaɨ*	May–July
Clusiaceae	*Garcinia macrophylla* Martius	*ʰtãwẽ ʰyaɨ*	July–August
Fabaceae	*Dipteryx* sp.	*wãi ʰyaɨ*	February–April
Humiricaceae	*Humiria balsamifera* Aubl.	*ʰtau ate* -	December–March
Mimosaceae	*Inga bourgoni* (Aublet) D.C.	*hãlĩ luwe ʰyaɨ*	February–April
Mimosaceae	*Inga edulis* Martius	*uli luwe ʰyaɨ*	January–May
Mimosaceae	*Inga* cf. *alba* (Sw) Willd.	*bãyekalõ luwe ʰyaɨ*	April–August
Mimosaceae	*Inga* sp.	*eʰkɔ luwe ʰyaɨ*	February–April
Moraceae	*Brosimum lactescens* (S. Moore) C. C. Berg	*ʰtɨ̃ʰtɨ̃wẽka ʰyaɨ*	April–June
Moraceae	*Helicostylis tomentosa* (P. & E.) Rusby	*wãiyõ walẽbã ʰyaɨ*	June–August
Moraceae	*Pseudolmedia laevigata* Tréc.	*hãlĩ wãiyõ ʰyaɨ*	January–March
Moraceae	*Pseudolmedia laevis* (R. & P.) Macbride	*uli wãiyõ ʰyaɨ*	January–March
Moraceae	*Sorocea muriculata* Miquel	*buʰta hele*	May–July
Polygonaceae	*Coccoloba fallax* Lindau	*ʰtoba ʰyaɨ*	January–May
Sapindaceae	*Matayba* sp.	*ahtikwɛ̃ ʰyaɨ*	July–August
Sapotaceae	*Chrysophyllum argenteum* Jacq.	*olẽkõ ʰyaɨ*	January–March
Sapotaceae	*Chrysophyllum sanguinolentum* (Pierre) Baehni	*tuwiyu ʰyaɨ*	January–March
Sapotaceae	*Ecclinusa guianensis* Eyma	*walikyɛ ʰyaɨ*	April–June
Sapotaceae	*Micropholis egensis* (A.D.C.) Pierre	*ʰkaile ʰyaɨ*	February–April
Vochysiaceae	*Qualea paraensis* Ducke	*ʰlẽbãle ʰyaɨ*	February–April

environment. The fruit is oval, about 3–5 cm long, with yellow-reddish skin when ripe and a yellow, juicy, gummy pulp. *walikye* has a sweet, tasty fruit, and, as with many Sapotaceae, the trunk and fruit exude sticky, milky latex. The fruiting period is relatively short, but the fruit is abundant enough to motivate extensive collecting expeditions among the Hotï.

In May 1998 we observed an ecologically and socially dynamic expedition to collect this fruit at Caño Iguana. A total of 37 individuals (nearly 25% of the village population) participated actively in the event, making it the largest communal economic activity we saw or heard about during our six months of cumulative residence in the community. Cognatically as well as affinally related groups (including infants and elderly), consisting of two to seven persons, participated in the harvest, which took place at about a two-hour walk from the community. Some groups built small shelters and spent a few days there, but most people went just for the day. While some men and boys climbed up several trees and cut down the fruit-laden branches, others gathered fruits that had already fallen to the ground or after the branches had crashed down. A few large trees were felled with an ax to gain access to the fruit, even though it appeared that they could have been climbed. Most of the fruit was consumed on the spot, although several groups managed to take one or more basketloads back to the village.

What are the ecological consequences of this intensive pattern of *walikye* exploitation? Table 3.2 shows the census figures for this plant species in the 1-ha ethnobotanical plots we surveyed in the four study communities. Where *Ecclinusa guianensis* is present (at Iguana, Majagua, and Kayamá), the comparative absolute densities of this species in high forest range from 1 to 7 individuals per hectare, whereas the absolute frequencies are between 1 and 6 (in a sampling universe of 25 quadrats per settlement). By contrast, the density and frequency of *E. guianensis* in the collection zone referred to above was much greater. There we counted 47 large individuals (>10 cm dbh) of *E. guianensis* in an area we estimated to cover approximately 1 ha. This species appeared to be the most dominant arboreal species in that patch of forest. Furthermore, we also saw numerous smaller individuals of the same species in the area. The diversity of heights and thicknesses of *E. guianensis* stems tells us that the local population is extremely well balanced according to age group and is a key indicator of successful colonization. Another significant feature of the *walikye* harvest zone was the conspicuous presence of indicators of past human disturbance, in the form of old tree stumps and logs from cut-down trees.

Human management through harvesting practices over long periods of time seems to be the most reasonable explanation for the high density of *Ecclinusa guianensis* in particular patches of forest. Even if the Hotï do not act consciously, their collection techniques appear to foster the dominance and maintenance of this species locally. Pertinent ecological consequences of these techniques include: enhancement of local seed banks of this species while collecting the fruit, since the Hotï eat some of it on the spot and let the seeds fall to the ground; regional dispersal of seeds, since a significant portion of the harvest is transported back to the settlement and since some of the fruit is eaten during the walk home and the seeds dropped along the trail while the rest are thrown away at the settlement site; creation of gaps by felling larger, mature individuals, which eventually allows the development of new individuals from the same or other species; liberation of nutrients as a result of cutting down branches or tree trunks, which benefits the plants around them; and stimulation of new branch growth through the selective removal of branches laden with fruit.

Analogous consequences can be observed in other arboreal fruits collected by the Hotï, such as *tuwiyu* ^{h}*yaɨ* (*Chrysophyllum sanguinolentum*) and ^{h}*tau ate* ^{h}*yaɨ* (*Humiria balsamifera*). The fructification of both species takes place in the dry season (January–March) and stimulates overnight camping trips. Both species normally grow on the

Table 3.2
Synecological values for *Ecclinusa guianensis* in the four 1-hectare Hotï forest plots

Location	Absolute density[a]	Relative density[b]	Absolute frequency[c]	Relative frequency[d]	Relative dominance[e]	Importance value[f]
Caño Iguana	5	1.41	4	1.40	1.38	4.19
Caño Majagua	1	0.18	1	0.27	0.05	0.50
San José de Kayamá	7	1.30	6	1.51	1.85	4.66
Caño Mosquito	0	0	0	0	0	0

[a] Number of individuals of the species occurring in the plot.
[b] Number of individuals of the species divided by the number of all individuals in the plot.
[c] Number of subplots (400 m^2) in which one or more individuals of the species occur.
[d] Absolute frequency of the species divided by the total absolute frequency of all species.
[e] Aggregate basal area of the species divided by total basal area of all species.
[f] Sum of the relative density, relative frequency, and relative dominance.

stony ridge tops in premontane and montane forests (>400 m asl), and their presence in lower-lying regions is explained by the Hotï as the exceptional result of the dispersal actions of people or animals. *ʰtau ate* is especially abundant in the vicinity of Caño Iguana and is the object of frequent collection expeditions. The fruit is oval, measuring about 1–2 cm in length, is green-to-violet black in color, and has a sweet pulp that surrounds a single seed. When ripe and ready to be eaten, it falls naturally to the ground. In this case, the main disturbance behavior of the Hotï is simply to disperse the seeds at least as far as their settlement or campsites.

Exploitation and Manipulation of Palms

Without a doubt, palms hold a special place in the resource-management strategies of lowland South American cultures (cf. Balick 1984). Recent studies conducted in the Colombian Amazon show that palms have been used by humans continuously since the early Holocene (Morcote-Ríos et al. 1998). It is rare to find a contemporary rural group that is not dependent on palms in some way. The Hotï are no exception to this pattern. As shown in Table 3.3, 10 species of palms are extensively exploited for a wide variety of reasons: food, human medicine, hunting medicine, fish bait, house building materials (thatch, wall planks), furniture (tables, benches, shelves), basketry and mats, ceremonial dress, hunting tools (lance, bow, blowgun dart, shooting pellet), household utensils (broom, food container, food grater, needle), fire drill, and cotton spindle. The contribution of palms to the Hotï diet is quite significant. The collection of four species of palm fruit (*Attalea maripa, A. macrolepis, Oenocarpus bacaba,* and *Bactris gasipaes*) has been cited as a main motivation for camping trips. The *cucurito* palm, *ulu hi* (*A. maripa*), is an especially important food source, constituting a staple food during the harvesting period (May–July). Perhaps for this reason, the Hotï regard it as the first plant created on earth, deep in the mythological past. Trekking and camping behavior during the rainy season is largely determined and regulated by opportunities to

Table 3.3
The most common palms exploited and managed by the Hotï

Hotï name	Latin name	Uses
hãlĩ ba	*Astrocaryum gynacasthum* Mart.	Food, bows, fish bait
awaʰto	*Attalea macrolepis* Mart.	Food, house-building material, basketry, darts, food containers
ulu	*Attalea maripa* (Aubl.) Mart.	Food, basketry, darts, house-building material, food containers, combustible material
ʰkolowa	*Attalea* sp.	Food, house-building material, darts, food containers
lẽbã	*Bactris gasipaes* Mart.	Food, fish bait
waʰwɨ	*Euterpe precatoria* Wallace	House-building material, hunting material, combustible material, fish bait
bãlaya	*Geonoma deversa* (Pait.) Kunth	Food, house-building material
tuli	*Mauritia flexuosa* L.f.	Food, basketry, house-building material, ceremonial dress
hãlĩ baʰte	*Oenocarpus bacaba* Mart.	Food, basketry, house-building material
bũli	*Socratea exorrhiza* (Mart.) Wendl.	Food, food graters, fish bait, basketry, house-building material, fish traps, combustible material

harvest *ulu* fruit. Thus it is common to find camp settlements next to *ulu* stands. The fruit is oval, about 3 cm long, with a leathery, ochre-brown skin that covers the soft, whitish pulp. It can be eaten raw or cooked in the form of juices (*ʰyu*) or soups (*ʰtaʰta*). The *seje* palm, *hãlĩ baʰte hĩ* (*O. bacaba*), is another keystone species for Hotï subsistence, providing edible fruit as well as edible weevil larvae (see below). The fruit is round, about 1 cm in diameter, with a purple-black skin and oily, edible mesocarp that is boiled and eaten or drunk as a beverage. This erect palm is found mostly on hill slopes and well-drained spots.

The conspicuous presence of palms throughout the Maigualida landscape is attested to by the data presented in Table 3.4. The dominance of palms is greatest in the riverine forests at Caño Majagua. Here the abundance of *Oenocarpus bacaba* is particularly striking, reaching an absolute density of 59 individuals per hectare (more than 10% of all individuals in the plot).[5] The presence of *Attalea maripa* is also substantial, showing absolute densities of 6 or 7 individuals per hectare in three of the four forest plots. Such densities appear meaningful when viewed in the context of the low densities recorded for most other species found within the plots. The vast majority of taxa on these plots are represented by five or fewer individuals: for Caño Mosquito, 89%; for Caño Majagua, 88%; and for Kayamá, 86%. In short, *A. maripa* is one of the most abundant plant species found in three of the four study sites.

The existence of palm-dominant forests in other Amazonian areas has been associated directly with human management (Balée 1988; Anderson et al. 1991; Heinen & Lizarralde

Table 3.4

Synecological values for palm species in the four 1-hectare Hotï forest plots

Species, location	Absolute density	Relative density	Absolute frequency	Relative frequency	Relative dominance	Import val
Astrocaryum gynacasthum						
Caño Mosquito	1	0.18	1	0.25	0.03	0.4
Attalea maripa						
Caño Mosquito	7	1.26	5	1.31	1.63	4.2
Caño Majagua	6	1.07	5	1.40	7.37	9.8
San José de Kayamá	7	1.30	3	0.76	3.88	5.9
Attalea macrolepis						
San José de Kayamá	1	0.19	1	0.25	0.72	1.1
Attalea sp.						
Caño Majagua	6	1.07	1	0.28	5.69	7.0
Euterpe precatoria						
Caño Majagua	10	1.78	4	1.12	0.44	3.3
Canõ Iguana	3	0.85	3	1.08	0.16	2.0
San José de Kayamá	12	2.23	8	2.02	0.81	5.0
Oenocarpus bacaba						
Caño Mosquito	2	0.36	2	0.52	0.27	1.1
Caño Majagua	59	10.48	17	4.75	6.46	21.6
Caño Iguana	1	0.28	1	0.36	0.04	0.6
Socratea exorrhiza						
Caño Mosquito	12	2.16	10	2.62	0.43	5.2
Caño Majagua	29	5.15	11	2.98	0.99	9.1
Caño Iguana	8	2.25	7	2.53	0.58	5.3
San José de Kayamá	3	0.56	2	0.51	0.14	1.2

1991; Heinen et al. 1995). This finding provokes the question, To what extent is the abundance of the palm species mentioned above the product of human intervention? The main kinds of human disturbance of possible relevance to this example include fire, felling, transport, and cooking. Both *Oenocarpus bacaba* and *Attalea maripa* are fire-resistant species. The Majagua plot was partially burned during the dry season of 1998 in a fire that escaped from a nearby swidden plot. Twenty-three *O. bacaba* individuals were severely burned and indeed appeared to be dead upon initial observation. Returning to the site five months later, we were astonished to find that every one of them had recuperated successfully. The resilience of *O. bacaba* shown in this case may have to do with the fact that it has a higher water input near its stem than do open-area rainfall or dicotyledonous tree species (Schroth et al. 1999). *Attalea maripa* is similarly resistant, for it often survives the initial burning of a new garden, making this pioneer species a long-lived one. In this regard, it is interesting to note that we observed numerous *ulu* monodominant patches, called *ulu čo* (*A. maripa* forest), during a long trek we made into the headwaters of the Cuchivero River, which is the traditional heartland of Hotï settlement. Our Hotï collaborators explained that some of the *ulu čo* are clearly the living remains of past garden and settlement sites.[6] In other cases, the *ulu čo* were found on ridge tops, where gardens are not usually located. However, we also noticed that some forest patches on ridge tops had been burned, and our collaborators confirmed that people initiated these fires during the height of the dry season. It is therefore possible that the ridge-top *ulu čo* were established through human-induced seed depositions advantaged by human-set fires.[7] In sum, the generous endowment of palms such as *O. bacaba* and *A. maripa* throughout the Maigualida landscape may be interpreted as supporting the claim that "fire is everywhere in the Amazonian natural history" (Pyne 1998: 88).

To harvest edible palm fruit or fronds for house thatching, the collector normally climbs the trunk (or an adjacent one) and cuts down the whole bunch or the leaves, as the case may be. But on certain occasions the collector chops down the entire palm tree in order to harvest these products, a practice that appears to be destructive. However, this practice is usually reserved for very mature or senescent individuals that are past their fertility peak or no longer producing fruit. Our Hotï collaborators stated that individuals of *Oenocarpus bacaba* and *Attalea maripa* are cut down when they are "too tall and old" in order to make way for younger individuals that will then "grow faster and stronger." These perceptions are in fact consistent with the information found in the scientific literature, where it has been reported that *O. bacaba* individuals approaching senescence exhibit a reduction in growth rate, reproductive potential, and biomass generation (Balick 1988: 129). Moreover, although the senescent palm has fewer leaves, the open structure and horizontal orientation of the crown shade the plants growing underneath it. Therefore, cutting down the whole plant not only releases nutrients but also allows more direct insolation, which benefits seedlings, saplings, and young individuals. In this sense, the selective removal of unproductive individuals facilitates the development of newly productive ones, which also benefit from the nutrients released by the fallen trunk. As a result, the *O. bacaba* stands that we observed were made up of individuals of diverse sizes, ages, and phenological stages. Unfortunately, in the available literature there are no estimates of ages for the life expectancy and life stages of *Oenocarpus* sp. (Balick 1988: 129), due to high plasticity and the inability to make correlations between size and age of this species (Balick 1986: 77), although leaf morphology is considered to be a good indicator of life stage in palms (Balick 1988: 130).

The Hotï may also influence the distribution and regeneration of palms as a consequence of harvesting, handling, and consuming their fruit. The collection and transport of large quantities of fruit to settlements and campsites implies the formation of large seed banks in particular locations. For example, in May 1996 we visited the campsite of a Hotï nuclear family next to an *ulu* grove in the Upper Cuchivero region. The family had been staying at this camp for the past three to four weeks and reported that *ulu* fruit was the principal food they had consumed during that time. We observed that a few tall *ulu* trees had been felled recently and that huge piles of shells and seeds of the freshly eaten fruit were all around the small clearing where lean-tos had been built. The processing of palm fruit, especially its cooking, may also hasten germination and growth. This is certainly the case with *Oenocarpus bacaba,* for which soaking the seeds in warm water or boiling them prior to eating increases the global percentage of seed germination from 42% to 98% (Balick 1988: 122–123). With *Attalea maripa,* the hard shells are sometimes cracked open with a rock in order to look for larvae (*ulu abĩye*), which are eaten or used as fish bait. In sum, Hotï strategies of exploiting and manipulating palms influence their distribution, density, and regeneration patterns.

Management of Palm Grubs

Palm grubs, corresponding to the larvae of two weevil species, *hãlĩ ba*^{h}*tebohã* and *uli ba*^{h}*tebohã* (*Rhinostomus barbirostris* and *Rhynchophorus palmarum,* respectively, Curculionidae, Order Coleoptera), are very common in the Hotï diet. Although we initially considered the weevil larvae wild resources, we soon became aware that the Hotï consciously manipulate their production and reproduction through a form of management that might be considered pseudo-domestication. Their management strategy relies on knowledge and exploitation of the relationship between these coleopteran insects and a few palm species, especially *hãlĩ ba*^{h}*te hĩ* (*Oenocarpus bacaba*).

Using their knowledge of the specific reproductive ethologies of *Rhinostomus barbirostris* and *Rhynchophorus palmarum,* the Hotï re-create the environmental conditions needed to raise them. The female weevil deposits her eggs in the decaying pith of fallen stems of *Oenocarpus bacaba* and, to a lesser extent, of *O. bataua, Attalea maripa,* and *A. macrolepis.* The wood of these palm species is very fibrous yet soft to medium in hardness and decomposes relatively rapidly after the trunk falls down. As such, it provides the ideal incubation and nurturing medium for the developing weevils. So the Hotï cut the palms down in order to provide extra breeding grounds for the weevils. This practice is mostly reserved for *O. bacaba,* because the Hotï regard the other palm species as less reliable sources of weevil larvae production, but they occasionally fell the other species for this purpose as well. The stems selected for felling correspond to individuals approaching senescence and hence past their most productive stage. In fact, the Hotï assert that if a younger, more fertile individual is felled, the wood will not decompose fast enough and the weevils will not lay their eggs there. This may be due to the fact that the pregnant weevils are instinctively repulsed from laying their eggs in the logs of younger palms or that the slower decomposition rate of the latter does not allow the eggs and larvae to prosper; a more definitive answer will require further investigation. In any case, felling the older, larger, and taller individuals creates light gaps that benefit younger, smaller, and more fertile individuals (see the previous section).

We found that the harvesting of palm-grown weevil larvae was practiced most frequently at the Majagua and Mosquito communities, precisely where the human population density is lowest.[8] In these areas, the local landscape is sprinkled with small clearings or sometimes patches in closed canopy forest (20–40 m), where two to four felled trunks of *Oenocarpus bacaba* lie. The Hotï fell the palm stems and then leave them for a few months, thus allowing natural decomposition to begin, creating ideal conditions for the weevil larvae, which feed on the fibrous wood of the palm. After three to four months, the person who felled the stem will return to the site, usually in the company of his spouse and/or children, in order to harvest the larvae and nymphs. The log is split open with an ax, and the inner fiber is sliced up with a knife and then meticulously sifted by hand for edible larvae. A large portion of the larvae are eaten raw on the spot; the rest are taken home and cooked in soups or wrapped in leaves and then roasted. If one waits any longer, the larvae will have become too mature, too adult-like, to be considered edible and thus, from the Hotï perspective, represent a loss of food resources.[9] However, it is not uncommon to find that some of the eggs have already reached the adult stage by the harvest date; the collectors generally ignore these individuals. Furthermore, it is also not uncommon for the collector to leave some ostensibly edible larvae behind. According to our Hotï collaborators, a large number of larvae and adults normally die trapped inside the palm trunk, if not liberated by the harvest procedure, so in that sense the harvest may actually provide a positive input to the local weevil population. In any case, the fixed period of larvae maturation and the natural mortality levels impose a fairly limited window of harvesting opportunity, which we interpret to mean that the collection outings are consciously planned from the moment of felling.

The harvest and consumption of weevil larvae occurs year-round, availability depending mainly on the level of human management. In the Majagua and Mosquito communities the larvae are harvested an average of three times per week. So it is fair to say that this resource makes a very substantial contribution to the Hotï diet, being especially rich in protein, fat, and carbohydrate content.[10] Moreover, it is a very reliable source of protein and fat that can be used to offset periodic shortages of fish or game.

Obviously, the management of palm larvae described here goes beyond simple foraging of natural resources and implies pseudo-domestication, by which we mean creating an interdependence among animal, plant, and human populations. The Hotï practice of felling older palm trees and harvesting weevil larvae from them has a direct influence on the abundance and regeneration of both the animal and plant species involved. At the same time, the subsistence economy of the Hotï is heavily dependent on the nutritional and other material benefits of the weevils as well as the palms. On the social plane, the *Oenocarpus bacaba* stands constitute one of the few types of natural resource that are invested with a certain "property value" by the Hotï. The stands are recognized to have owners, to whom prior rights of exploitation are accorded, and we have even heard of cases in which the stands were inherited from a previous owner. Indeed, the palm stands, along with other strategic resources (e.g., hunting trails, old gardens), seem to provide the material basis for territorial behavior among the Hotï, which in turn conditions various aspects of their settlement pattern and social relations (cf. Storrie 1999).

Gap Cultivation

Another kind of creative disturbance behavior performed by the Hotï entails the cultivation of certain crop plants in natural gaps in the high forest along their walking trails. We observed two kinds of gaps being utilized in this fashion: natural tree-fall clearings and seasonally flooded open areas along creeks. The Hotï refer to these resource areas as *bãlã balo* (trail garden). Like those found among the Kayapó (Posey 1983a, 1984) and Piaroa (Zent 1992: 168–169), these trail gardens provide people with extra food supplies that can be used during their frequent trekking activities. The trail gardens are usually far from the settlements and, according to the Hotï, were much more extensive a generation ago. Some informants report that the bulk of their ancestors' diet consisted of palm fruits, wild game, honey, and cultivated plants from the trail gardens. Most of the cultigens planted in trail gardens are relatively short-lived plants that are favorite food items and also release nutrients when they die, such as maize, manioc, sweet potato, yam, ocumo, and sugarcane. Other cultivars, such as banana or plantain and papaya, can survive longer,[11] although they eventually disappear when the surrounding trees overtake the clearing and close the canopy.

The peach palm, *lẽbã hĩ* (*Bactris gasipaes*) is also planted in trail gardens, but its exploitation involves an interesting case of human introduction of a natural resource that apparently is being maintained due to its mutualistic interaction with the wild fauna of the region. According to Schultes (1994: 203), the peach palm is "unknown in the wild" but is "widely cultivated throughout tropical America." However, the majority of the harvests we observed and recorded in the field came from "wild" stands, according to the people who collected the fruit. They claimed that the agouti (*Dasyprocta leporina*) is responsible for the abundant proliferation of this species. Agoutis not only eat the fruit but also bury the seeds in the ground, thus in effect planting new trees and stands.

Honey Extraction

One of the most distinctive features of the Hotï subsistence system is their avid procurement of honey. Although the Hotï spend a good deal of time year-round searching for new nests of *bãilã* "bees" (Meliponinae, Apinae), honey collection peaks during the dry season (December–March), and honey becomes one of the most frequently eaten foods. On dry-season treks, when access to cultivated foods is diminished, honey serves as the staple of dietary carbohydrates. In addition to honey, *bãilã helã,* the Hotï consume certain types of bee pollen, *bãilã bũʰkɛhu,* and pupae, *bãila ilĩbõ,* and also gather the wax, *bãilã õ,* which is used to manufacture torches, *bãilã bõkõ.*

The Hotï possess a vast and complex knowledge of Hymenoptera biology, ethology, ecology, and geography, and we are only able to skim the surface of it at this time. We recorded 23 folk species of *bãilã.* However, since no bee specimens were collected (due mainly to lack of legal permits to do so), the correspondence of folk species with scientific species cannot be determined. The natural history of each species is well known. Walking through the forest, our Hotï companions would often point to the presence of different bees and mention something about their habits or habitats. The perception of bees in any context usually elicits a responsive action or at least a comment. The Hotï constantly monitor their environment for movements of bees and locations of their colonies, and this information goes to update their cognitive cartography of wild resources. Newly spotted beehives are quickly assessed with regard to size, potential amount of harvestable honey, and best time to harvest the product. In searching for honey, the Hotï often make use of their knowledge of the tree species in which the bees prefer to nest. Our data show that at least a dozen local bee taxa are associated with specific trees, as shown in Table 3.5. This knowledge suggests potential interspecific co-evolutionary processes that deserve further study, since the biology and ecology of native Amazonian bees are poorly understood by Western science.

Honey-collecting trips are flexible events that can last a few minutes or several days. In the latter case, camp settlements are established and occupied until all the honey in the local area has been depleted. Once the hive has been spotted, the species identified, and the potential productivity and accessibility sized up, the extraction process commences quickly and usually involves one of the following techniques: climbing the tree where the nest is situated or a nearby tree, felling the tree containing the nest, or chopping a hole in the tree trunk. In the first technique the nest is attached externally to the trunk or tree branch and hence is directly accessible. If the bees are an aggressive, stinging species, a makeshift torch will be assembled (usually from dry palm leaves; see Table 3.3) and lit with fire before the collector approaches the nest too closely. Using the torch, the collector will apply smoke to ward off the enraged bees or even partially burn the nest itself in order to kill some of the warriors. When the threat of being stung has been diminished or removed, the collector will use a knife or a pole to knock the nest down to the ground, where it can be easily harvested. However, the vast majority of bee species exploited by the Hotï do not sting, so usually there is no need to apply fire. In the second scenario, the tree may be too tall, access too difficult for the tree to be climbed easily, or the nest located inside the trunk, so the collector simply chops down the whole tree. If the hive is situated in a cavity inside the trunk and the entrance hole is too small for a person to gain access to it, the collector must chop out one or more openings to reach the honey.

Given the great diversity of bees and the complexity of their interspecific interactions, assessing the ecological impacts of honey extraction obviously is not a quick and easy exercise. Furthermore, we did not include apian biology and ecology as a focal subject in our original research design and therefore have not investigated this topic in as detailed and rigorous fashion as it deserves. However, based on the testimony of our Hotï collaborators, who do possess long-term, intimate knowledge of bees, as well as our own multiple observations of honey-harvesting procedures, we are able to offer some tentative conclusions about the global impacts of this subsistence activity on the surrounding environment.

We are concerned primarily with assessing the implications of harvesting practices in regard to the conservation of bee populations and the plants they pollinate and interact with. In some of the extraction events we witnessed, the collectors seemed to leave small portions of the nest behind, a practice that has been attributed elsewhere (Posey 1983b) to conserving the viability of the colony. By contrast, in other harvest events it appeared as though all of the available honey, pollen, pupa, combs, and wax were taken away. In any case, the queen, upon whom the survival of the colony depends, is generally not abducted or killed during the harvest. According to our Hotï consultants, the act of harvesting a nest does not doom the colony but instead incites the swarm to abandon the site and reestablish the nest in a nearby spot. If this is true, the predatory behavior of people functions as a catalyst to the migration and renovation of

Table 3.5
Hotï bee-tree association

Bee taxon	Nest site (tree species, etc.)
uli bãilã	*yewi bãu ʰyaɨ* [*Protium tenuifolium* (Engler) Engler]
luelãtɨ bãilã	*ʰtubĩlu ʰyaɨ* [*Terminalia amazonia* (Gmel) Exell] *bũʰkɨkɨ ʰyaɨ* [*Jacaranda copaia* (Aubl.) D. Don] *hãlĩ hõlõ ʰyaɨ* [*Eschweilera parvifolia* Mart. ex A. P. D. C.] *bãlĩo ʰyaɨ* (*Pterocarpus rohrii* Vahl)
ʰwodekai bãilã	*hãlĩ hõlõ ʰyaɨ* [*Eschweilera aff. pedicellata* (Rich.) Mori] *bũʰkɨkɨ ʰyaɨ* [*Jacaranda copaia* (Aubl.) D. Don] *alawilĩ ʰyaɨ* [*Ouratea castaneifolia* (DC.) Engler] *yewi bãu ʰyaɨ* (*Protium* sp.) *ʰkaba ʰyaɨ* [*Tachigali guianensis* (Benth.) Zarucchi & Herend] *hãlĩ bãu ʰyaɨ* [*Progium sagotianum* Marchand]
baʰta bãilã	*ʰwĩʰwio ʰyaɨ, balo ʰwaĩʰka ʰyaɨ* [*Guazuma ulmifolia* Lam.]
ɨlɛtia bãilã	*hãlĩ hõlõ ʰyaɨ* [*Eschweilera parvifolia* Mart. ex A. P. D. C.] *alawilĩ ʰyaɨ* [*Ouratea castaneifolia* (DC.) Engler] *bũ tawɨ*
kwailɛ̃wa bãilã	*ʰkawɨle ʰyaɨ* (*Erisma uncinatum* Ward.) *bãbãi ʰyaɨ* (*Ludwigia* sp.) *hãlĩ bãu ʰyaɨ* (*Protium* sp.)
ʰkwaʰlaho bãilã	*bũʰkɨkɨ ʰyaɨ* [*Jacaranda copaia* (Aubl.) D. Don] *hãlĩ waĩyo ʰyaɨ* [*Pseudolmedia laevis* (R. & P.) Macbride]
aʰtuwehai bãilã	*lĩʰtawɨ, bũʰkɨkɨ ʰyaɨ* [*Jacaranda copaia* (Aubl.) D. Don]
bũ bãilã	*uli laudau ʰyaɨ* (*Sloanea* sp.) *ʰtawɨ bokwa* [inside fallen log]
čolobã bãilã	*uli tabali ʰyaɨ* [*Ceiba pentandra* (L.) Gaertn.]
lẽikwa bãilã	*lẽikwa bẽ dekaibata* [subterranean]
yewi bãilã	*alawilĩ ʰyaɨ* [*Ouratea castaneifolia* (DC.) Engler]
iyɛ̃ʰtela bãilã	*hãlĩ waĩyo ʰyaɨ* (*Pseudolmedia* sp.)
hitɨ bãilã	*waĩʰtukwa* (inside termite nest)

nests. Such movement may be beneficial from a production standpoint because it limits the buildup of parasites or fungi in the nest over time, which might weaken the colony, or because it precludes invasion by the aggressive, nonnative Africanized bees. Moreover, the Hotï claim that the building of a new nest results in an increase of nectar and pollen-gathering activity by the bees, due to the imperative of restoring the colony's food supply. In other words, the bees work harder when they have to establish a new nest. Having to collect more nectar and pollen, the bees are more apt to pollinate a greater number of flowers. The establishment of a new nest does not take very long, and the same colony of bees subject to human predation will be resettled, replenished, and ready to be harvested again in one or two years, depending on the species. By this account, Hotï manage the bee populations and the honey they produce as if they were renewable resources, which may explain why this activity is sustainable at a certain low level of predatory pressure. In sum, Hotï honey-extraction practice may be regarded as ecologically conservative, because it does not imply the inevitable extinction of the colony, it accelerates the movement of bees within the local environment, it stimulates an increase of nectar gathering and hence pollination, and it is renewable after a one- or two-year wait.

The rather common practice of felling trees to facilitate the extraction of honey also leaves its mark on the landscape. The frequent creation of gaps in closed-canopy forest simulates small, natural disturbance events that trigger multiple dynamics in the local biotic communities and populations. It should be stressed that most of the felled trees associated with the collection of honey are old (partially or mostly hollow) or experiencing drying. Creating gaps where old trees once stood has the effect of speeding up the turnover of forest species and patches, which in turn contributes to the increase of local biodiversity, if only by creating more and diverse microhabitats. Along Hotï trails it is common to observe abundant traces of past honey-extracting episodes. Just to give an idea how intensive this impact can be, three of the large tagged trees in the 1-ha forest plot at Caño Iguana were felled for the purpose of honey gathering during our three-year research period, thereby opening up light gaps that were quickly colonized by other plant types (i.e., Cecropiaceae, Urticaceae, Piperaceae). A simple extrapolation of these results leads to the conclusion that one tree per hectare per year is felled by Hotï honey extractors. The real level of impact will, of course, vary according to settlement size and permanence and other variables, but it should nevertheless be clear that the very frequent creation of gaps by human honey gatherers must be taken into account when considering the variables that determine the floristic composition of the Maigualida region. However, more systematic studies of honey extraction and gap creation by the Hotï are needed in order to define just what their role is.

Spatial-Temporal Context of Hotï Disturbance Behavior

The preceding description of Hotï disturbance behaviors and their environmental effects provides an essentially synchronic perspective of the possible anthropogenic creation of the Sierra de Maigualida landscape, necessitated in part by the space and time limitations of our field research project. However, especially given the small scale of disturbances represented in these examples, a diachronic perspective is obviously needed in order to grasp the full magnitude of human modification of the natural environment. Thus it is important to consider briefly the spatial-temporal context in which these behaviors occur.

The cultural and environmental history of the Hotï and the Sierra de Maigualida region they inhabit is a big question mark. Because contact with the Western world is barely a generation old, except for vague allusions there is no mention of the history in the literature prior to that time. No archaeological research has been conducted so far, so we have no idea about the antiquity or evolution of human settlement in the region.

Phytoecological studies in the Maigualida are incipient and confined mostly to the high-altitude, *tepuy* summit zones (>2000 m asl) (O. Huber pers. comm.). The ethnographic research realized since contact was established, beginning with missionaries', explorers', and journalists' accounts in the early 1970s and culminating with our own research at the end of the 1990s, points to considerable sociocultural, technological, and environmental changes during this period, which have affected some but not all communities (for more information, see S. Zent & E. L. Zent 2004). Such changes include migration to new habitats, larger and more permanent settlements, economic reorientations (notably greater emphasis on agriculture and fishing in some communities), and generalized access to Western trade goods. In regard to the last, it is important to note that the acquisition of steel axes may have facilitated tree felling, certainly one of the key forms of environmental disturbance by the Hotï. We therefore caution that the types of disturbance behavior reported in this chapter, and the implied environmental effects thereof, cannot be extrapolated in a simple and straightforward way into the past. Basically, the only information about the past that is available to us comes from the Hotï themselves. According to their oral history, the circum-Maigualida region, especially the upper Cuchivero, is their traditional homeland, and many more people lived there in the past than do today. Furthermore, they believe that the precontact Hotï did more trekking, relied to a greater extent on palm fruits, maintained more extensive networks of trail gardens, exploited palm grubs and honey at least as much as they do today, had very few steel tools, and made greater use of fire to fell trees and clear unwanted vegetation. A more complete diachronic picture of human impact on the Maigualida environment will depend on the expansion of research in this fascinating region. We think that several research lines could help to bring into focus this clouded history: archaeoethnobotany, paleoecology, patch dynamics, botanical inventory, community floristics, faunal censuses, and human-Hymenoptera-plant relationships.

Discussion

Our Hotï case study provides various examples of how the subsistence activities of a foraging-oriented, indigenous human population in the tropical forest exerts a creative impact on the composition, diversity, configuration, and functional integration of the biological habitat. Although the precise history of this region and its people are not well known, we consider it likely that such impacts, compounded in time and space through long-term occupation of the Maigualida region, were indeed responsible for shaping and defining to a meaningful extent the floristic structure and composition of significant fragments of the forest. In other words, we interpret the data collected so far as supporting the hypothesis that human activities constitute a central variable in the formation and conservation of this ecosystem. This conclusion and the facts that support it contain important implications, both for the theoretical understanding of the generative relationship between small-scale, low-technology, foraging-dependent Amazonian Indians and the culturalized forests they inhabit and for the practical applicability and design of a biocultural conservation strategy in this environment.

Although high biotic diversity and speciation in the Tropics have not been fully explained (Pianka 1966), one appealing explanation perceives them as a function of the interaction of complex interspecific relationships, such as competition and predation, with disturbance events (Waide & Lugo 1992: 176; cf. Sousa 1984). Many of the traditional subsistence activities of native Amazonian human populations, such as those outlined in this chapter, encompass both processes simultaneously—that is, they are disturbance events interwoven with predatory and dispersal practices akin to those of nonhuman species. If, as reported, between 75% and 95% of tropical forest seeds are dispersed by animals (Janzen & Vázquez 1991; Julliot 1997; Pacheco & Simonetti 2000), then we wonder, What is the proportional responsibility of human dispersal and manipulation of plant reproductive material for species distributions and community composition patterns? In the first place, humans are more mobile and are capable of carrying larger amounts of fruits than are most other seed-dispersing animals.[12] Second, we contend that the seed handling by people has greater potential ecological impact than does comparable activity by nonhuman animals precisely because human predation and dispersing activities often occur in direct or indirect association with disturbance actions such as tree felling, fire setting, and faunal harvests. A good example of the interconnectedness and amplifying effect of different disturbance actions is provided by considering a series of disturbance events, starting with a fruit harvest and culminating in the founding of a base settlement, that we were able to monitor over the course of two years in the lower Mosquito region. First, felling two old *ulu* palm trees for the immediate purpose of harvesting the fruit created a small clearing. A temporary settlement was established nearby, and most of the fruit was eaten and then discarded at this campsite. The site was revisited a few months later, and the trunk was split open in order to harvest palm grubs. Sometime later a few individuals of maize, papaya, squash, and cotton were planted in the clearing, effectively establishing a trail garden. About six months after that, a more permanent dwelling was constructed and a swidden plot felled, burned, and planted in the same general location. Juvenile *Attalea maripa* individuals were observed to be one of the most common weed plants in the garden.

One of the most significant forms of ecological disturbance attributed to forest-dwelling humans is the creation and manipulation of gaps in the forest cover. Gaps are considered central to the "structure and function" as well as to the maintenance of "high tree species richness" in tropical forests (Brokaw 1985: 54). Studies carried out in different forest ecosystems around the world have shown that many species are dependent on gaps or open grounds (Sutherland 1998: 210). For example, some emergent tree species "may depend on growth in a gap to reach maturity" (p. 54). A big difference in the way gaps are created and succession instigated by humans versus natural processes is that the former cannot be described as a random or stochastic process. The Hotï case should well illustrate that gap making by people is biased toward the extinction of less useful plants and the immigration or succession of more useful plants. Posey (1998: 105) reports analogous behavior among the Kayapó, noting that gap management produces "resource islands" for humans and game animals. The research results reported for other traditional Amazonian groups (Denevan & Padoch

1987; Irvine 1989; Balée 1993; Zent 1995, 1997) also confirm the "utilitarian bias" of gap management by people.

The ecological impact of human disturbance behaviors can be analyzed as a function of three principal variables describing disturbance events: scale, frequency, and intensity (cf. Petraitis et al. 1989; Hoopes & Harrison 1998; Sutherland 1998). Scale, also referred to as size or magnitude, is defined as the amount of area or resources disrupted. Most of the disturbance areas created and/or manipulated by the Hotï fall in the range of very small (single tree falls, usually less than 200 m²), small (multiple tree falls, usually less than 0.5 ha), and medium (swidden plot felling and burning, usually less than 1 ha), the scale differences being correlated with different regenerative mechanisms as well as different recovery times (cf. Uhl 1983). Frequency, or periodicity, is the number of events per unit of time. The relevant literature mentions three basic categories for describing the frequency of ecological disturbances: high-frequency events (short recurrence intervals of 3–30 years), where the environments experiencing them are usually dominated by early succession species; low-frequency events (long recurrence intervals of 50–200 years), characteristic of environments dominated by late succession species; and medium-frequency events (recurrence intervals of 30–50 years), associated with environments having the greatest mix of early-succession and late-succession species and hence the highest levels of biodiversity (cf. Barbour et al. 1987; Pickett et al. 1992; Odum 1993; Meffe & Carroll 1994). It is significant to note that Hotï shifting cultivation practices (i.e., swidden-fallow cycles) and intermittent exploitation of hunting-collecting territories in conjunction with intergenerational settlement shifts correspond to the 30–50-year interval range. Intensity refers to the severity or lasting impact of the disturbance event in terms of the time and energy costs of recovery (sometimes conceived as scale × frequency). Two intensity levels are usually distinguished: low intensity, which are localized and reversible, without large energy investments (stimulating colonization by early-succession species, eliminating some but not all seeds of late-succession species, and prolonging the succession process); and high intensity, which are extensive and require high energy investments to return to the original state (i.e., implying drastic structural and compositional changes). The disturbances observed among the Hotï tend to be of low intensity. There is no evidence to suggest that they are responsible for intensive modifications of habitat, such as the conversion of forest to savanna, although this possibility needs to be investigated in a few restricted locations, such as San José de Kayamá, where some patches of savanna are found.[13] In sum, the case of Hotï disturbance behavior, analyzed here in terms of scale, frequency, and intensity, confirms the view that relatively unacculturated, indigenous peoples act as small-to-intermediate-impact disturbance agents in the Amazonian landscape. This role is consistent with the enhancement and maintenance of biodiversity levels and therefore should be taken into account when developing models for explaining the extraordinary richness of the tropical forest biota (Goldammer 1992; Balée 1998a).

It should be clear from our discussion that the Hotï and other Amazonian Indians constitute an important component in the dynamic structure and function of Amazonian ecosystems as we know them today. What is the implication of this finding for biological conservation planning? First, it is necessary to recognize that these people do not always conform to Western standards, ideals, or images of environmentally friendly behavior. The research results reported in this chapter do not support indiscriminate acceptance of the ecologically noble savage hypothesis, which supposes that the exploitative habits of indigenous peoples are perfectly adjusted to natural resource constraints and, therefore, that a state of environmental equilibrium prevails (cf. Redford 1991; Kidder 1998; Grande 1999). Rather, from our perspective these people are perceived as conscious and purposeful agents of ecological disturbance who seek to effectuate environmental change by manipulating biological and cultural processes in order to satisfy a resource production goal. However, we also emphasize the ethical neutrality of these results: disturbance behavior by humans neither necessarily induces environmental degradation or species extinction nor suggests that the potential increase of biodiversity due to human management will improve biospheric conditions (cf. Balée 1998a). But such behavior, independent of its valuation by contemporary environmental ideologies, is nevertheless meaningful for the biological composition and ecological organization of contemporary Amazonian ecosystems.

Although we have highlighted the role of Amazonian Indians as ecological disturbance agents, we should also reiterate that they have occupied this region for a very long time, and it would be highly inaccurate to accuse them of having caused the large-scale degradation of their habitat or the erosion of its immense biological wealth.[14] Obviously, the Indians have an intimate knowledge of their biotic surroundings, are skillful manipulators of ecological relationships and processes, and tend to pursue sustainable resource-management practices. The results of the present study show that forager groups such as the Hotï are small- and intermediate-disturbance providers, services that emulate natural processes and actually enhance biodiversity. Accordingly, we conclude that these people have much to contribute to biocultural conservation programs: their ethnoecological beliefs and practices provide models for a new environmental ethic; their subsistence behaviors supply vital inputs for ecosystem maintenance; and the people themselves constitute strategic human resources for implementing development initiatives. In this regard, perhaps the main threat to the compatible coexistence of indigenous human populations and the rich biotic diversity characteristic of their homelands stems from the rapid and drastic changes in social and economic lifestyle that these people are currently experiencing. Among the Hotï, as among other groups (cf. Mansutti 1988; S. Zent 1999), there is a definite trend toward larger and more stationary settlements, and consequently one can observe wild-resource depletions in the vicinity of these areas. Another problematic trend involves the growing use and dependence on Western goods, whereby former luxuries have become necessities, stimulating surplus production as well as the development of more consumer-oriented ideologies.[15] The great challenge for integrated biological-cultural conservation initiatives is to strike a balance between the modernizing socioeconomic institutions and the maintenance of biodiversity and ecosystem functioning. It is simply not realistic to expect that the Indians' lifestyles can be frozen in time or prevented from changing. In any event, we are once again reminded that the relationship between Amazonian Indians and the ecosystems they inhabit is a dynamic one. Strategies for biocultural conservation must embrace this essential truth.

Acknowledgments

Thanks are due to Melissa Melby, Lavon Smith, Luisa Maffi, Tom Carlson, Bill Balée, Charles Peters, Sally Kerr, Erika Wagner, and Rafael Gassón for their editorial and critical commentary on earlier versions of this chapter. Leticia Marius assisted in preparing the map. The fieldwork was made possible by financial support from the Consejo Nacional de Investigaciones Científicas y Tecnológicas, the National Science Foundation, the Wenner-Gren Foundation for Anthropological Research, and the Instituto Venezolano de Investigaciones Científicas. Herbarium support was provided by the Herbario Victor Manuel Ovalles, the Herbario Nacional de Venezuela, and the Herbario Julian Steyermark del Estado Amazonas.

Notes

1. Overlapping our research, Robert Storrie, a doctoral student from Manchester University, conducted fieldwork with the Hotï from 1994 to 1997. His doctoral dissertation (Storrie 1999) focuses on kinship, cosmology, and personhood.

2. Both resource types are intimately related in knowledge as well as practice. Thus hunting expeditions are often organized according to knowledge of the feeding behaviors of different game animals, the fruiting schedules of the plants they eat, and the localities where stands of these plants are found.

3. It is probable that steel tools were acquired by the Hotï relatively recently (see Coppens 1983) and, therefore, that shifting cultivation was less intensively practiced before the modern (i.e., postcontact) period (cf. Denevan 1992b). The corollary of this likely circumstance is that reliance on foraging was even greater in the recent past than it is today.

4. A similar argument is made by Fragoso (1997), who considers that the small-scale seed-dispersing behaviors of the lowland tapir influence the landscape-scale diversity and patchiness of vegetation in a Brazilian rain forest.

5. *Oenocarpus bacaba* has also been reported as the most abundant species in sample plots located in upland forests in the Brazilian Amazon (Ferreira & Prance 1998).

6. In some cases, our collaborators were able to recall who lived there.

7. Fragoso (1997) also observed numerous *Attalea maripa*–dominant stands in protected (i.e., uninhabited by humans) primary rain forest in northern Brazil. His research found that primary seed dispersal by tapirs and secondary dispersal by rodents, rather than edaphic factors, are mainly responsible for the clumped distribution of this species. Building on this insight, we argue that the *ulu čo* phenomenon in the forests of the Maigualida region, which have long been occupied by humans, is best explained as the result of human transport and handling of seed material, along with the removal of competing species through fire application.

8. This would seem to indicate that the local populations of *Oenocarpus bataba, Rhinostomus barbirostris,* and *Rhynchophorus palmarum* have suffered some degree of depletion in the vicinity of the more nucleated, sedentary communities (probably due to greater exploitation pressure) and perhaps also that this form of resource management is sustainable only under the traditionally small, dispersed, and mobile settlement pattern.

9. We did witness a few exceptions to this norm, however. Members of one of the settlement units in the Mosquito region ate adult weevils, but not many of them.

10. Nutritional analysis of *Rhynchophorus palmarum* larva shows that its caloric value is 165 Kcal / 100 g; its protein content, 7.33 g / 100 g, which is superior to that of milk (3.1 g / 100

g); its fat content, 10.9 g / 100 g; and its carbohydrate content, 9.4 g / 100 g. It is also rich in vitamin A (equal to 85.0 μg retinol) and vitamin E (9.82 mg / 100 g fresh weight of alfa tocoferol) and contains 11 essential amino acids (Cerda et al. in press).

11. We encountered a fruit-bearing *Musa* stand in a trail garden in the Upper Cuchivero region that our Hotï guide claimed was at least 10 years old.

12. Fragoso and Huffman (2000) refer to essentially the same capabilities in arguing the particular importance of tapirs for plant diversity and distribution patterns in Amazonian rain forests.

13. We did observe the Hotï at Kayamá burning the savanna grass during the late dry season, motivated by the stated desire to "clean up" the savanna and thus facilitate walking and permit better visibility for hazards such as snakes.

14. Some scholars believe that indigenous human populations are partly responsible for the conversion of forest to savanna habitat over long time periods in the Gran Sabana (Fölster 1992) and the Sierra Parima (Huber et al. 1984; Smole 1989) regions of southern Venezuela, but these ideas are still somewhat controversial. In any case, the environmental modifications that occurred in these areas do not come close in scale or gravity to the devastating deforestation caused by the expansion of national and international social and economic frontiers into Indian-occupied lands in the last 30–40 years (cf. Fearnside 1987; Moran 1993).

15. For the nonmission Hotï, captive wild animals (mainly monkeys, parrots, parakeets, and toucans) have become a most valuable commodity, which they trade downriver to other Indians (Yabarana, Piaroa, E'ñepa) in exchange for cash or Western manufactured goods. However, most animals die in captivity before an exchange can be made. We witnessed five infant or juvenile spider monkeys (*Ateles belzebuth*) taken alive in a two-week period in the Mosquito region. Not one survived long enough to be taken away and traded.

Literature Cited

Anderson, A. B. & E. M. Ioris. 1992. The logic of extraction: Resource management and income generation by extractive producers in the Amazon. Pp. 175–199 *in* K. H. Redford & C. Padoch (eds.), Conservation of Neotropical Forests: Working from Traditional Resource Use. New York: Columbia University Press.

——— & D. A. Posey. 1989. Management of a tropical scrub savanna by the Gorotire Kayapó of Brazil. Pp. 159–173 *in* D. A. Posey & W. Balée (eds.), Resource Management in Amazonia: Indigenous and Folk Strategies. Advances in Economic Botany, 7. Bronx: New York Botanical Garden.

———, P. H. May & M. Balick. 1991. The Subsidy from Nature: Palm Forests, Peasantry, and Development on the Amazon Frontier. New York: Columbia University Press.

Balée, W. 1988. Indigenous adaptation to Amazonian palm forests. Principes 32(2): 47–54.

———. 1989. The culture of Amazonian forests. Pp. 1–21 *in* D. A. Posey & W. Balée (eds.), Resource Management in Amazonia: Indigenous and Folk Strategies. Advances in Economic Botany, 7. Bronx: New York Botanical Garden.

———. 1992. People of the fallow: A historical ecology of foraging in lowland South America. Pp. 35–57 *in* K. H. Redford & C. Padoch (eds.), Conservation of Neotropical Forests: Working from Traditional Resource Use. New York: Columbia University Press.

———. 1993. Indigenous transformation of Amazonian forests. L'Homme 126–128: 231–254.

———. 1994. Footprints of the Forest: Ka'apor Ethnobotany—The Historical Ecology of Plant Utilization by an Amazonian People. New York: Columbia University Press.

———. 1995. Historical ecology of Amazonia. Pp. 97–110 *in* L. Sponsel (ed.), Indigenous Peoples and the Future of Amazonia. Tucson: University of Arizona Press.

———. 1998a. Introduction. Pp. 4–12 *in* W. Balée (ed.), Advances in Historical Ecology. New York: Columbia University Press.
——— (ed.). 1998b. Advances in Historical Ecology. New York: Columbia University Press.
Balick, M. J. 1984. Ethnobotany of palms in the Neotropics. Pp. 9–23 *in* G. T. Prance & J. A. Kallunki (eds.), Ethnobotany in the Neotropics. Advances in Economic Botany, 1. Bronx: New York Botanical Garden.
———. 1986. Systematics and Economic Botany of the Oenocarpus-Jessenia (Palmae) Complex. Advances in Economic Botany, 3. Bronx: New York Botanical Garden.
———. 1988. Jessenia and Oenocarpus: Neotropical Oil Palms Worthy of Domestication. Rome: Food and Agriculture Organization of the United Nations.
Barbour, M. G., J. H. Burk & W. D. Pitts. 1987. Terrestrial Plant Ecology. Ed. 2. Menlo Park, Calif.: Benjamin/Cummings Publishing Co.
Beckerman, S. 1993. Major patterns in indigenous Amazonian subsistence. Pp. 411–424 *in* C. M. Hladik, A. Hladik, O. F. Linares, H. Pagezy, A. Semple & M. Hadley (eds.), Tropical Forests, People and Food: Biocultural Interactions and Applications to Development. Man and the Biosphere Series, 13. Paris: UNESCO and the Parthenon Publishing Group.
———. 1994. Hunting and fishing in Amazonia: Hold the answers, what are the questions? Pp. 177–200 *in* A. C. Roosevelt (ed.), Amazonian Indians from Prehistory to the Present. Tucson: University of Arizona Press.
Bodin, R. 1969. Finding brown gold. Brown Gold 27(12): 7,9.
Botkin, D. B. & M. J. Sobel. 1975. Stability in time-varying ecosystems. American Naturalist 109: 625–646.
Bou, J. 1970. Found . . . but lost. Brown Gold 28(3): 6–7.
Brokaw, N. V. 1985. Tree falls, regrowth, and community structure in tropical forests. Pp. 53–69 *in* S. T. A. Pickett & P. S. White (eds.), The Ecology of Natural Disturbance and Patch Dynamics. New York: Academic Press.
Cavelier, I., C. Rodriguez, L. F. Herrera, G. Morcote & S. Mora. 1995. No solo de caza vive el hombre: Ocupación del bosque amazónico, holoceno temprano. Pp. 27–44 *in* I. Cavelier & S. Mora (eds.), Ambito y ocupaciones tempranas de la América tropical. Bogotá: Instituto Colombiano de Antropología, COLCULTURA, Fundación Erigaie.
Cerda, H., R. Martínez, N. Briceño, L. Pizzoferrato, D. Hermoso & M. Paoletti. In press. Cría, análisis nutricional y sensorial del picudo del cocotero *Rhynchophorus Palmarum* (Coleoptera: Curculionidae), insecto de la dieta tradicional indígena amazónica. Ecotrópicos.
Clark, K. & C. Uhl. 1987. Farming, fishing, and fire in the history of the upper Rio Negro region of Venezuela. Human Ecology 15(1): 1–26.
Clement, C. R. 1989. A center of crop genetic diversity in western Amazonia. BioScience 39: 624–631.
———. 1999a. 1492 and the loss of Amazonian crop genetic resources, I: The relation between domestication and human population decline. Economic Botany 53(2): 188–202.
———. 1999b. 1492 and the loss of Amazonian crop genetic resources, II: Crop biogeography at contact. Economic Botany 53(2): 203–216.
Colinvaux, P. A. 1989. The past and future Amazon. Scientific American, May, 68–74.
Connell, J. H. 1978. Diversity in tropical rain forests and coral reefs. Science 199: 1302–1310.
Coppens, W. 1975. Contribución al estudio de las actividades de subsistencia de los Hotis del río Kaima. Boletín Indigenista Venezolano, n.s. 16(12): 65–78.
———. 1983. Los Hoti. Pp. 2: 243–302 *in* W. Coppens (ed.), Los aborígenes de Venezuela. Caracas: Fundación La Salle / Monte Avila Editores.
——— & P. Mitrani. 1974. Les Indiens Hoti: Compte rendu de missions. L'Homme 14(3–4): 131–142.

Corradini, H. 1973. Los indios chicanos. Venezuela Misionera 35(405): 6–9; 35(406): 42–44; 35(407): 88–91.
Crumley, C. (ed.). 1994. Historical Ecology. Santa Fe, N.Mex.: School of American Research Press.
Cruxent, J. M. 1961. Los Shikana del río Kaima. Unpublished manuscript.
Denevan, W. M. 1992a. The pristine myth: The landscape of the Americas in 1492. Annals of the Association of American Geographers 82(3): 369–385.
———. 1992b. Stone vs. metal axes: The ambiguity of shifting cultivation in prehistoric Amazonia. Journal of the Steward Anthropological Society 20(1–2): 153–165.
——— & C. Padoch (eds.). 1987. Swidden-Fallow Agroforestry in the Peruvian Amazon. Advances in Economic Botany, 5. Bronx: New York Botanical Garden.
Descola, P. 1996. In the Society of Nature: A Native Ecology in Amazonia. N. Scott, trans. Cambridge and New York: Cambridge University Press.
Durbin, M. 1977. A survey of the Carib language family. Pp. 23–38 *in* E. Basso (ed.), Carib-Speaking Indians: Culture, Society and Language. Tucson: University of Arizona Press.
Dye, P. 1970. Yowana contact. Brown Gold 28(8): 3–5, 15.
Eden, M. J., W. Bray, L. Herrera & C. McEwan. 1984. Terra preta soils and their archaeological context in the Caquetá Basin of southeast Colombia. American Antiquity 49(1): 125–140.
Eibl-Eibesfeldt, I. 1973. Die Waruwádu (Yuwana), ein kurzlich entdeckter, noch unerforschter Indianerstamm Venezuelas. Anthropos 68(1–2): 137–144.
Eisenberg, J. F. 1989. Mammals of the Neotropics, vol. 1: The Northern Neotropics: Panama, Colombia, Venezuela, Guyana, Suriname, French Guiana. Chicago: University of Chicago Press.
Fearnside, P. 1987. Deforestation and international economic development projects in Brazilian Amazonia. Conservation Biology 1(3): 214–221.
Ferreira, L. V. & G. T. Prance. 1998. Species richness and floristic composition in four hectares in the Jau National Park in upland forests in Central Amazonia. Biodiversity and Conservation 7(10): 1349–1364.
Fölster, H. 1992. Holocene autochtonous forest degradation in southeast Venezuela. Pp. 25–44 *in* J. G. Goldammer (ed.), Tropical Forests in Transition: Ecology of Natural and Anthropogenic Disturbance Processes. Basel, Switzerland, and Boston: Birkhäuser Verlag.
Fragoso, J. M. V. 1997. Tapir-generated seed shadows: Scale-dependent patchiness in the Amazon rain forest. Journal of Ecology 85: 519–529.
——— & J. M. Huffman. 2000. Seed-dispersal and seedling recruitment patterns by the last Neotropical megafaunal element in Amazonia, the tapir. Journal of Tropical Ecology 16: 369–385.
Goldammer, J. G. 1992. Tropical forests in transition: Ecology of natural and antropogenic disturbance processes—An introduction. Pp. 1–16 *in* J. G. Goldammer (ed.), Tropical Forests in Transition: Ecology of Natural and Anthropogenic Disturbance Processes. Basel, Switzerland, and Boston: Birkhäuser Verlag.
Gorzula, S. & J. Celsa S. 1998. Contribution to the herpetofauna of the Venezuelan Guayana, I: A data base. Scientia Guaianae, 8. Caracas: REFOLIT.
Gragson, T. L. 1995. Pumé exploitation of *Mauritia flexuosa* (Palmae) in the Llanos of Venezuela. Journal of Ethnobiology 15(2): 177–188.
Grande, S. M. A. 1999. Beyond the ecologically noble savage: Deconstructing the white man's Indian. Environmental Ethics 3(21): 307–320.
Gremone, C., F. Cervigón, S. Gorzula, G. Medina & D. Novoa. 1986. Fauna de Venezuela: Vertebrados. Caracas: Editorial Biosfera.
Guarisma P., V. 1974. Los Hoti: Introducción etno-lingüística. Thesis, Universidad Central de Venezuela.

——— & W. Coppens. 1978. Vocabulario Hoti. Antropológica 49: 3–27.

Haffer, J. 1969. Speciation in Amazonian forest birds. Science 165: 131–137.

Heinen, H. D. & R. Lizarralde. 1991. El ecosistema morichalero del delta Orinoco y su utilización humana. Pp. 481–526 *in* J. J. San José & J. Celecia (eds.), Enfoques de ecología humana aplicados a los sistemas agrícolas tradicionales del trópico americano. Caracas: CIET/ UNESCO.

———, J. J. San José, H. Caballero A. & R. Montes. 1995. Subsistence activities of the Warao Indians and anthropogenic changes in the Orinoco Delta vegetation. Pp. 312–334 *in* H. D. Heinen, J. J. San José & H. Caballero A. (eds.), Naturaleza y ecología humana en el neotrópico. Scientia Guaianae, 5. Caracas: REFOLIT.

Henley, P., M-C. Mattei-Müller & H. Reid. 1994–1996. Cultural and linguistic affinities of the foraging people of northern Amazonia: A new perspective. Antropológica 83: 3–37.

Hoopes, M. F. & S. Harrison. 1998. Metapopulation, source-sink and disturbance dynamics. Pp. 125–151 *in* W. Sutherland (ed.), Conservation Science and Action. Oxford: Blackwell Science.

Huber, O. & C. Alarcón. 1988. Mapa de vegetación de Venezuela. 1:2.000.000. Caracas: Ministerio del Ambiente y de los Recursos Naturales Renovables / The Nature Conservancy.

———, J. A. Steyermark, G. T. Prance & C. Alés. 1984. The vegetation of the Sierra Parima, Venezuela-Brazil: Some results of recent exploration. Brittonia 36(2): 104–139.

Irvine, D. 1989. Succession management and resource distribution in an Amazonian rain forest. Pp. 223–237 *in* D. A. Posey & W. Balée (eds.), Resource Management in Amazonia: Indigenous and Folk Strategies. Advances in Economic Botany, 7. Bronx: New York Botanical Garden.

Jangoux, J. 1971. Observations on the Hoti Indians of Venezuela. Unpublished manuscript.

Janzen, D. & C. Vázquez Y. 1991. Aspects of tropical seed ecology of relevance to tropical forested wildlands. Pp. 137–154 *in* A. Gómez-Pompa, T. C. Whitmore & M. Hadley (eds.), Rain Forest Regeneration and Management. Man and the Biosphere Series, 6. Paris: Unesco.

Julliot, C. 1997. Impact of seed dispersal by red howler monkeys *Alouatta seniculus* on the seedling population in the understorey of tropical rain forest. Journal of Ecology 85: 431–440.

Junk, W. J. 1995. Human impact on neotropical wetlands: Historical evidence, actual status, and perspectives. Pp. 299–311 *in* H. D. Heinen, J. J. San José & H. Caballero A. (eds.), Naturaleza y ecología humana en el neotrópico. Scientia Guaianae, 5. Caracas: REFOLIT.

Kaplan, H. & K. Kopischke. 1992. Resource use, traditional technology, and change among native peoples of lowland South America. Pp. 83–107 *in* K. H. Redford & C. Padoch (eds.), Conservation of Neotropical Forests: Working from Traditional Resource Use. New York: Columbia University Press.

Kaufmann, J. B. & C. Uhl. 1990. Interactions of anthropogenic activities, fire and rain forests in the Amazon Basin. Pp. 117–134 *in* J. G. Goldammer (ed.), Fire in the Tropical Biota: Ecosystem Processes and Global Challenges. Berlin and New York: Springer-Verlag.

Keogh, F. K. 1995. Where rocks grow and God has shoes: Reflections and shifting realities in the Venezuelan Amazon. Ph.D. diss., University of Michigan.

Kidder, T. 1998. The rat that ate Louisiana: Aspects of historical ecology in the Mississippi River Delta. Pp. 141–168 *in* W. Balée (ed.), Advances in historical ecology. New York: Columbia University Press.

Linares, O. L. 1998. Mamíferos de Venezuela. Caracas: Sociedad Conservacionista Audubon de Venezuela.

Mansutti, A. 1988. Pueblos, comunidades y fondos: Los patrones de asentamiento Uwotjuja. Antropológica 69: 3–36.

MARNR–ORSTOM. 1988. Atlas del Inventario de Tierras del Territorio Federal Amazonas. Caracas: MARNR–DGSIIA.

Meffe, G. & C. R. Carroll. 1994. Principles of Conservation Biology. Sunderland, Mass.: Sinauer Associates.

Meggers, B. J. 1987. The early history of man in Amazonia. Pp. 151–174 *in* T. C. Whitmore & G. T. Prance (eds.), Biogeography and Quaternary History in Tropical America. Oxford: Clarendon Press.

———. 1994. Archeological evidence for the impact of mega-Niño events on Amazonia during the past two millennia. Climatic Change 28: 321–338.

———. 1995. Judging the future by the past: The impact of environmental instability on prehistoric Amazonian populations. Pp. 15–43 *in* L. Sponsel (ed.), Indigenous Peoples and the Future of Amazonia. Tucson: University of Arizona Press.

Milton, K. 1991. Comparative aspects of diet in Amazonian forest-dwellers. Philosophical Transactions of the Royal Society of London, Ser. B 334: 253–263.

Moran, E. 1993. Deforestation and land use in the Brazilian Amazon. Human Ecology 21(1): 1–21.

Morcote-Ríos, G., G. Cabrera-Becerra, D. Mahecha-Rubio, C. E. Franky-Calvo & I. Cavelier-F. 1998. Management of palms by groups of hunter-gatherers from the Colombian Amazon region. Caldasia 20(1): 57–74.

OCEI. 1985. Censo indígena de Venezuela 1982. Caracas: Oficina Central de Estadística e Informática.

———. 1993. Censo indígena de Venezuela 1992. Caracas: Oficina Central de Estadística e Informática.

Odum, E. 1993. Ecology and Our Endangered Life-Support Systems. Sunderland, Mass.: Sinauer Associates.

Pacheco, L. F. & J. A. Simonetti. 2000. Genetic structure of a mimosoid tree deprived of its seed disperser, the spider monkey. Conservation Biology 14(6): 1766–1775.

Padoch, C. & W. De Jong. 1992. Diversity, variation, and change in *ribereño* agriculture. Pp. 158–174 *in* K. H. Redford & C. Padoch (eds.), Conservation of Neotropical Forests: Working from Traditional Resource Use. New York: Columbia University Press.

Petraitis, P. S., R. E. Latham & R. A. Niesenbaum. 1989. Maintenance of species diversity by disturbance. Quarterly Review of Biology 64: 393–418.

Phelps, W. H., Jr. & R. M. de Schauensee. 1978. Una guía de las aves de Venezuela. Princeton, N.J.: Princeton University Press.

Phillips, O. L., P. Hall, A. H. Gentry, S. A. Sawyer & R. Vasquez. 1994. Dynamics and species richness of tropical rain forests. Proceedings of the National Academy of Sciences 91: 2805–2809.

Pianka, E. 1966. Latitudinal gradients in species diversity: A review of concepts. American Naturalist 100: 33–46.

Pickett, S. T., V. T. Parker & P. L. Fiedler. 1992. The new paradigm in ecology: Implications for conservation biology above the species level. Pp. 65–88 *in* P. L. Fiedler & S. K. Jain (eds.), Conservation Biology: The Theory and Practice of Nature Conservation, Preservation, and Management. New York: Chapman & Hall.

Posey, D. A. 1983a. Indigenous ecological knowledge and development of the Amazon. Pp. 225–257 *in* E. F. Moran (ed.), The Dilemma of Amazonian Development. Boulder, Colo.: Westview Press.

———. 1983b. Folk apiculture of the Kayapó Indians of Brazil. Biotrópica 15(2): 154–158.

———. 1984. A preliminary report on diversified management of tropical forest by Kayapó Indians of the Brazilian Amazon. Pp. 112–126 *in* G. T. Prance & J. A. Kallunki (eds.),

Ethnobotany in the Neotropics. Advances in Economic Botany, 1. Bronx: New York Botanical Garden.
———. 1988. Kayapó Indian Natural-Resource Management. Pp. 89–90 *in* J. S. Denslow & C. Padoch (eds.), People of the Tropical Rain Forest. Berkeley: University of California Press.
———. 1994. Environmental and social implications of pre- and postcontact situations on Brazilian Indians: The Kayapó and a new Amazonian synthesis. Pp. 271–286 *in* A. C. Roosevelt (ed.), Amazonian Indians from Prehistory to the Present. Tucson: University of Arizona Press.
———. 1998. Diachronic ecotones and anthropogenic landscapes in Amazonia: Contesting the consciousness of conservation. Pp. 104–118 *in* W. Balée (ed.), Advances in Historical Ecology. New York: Columbia University Press.
Prance, G. T. 1978. The origin and evolution of the Amazon flora. Interciencia 3(4): 207–222.
——— (ed.). 1982. Biological Diversification in the Tropics. New York: Columbia University Press.
Pyne, S. 1998. Forged in fire: History, land and anthropogenic fire. Pp. 64–103 *in* W. Balée (ed.), Advances in Historical Ecology. New York: Columbia University Press.
Redford, K. H. 1991. The ecologically noble savage. Cultural Survival Quarterly 15(1): 46–48.
Richards, P. W. 1952. Tropical rain forest. Cambridge: Cambridge University Press.
Rival, L. 1998. Domestication as a historical and symbolic process: Wild gardens and cultivated forests in the Ecuadorean Amazon. Pp. 232–250 *in* W. Balée (ed.), Advances in Historical Ecology. New York: Columbia University Press.
Roosevelt, A. C. 1992. Secrets of the forest. The Sciences 32(6): 22–28.
Salick, J. 1989. Ecological basis of Amuesha agriculture, Peruvian Upper Amazon. Pp. 189–212 *in* D. A. Posey & W. Balée (eds.), Resource Management in Amazonia: Indigenous and Folk Strategies. Advances in Economic Botany, 7. Bronx: New York Botanical Garden.
———. 1992. Amuesha forest use and management: An integration of indigenous use and natural forest management. Pp. 305–332 *in* K. H. Redford & C. Padoch (eds.), Conservation of Neotropical Forests: Working from Traditional Resource Use. New York: Columbia University Press.
——— & M. Lundberg. 1990. Variation and change in Amuesha agriculture in the Peruvian Upper Amazon. Pp. 199–223 *in* G. T. Prance & M. J. Balick (eds.), New Directions in the Study of Plants and People: Research Contributions from the Institute of Economic Botany. Advances in Economic Botany, 8. Bronx: New York Botanical Garden.
Sánchez, P. & K. Jaffé. 1992. Rutas de migraciones humanas precolombinas a la Amazonia sugeridas por la distribución del cacao. Interciencia 17(1): 28–34.
Sanford, R. L., J. Saldarriaga, K. E. Clark, C. Uhl & R. Herrera. 1985. Amazon rain-forest fires. Science 227: 53–55.
Schroth, G., L. F. da Silva, M.-A. Wolf, W. G. Teixeira & W. Zech. 1999. Distribution of throughfall and stemflow in multi-strata agroforestry, perennial monoculture, fallow and primary forest in central Amazonia, Brazil. Hydrological Processes 13(10): 1423–1436.
Schubert, C. 1988. Climatic changes during the last glacial maximum in northern South America and the Caribbean: A review. Interciencia 13(3): 128–137.
Schultes, R. E. 1994. The importance of ethnobotany in environmental conservation. American Journal of Economics and Sociology 53(2): 202–206.
Smith, N. J. H. 1980. Anthrosols and human carrying capacity in Amazonia. Annals of the Association of American Geographers 70: 553–566.
Smole, W. 1989. Yanomama horticulture in the Parima Highlands of Venezuela and Brazil. Pp. 115–128 *in* D. A. Posey & W. Balée (eds.), Resource Management in Amazonia: Indigenous and Folk Strategies. Advances in Economic Botany, 7. Bronx: New York Botanical Garden.

Sousa, W. P. 1984. The role of disturbance in natural communities. Annual Review of Ecology and Systematics 15: 353–391.

Sponsel, L. E. 1992. The environmental history of Amazonia: Natural and human disturbances, and the ecological transition. Pp. 233–251 *in* H. K. Steen & R. P. Tucker (eds.), Changing Tropical Forests: Historical Perspectives on Today's Challenges in Central and South America. Durham, N.C.: Forest History Society.

Sprugel, D. 1991. Disturbance, equilibrium, and environmental variability: What is 'natural' vegetation in a changing environment? Biological Conservation 58: 1–18.

Stahl, P. W. 1996. Holocene biodiversity: An archaeological perspective from the Americas. Annual Review of Anthropology 25: 105–126.

Stocks, A. 1983. Cocamilla fishing: Patch modification and environmental buffering in the Amazon várzea. Pp. 239–267 *in* R. B. Hames & W. T. Vickers (eds.), Adaptive Responses of Native Amazonians. New York: Academic Press.

Stone, D. (ed.). 1984. Pre-Colombian Plant Migrations. Papers of the Peabody Museum of Archaeology and Ethnology, 76. Cambridge: Harvard University Press.

Storrie, R. 1999. Being human: Personhood, cosmology and subsistence for the Hotï of Venezuelan Guiana. Ph.D. diss., University of Manchester.

Sutherland, W. 1998. Managing habitats and species. Pp. 202–219 *in* W. Sutherland (ed.), Conservation Science and Action. Oxford: Blackwell Science.

Uhl, C. 1983. You can keep a good forest down. Natural History, April, 71–79.

Unruh, J. & J. B. Alcorn. 1987. Relative dominance of the useful component in young managed fallows at Brillo Nuevo. Pp. 47–52 *in* W. M. Denevan & C. Padoch (eds.), Swidden-Fallow Agroforestry in the Peruvian Amazon. Advances in Economic Botany, 5. Bronx: New York Botanical Garden.

——— & S. Flores P. 1987. Relative abundance of the useful component in old managed fallows at Brillo Nuevo. Pp. 67–73 *in* W. M. Denevan & C. Padoch (eds.), Swidden-Fallow Agroforestry in the Peruvian Amazon. Advances in Economic Botany, 5. Bronx: New York Botanical Garden.

Vanzolini, P. E. 1973. Paleoclimates, relief and species multiplications in equatorial forest. Pp. 255–258 *in* B. J. Meggers, E. S. Ayensu & W. D. Duckworth (eds.), Topical Forest Ecosystems in Africa and South America: A Comparative Review. Washington, D.C.: Smithsonian Institution Press.

Vilera, D. 1985. Introducción morfológica de la lengua Hoti. Thesis, Universidad Central de Venezuela.

Waide, R. B. & A. E. Lugo. 1992. A research perspective on disturbance and recovery of a tropical montane forest. Pp. 173–190 *in* J. G. Goldammer (ed.), Tropical Forests in Transition: Ecology of Natural and Anthropogenic Disturbance Processes. Basel, Switzerland, and Boston: Birkhäuser Verlag.

White, P. S. & S. T. A. Pickett. 1985. Natural disturbance and patch dynamics: An introduction. Pp. 3–13 *in* S. T. A. Pickett & P. S. White (eds.), The Ecology of Natural Disturbance and Patch Dynamics. New York: Academic Press.

Whitehead, N. L. 1998. Ecological history and historical ecology: Diachronic modeling versus historical explanation. Pp. 30–41 *in* W. Balée (ed.), Advances in historical ecology. New York: Columbia University Press.

Whitmore, T. C. & G. T. Prance (eds.). 1987. Biogeography and Quaternary History in Tropical America. Oxford: Clarendon Press.

Zent, E. L. 1998. A creative perspective of environmental impacts by native Amazonian human populations. Interciencia 23(4): 232–240.

———. 1999. Etnobotánica Hotï: Explorando las interacciones entre la flora y el ser humano del Amazonas venezolano. Ph.D. diss., University of Georgia.

Zent, S. 1992. Historical and ethnographic ecology of the Upper Cuao River Wõthihã: Clues for an interpretation of native Guianese social organization. Ph.D. diss., Columbia University.

———. 1995. Clasificación, explotación y composición de bosques secundarios en el Alto Río Cuao, Estado Amazonas, Venezuela. Pp. 79–113 *in* H. D. Heinen, J. J. San José & H. Caballero A. (eds.), Naturaleza y ecología humana en el neotrópico. Scientia Guaianae, 5. Caracas: REFOLIT.

———. 1997. Piaroa and the Cracidae: Game management under shifting cultivation. Pp. 177–194 *in* S. D. Strahl, S. Beaujon, D. M. Brokks, A. J. Begazo, G. Sedaghatkish & F. Olmos (eds.), The Cracidae: Their biology and conservation. Hong Kong: Hancock House.

———. 1999. The quandary of conserving ethnoecological knowledge: A Piaroa example. Pp. 90–124 *in* T. Gragson & B. Blount (eds.), Ethnoecology: Knowledge, Resources, and Rights. Athens: University of Georgia Press.

——— & E. L. Zent. 2004. Ethnobotanical convergence, divergence, and change among the Hotï of the Venezuelan Guayana. Pp. 36–77 *in* T. J. S. Carlson & L. Maffi (eds.), Ethnobotany and Conservation of Biocultural Diversity. Advances in Economic Botany, 15. Bronx: New York Botanical Garden Press.

———, E. L. Zent & L. Marius. 2001. Informe final del Proyecto de Investigación S1–96000554, Etnobotánica Quantitativa de los Indígenas Hotï de la Región Circum-Maigualida, Estados Amazonas y Bolívar, Venezuela. Altos de Pipe, Venezuela: Instituto Venezolano de Investigaciones Científicas.

4

Indigenous Knowledge and Conservation of the Rain Forest: Ethnobotany of the Barí of Venezuela

Manuel Lizarralde

As Maffi (2001) points out, we are facing a complex crisis of environmental deterioration and loss of indigenous knowledge. The situation is so alarming that even popular magazines are voicing this concern. For example, the August 1999 issue of *National Geographic Magazine* included an inset map showing global language loss throughout history and an article about the loss of languages and cultures around the world (Davis 1999). The rapid erosion of numerous natural ecosystems and traditional societies is a global tragedy (Johnston 1994; Reaka-Kudla et al. 1997). New directions are needed to address these pressing issues.

Research has provided ample evidence that indigenous peoples have extensive knowledge of their environment. This chapter adds to this line of research by focusing on Barí indigenous knowledge and use of the rain forest, on the cultural, social, economic, and ecological changes these tropical lowland South American people are undergoing, and on the potential uses of their ecological knowledge for forest-conservation efforts. The Barí are currently facing an environmental and cultural crisis and are themselves eagerly seeking solutions to these problems.

The Barí: Ethnographic Background

The Barí are an indigenous society living in the tropical rain forest of the southwestern part of the Lake Maracaibo basin in western Venezuela and northeastern Colombia (Figure 4.1). Traditionally they have been swidden agriculturists, growing sweet manioc and complementing this staple of their diet by fishing, hunting, and gathering wild products from the forest. In traditional Barí subsistence, agriculture supplied the bulk of their foodstuff, with manioc (*Manihot esculenta,* or *mashú,* in the Euphorbiaceae) providing "more than 90% of the calories in the diet" (Beckerman 1983: 283). Other important crops, like banana, sweet potato, yam, pineapple, avocado, and chili pepper are also added to their diet. Fishing provides 75% of the protein; hunting, 25% (Beckerman 1980, 1994; Beckerman & Sussenbach 1983). The Barí

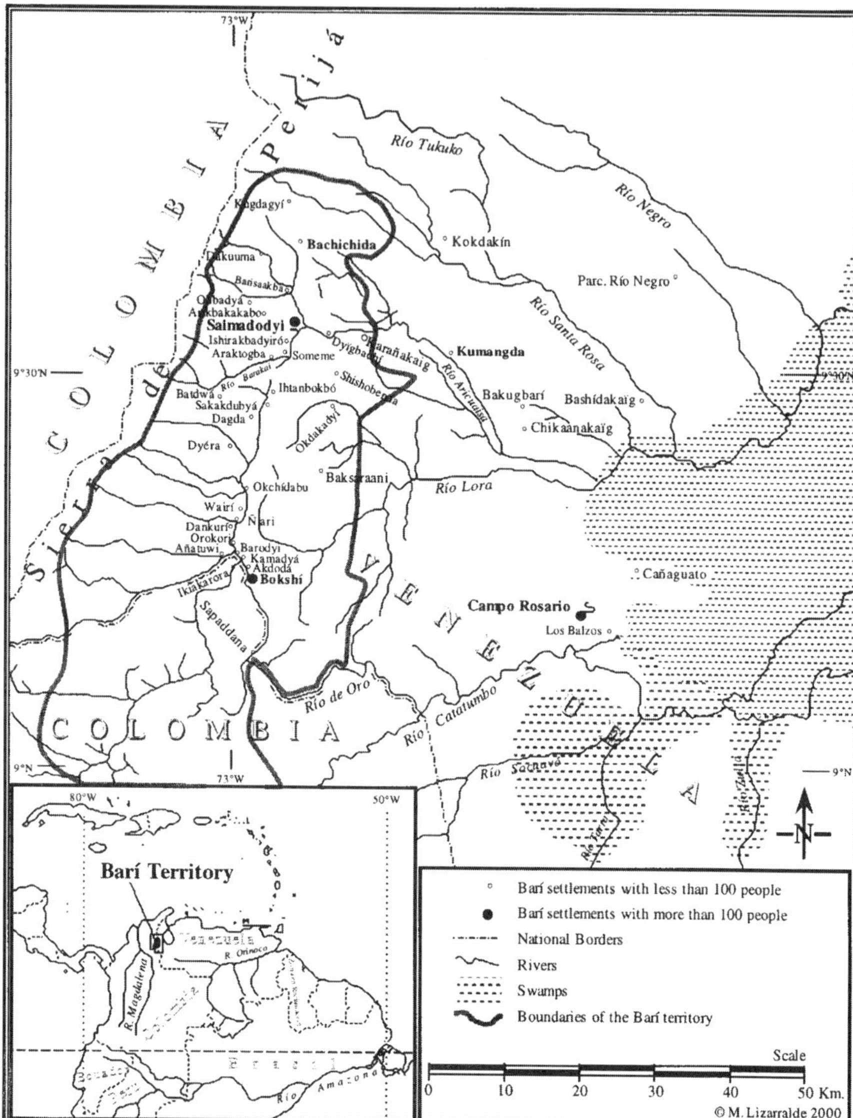

Figure 4.1. The Barí territory and its settlements.

also gather many kinds of forest products, principally palm grubs (*Rhynchophorus palmarum*), honey, palm fruits (mainly *Oenocarpus bataua,* or *arúu,* and *Attalea butyracea,* or *araktá,* in the Arecaceae), and many additional kinds of fruit from trees. Since the late 1960s the Barí have gradually entered the market economy, selling their labor, products, and lumber to buy food, medicine, and Western goods and to pay for the formal education of their children in boarding schools (Lizarralde 1991).

Today there are approximately 1850 Barí in Venezuela, spread among 37 villages and hamlets (OCEI 1985, 1993; R. Lizarralde pers. comm.). Approximately another 500 Barí live in two villages and a few hamlets across the Venezuela-Colombia border (Beckerman 1994; R. Lizarralde pers. comm). The Barí have experienced some dramatic changes in relation to their environment, especially their forest. They have lost 88% of their original territory since 1900 (Lizarralde & Beckerman 1982). Most of the remaining Barí territory lies inside a national park, the Parque Nacional Sierra de Perijá, which is still covered by rain forest. Almost all of the lost territory has been virtually deforested. With the loss of territory, the Barí population was reduced to 40% of its 1900 level. Lizarralde and Beckerman (1982) estimated the Barí population to be about 2000–2500 in 1900, which had decreased to 850 in 1964. Population numbers have been growing since then: From 1964 to 1982 the annual growth rate was 3.8%, doubling the population in 18 years (Lizarralde & Beckerman 1982). However, the Barí quality of life is decreasing, with malnutrition and illnesses (hepatitis, tuberculosis, and many kinds of intestinal parasites) as obvious side effects of social, economic, and ecological changes (Holmes & Scorza 1993). In addition, population growth is putting greater pressure on their reduced territory.

With this increase in population, the Barí are experiencing an accelerating process of Westernization. They are becoming more exposed to formal education and are consuming more Western processed products. With this change and the demand for more cash, they are becoming more engaged in producing cash crops such as rice, cacao, manioc, and banana. Numerous Barí are also raising cattle, although the future of this enterprise is uncertain due to the heavy toll it takes on the land in their territory. On the other hand, many Barí still maintain relatively traditional ways of life and largely retain their traditional knowledge.

Barí Ethnobotanical Knowledge

One of the most impressive features of Barí traditional knowledge is that a majority of people can identify and name almost all of the trees in their local forest using folk terms at different taxonomic ranks. In ethnobotanical interviews about plants, it is most relevant to elicit taxa at the more specific tanks, especially folk-generics. Folk-generic taxa are groups of organisms that are readily recognized by most members of a given society and are mostly named by single lexemes (cf. Berlin 1992). In this research I focused mainly on folk-generic taxa. However, as in other traditional societies, most knowledgeable Barí can also recognize and name taxa at the two lower taxonomic ranks (folk-specific and varietal).

Folk-generics and folk-specifics from many indigenous societies in the world generally show a high degree of correspondence to biological species as recognized in Western biotaxonomy (cf. Berlin et al. 1974; Bulmer 1974; Majnep & Bulmer 1977; Berlin 1981, 1992; Boom 1987; Prance et al. 1987; Balée 1989; Milliken et al. 1992, 1999). Most of the Barí folk-generic tree taxa correspond as Western scientific species of trees (70% of folk-generic taxa, or 58 out of the 82 trees that were identified to Western biotaxonomic species in the course of my research). The rest refer to genera or varieties. Since most of the folk-generic taxa are monotypic genera, these names tend to refer to species of trees (e.g., *bahku*, or *Cariniana pyriformis*, in the Lecythidaceae, and *báahdã*, or *Copaifera langsdorffii*, in the Caesalpinioideae). However, in the forest-plot interviews, several Barí informants employed folk-generic terms to refer to either Western biotaxonomic genera or varieties of biotaxonomic species. For example, many Barí used the term *ishiba* to include several species of the same genus, *Ficus*, in the Moraceae, a polytypic genus. The same can be noted for the term *kankarabaa*, used to include several species of another genus, *Inga*, in the Mimosoideae. This is quite common for polytypic genera. Other informants were able to use quite sophisticated terms that refer to varieties of species (e.g., *íshiraberi* and *baroo*, referring to different varieties of *Spondias mombin* L., in the Anacardiaceae).

In addition, each individual Barí has a large repertoire of tree names, ranging from approximately 175 names to possibly as many as 500. In this research I collected nearly 350 different names to which I linked 212 different folk-generics and folk-specifics found in the 4.83 ha of primary and old secondary forest in which I conducted my study. This section of forest includes 3162 trees 10 cm diameter at breast height (dbh) or higher. After checking all the terms on my list, one Barí claimed that it covered fewer than half of the trees in the forest. This suggests that my inventory of Barí trees is not yet complete. It appears that the more knowledgeable informants can label correctly virtually 100% of the trees in the area where they grew up, as in the case of a Barí elder who could name all of the trees over a 1-ha plot of forest. If an elder cannot name a given tree, it is usually because the tree is either an immature individual missing its diagnostic features or a species rarely encountered and therefore represented by very few individuals. Combining the results from the 20 informants I interviewed (which included nearly 17,000 naming events), the Barí were able to name 91.3% of the trees in the study (Figure 4.2). All 20 of the Barí informants interviewed were also able to correctly identify 76.8% of all trees at the level of generic taxa (where correct identification is understood as three or more people agreeing on the same term for the same tree).

Numerous studies have shown that the distribution of knowledge within human communities is not homogeneous (cf. Berlin 1992; Figure 4.3). Among the Barí, as with other indigenous peoples (cf. Berlin et al. 1974; Hays 1974; Ellen 1979; Alcorn 1984; Boster 1985; Weller 1987; Berlin 1992; S. Zent & E. L. Zent 2004), botanical knowledge varies greatly from person to person due to factors such as age, subsistence practice, gender, bilingualism (Barí and Spanish), exposure to formal education, and length of residency (Lizarralde 1997, 2001). The growing diffusion of Spanish and formal education is causing the loss of ethnobotanical knowledge and increasing the variation in this knowledge. In the Barí case, my research showed that people older than 39 years recognize a higher percentage of trees than do younger people (80.5% versus 63.0%; Figure 4.4), men recognize a higher percentage of trees than do women (83.0% versus 60.5%; Figure 4.4), and local people recognize a higher percentage of trees than do nonlocals (85.8% versus 69.8%; Figure 4.5).

The differences between men and women should not be taken as implying that Barí women are less knowledgeable overall because they know less about forest trees.

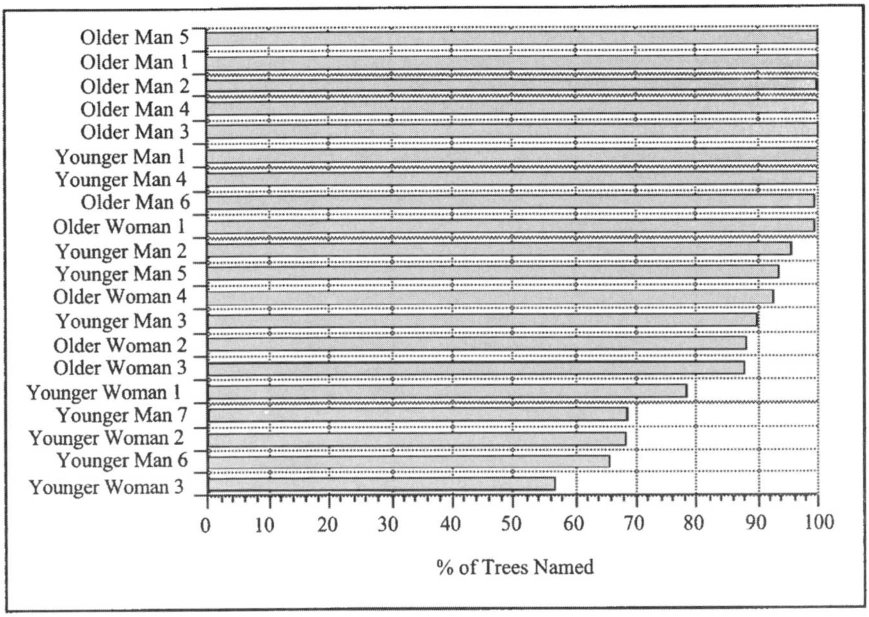

Figure 4.2. Numbers of trees Barí informants could name in all interviews in all plots (n=16,795).

The difference in knowledge is most likely produced by the fact that Barí men participate in activities that more frequently expose them to these rain-forest trees, which therefore breeds familiarity with the trees associated with their activities. Barí women are less exposed to the forest, but in other domains they may have greater plant knowledge than do men. For instance, among the Matsigenka people of Peru, women know more about medicinal plants associated with women's health and childbearing than do men (G. Shepard pers. comm. 1996). Therefore, since my research focused on tree knowledge, it is biased toward representing primarily men's knowledge among the Barí people. Also, in terms of the age factor, the rate of correct identification of trees would be much higher (96–100%) if one interviewed only the most knowledgeable informants, which are normally the older men.

The loss of traditional ethnobotanical knowledge caused by factors such as formal education, bilingualism (Barí and Spanish), and mode of subsistence is clearly observed when an older, monolingual Barí informant who does not have exposure to formal education and still practices a traditional type of subsistence (i.e., older man 5 in Figure 4.3) is compared with a fluently bilingual young man who has a high school diploma and practices a different type of subsistence (younger man 6). There is nearly a 40% difference in the ability of these two men to accurately identify forest trees, and this clearly represents a dramatic loss of traditional knowledge in one generation. Even more striking is the difference in accuracy of identification between older woman 1 and her daughter, younger woman 3: nearly 50%. Younger woman 3 has less knowledge

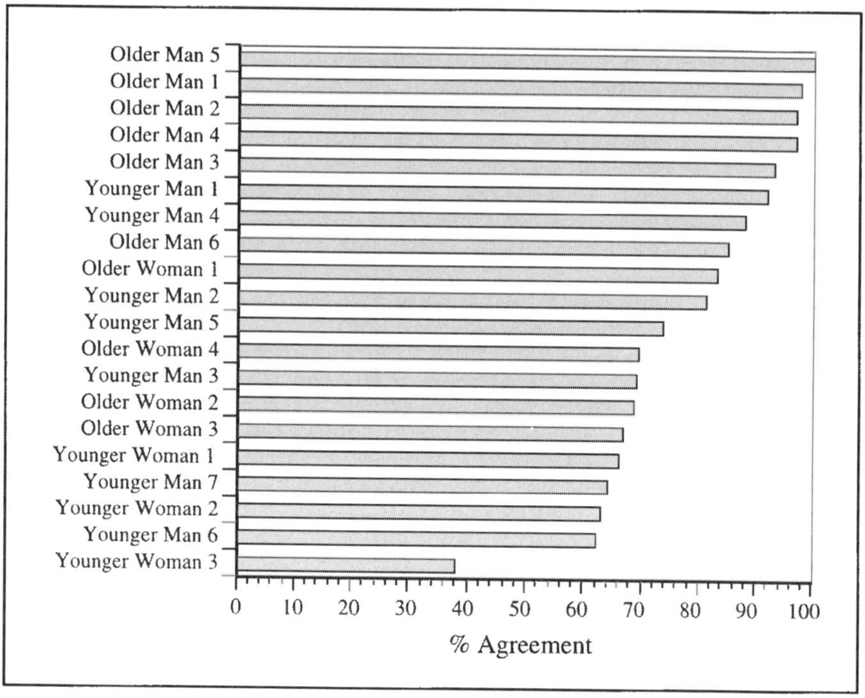

Figure 4.3. Percentages of agreement in naming trees by all Barí informants in all interviews ($n = 16{,}795$).

of rain-forest trees because she spent several years out of her community and in boarding schools. In fact, most young Barí are now exposed to Western education. It is clear that formal education, as well as living outside their traditional environment, can have a negative impact on younger Barí's ethnobotanical knowledge. The loss of this knowledge has serious implications for future conservation efforts. Before addressing this issue, a description of the potential uses of such knowledge is in order.

Barí Use of Folk-Generic Taxa of Trees in Plotted Forest

The data for the Barí's potential use of the forest are slightly below average for South American lowland indigenous peoples. Other indigenous peoples, such as the Ka'apor or Chacobo, have been found to have a higher percentage of uses for trees (Prance et al. 1987). The lower data for the Barí may be due to two relocations. First, the Barí appear to have moved to the western side of Lake Maracaibo shortly before 1622 (Beckerman 1978: 18–19). Second, although they used to live most of the year on the lower flooded land of the territory located east of the Serranía de Abusanqui (Figure 4.1), since 1960 most Barí have been living permanently in the valleys between the Serranía de Abusanqui and the Sierra de Perijá and between the villages of Bachichida and Bokshi, where this study was conducted. The ethnobotanical literature affirms

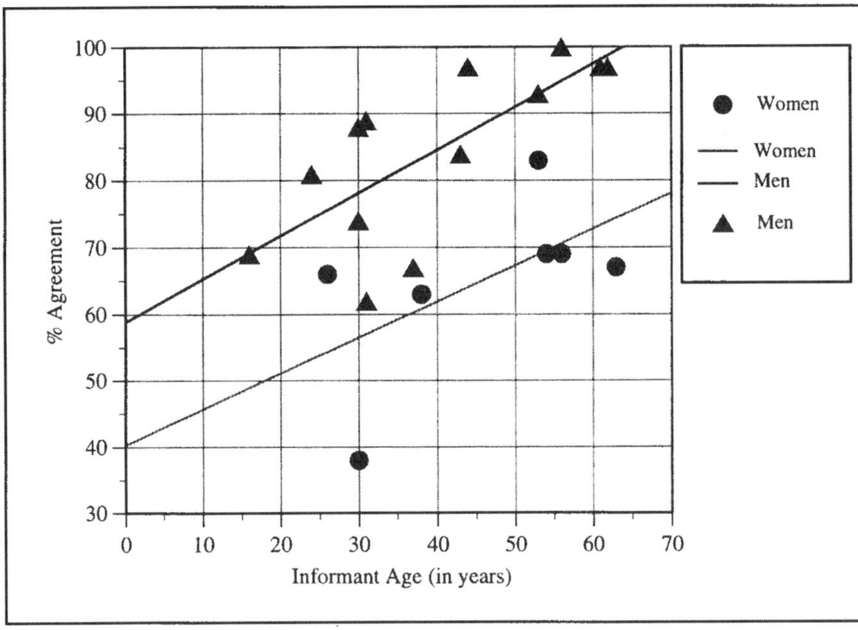

Figure 4.4. Scatter plot of age and agreement for Barí women ($n = 7$) and men ($n = 13$).

that societies with centuries of residence in a given environment tend to develop detailed empirical knowledge of the local flora (cf. Berlin 1992; Balick & Cox 1996: 7). In the Barí case, lack of a long historical residence in their current region, due to this recent migration and relocation of their settlements, would certainly imply reduced time available to the Barí to discover useful trees.

In this section, descriptions of species of trees that have potential uses for the Barí refer to trees that were plotted in the 4.83 ha of forest selected for this study. Of the 171 folk-generic taxa of trees on which information was obtained, the Barí have potential uses for 70.2% (120 taxa) for food, medicine, fuel, construction material, technology (material culture), and commercial timber (Figure 4.6).

Many Barí folk-generic taxa (69, or 32.5% of all taxa) have multiple uses (e.g., *ahka,* or *Sagotia racemosa,* in the Euphorbiaceae, has two types of uses, and *bahku,* or *Cariniana pyriformis,* in the Lecythidaceae, has five types of uses). Taking these overlaps into account, overall the Barí forest offers 120 (70.2%) folk-generic taxa for meeting the food, firewood, medicine, house construction, and material culture needs of their community.

For food, the Barí interviewed have potential uses for 52 (30.4%) folk-generics in the 4.83 ha studied. Palms (Arecaceae) are the most important plant family as a source of food in both frequency and bulk. The Barí use 11 species of palms for their fruits and 1 for palm hearts. Many palm fruits are available in large quantities, which are consumed by Barí households primarily from January to July. Of other trees, the most common sources of fruits found in the forest are *Pouteria anibaefolia* (*abogboo,* in the Sapotaceae),

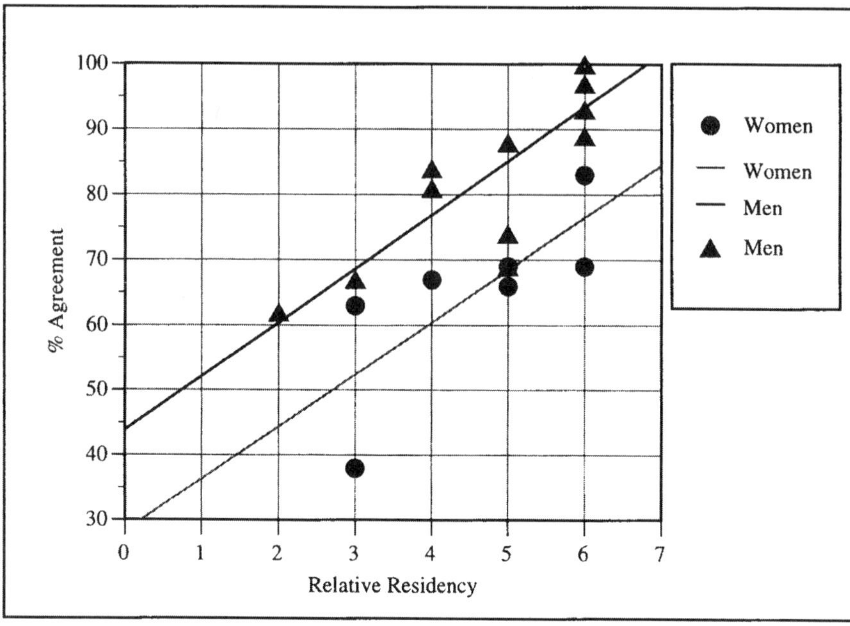

Figure 4.5. Scatter plot of relative residency and agreement for Barí women ($n = 7$) and men ($n = 13$). *Key:* 1 = moved to the territory recently; 2 = moved to the territory a few years ago; 3 = moved to the territory many years ago, or moved recently to an area that has trees similar to those in their ancestral area of residence; 4 = moved to the territory more than half their lifetime ago; 5 = moved from a similar territory many years ago, or older people who moved as teenagers and grew up in the area; 6 = spent all of their life in the territory.

Spondias mombin L. (*baroo,* in the Anacardiaceae), *Protium heptaphyllum* (*ishkugbaa,* in the Burseraceae), and many species of the *Inga* genus (in the Mimosoideae). Tree fruits contribute complex carbohydrates, fiber, and vitamins to the human diet.

The Barí prefer one particular tree species, *Sagotia racemosa* (*ahkaa,* in the Euphorbiaceae), for firewood, but they also regularly use other species (*Astronium graveolens,* or *tumma,* in the Anacardiaceae; *Dialium guianense,* or *ogchiri.* in the Caesalpinioideae; *Luehea seemanni,* or *burunbunkaa,* in the Tiliaceae; *Oxandra venezuelensis,* or *chirabu,* in the Annonaceae; and *Tabebuia chrysea, or karikã,* in the Bignoniaceae). The Barí claim that 93 folk-generic taxa have potential use as firewood. Firewood trees represent 54.4% of all the folk-generic taxa. *S. racemosa* is also used as a source of medicine.

For medicinal purposes, the use of trees is low, since most medicinal plants tend to be vines, herbs, or hemiepiphytes. There were only 250 medicinal trees among the trees plotted, representing 6 folk-generic taxa. These 6 folk-generic taxa constitute 3.5% of all the tree taxa plotted. Of these taxa, two species (*ahka,* or *Sagotia racemosa,* in the Euphorbiaceae, and *bahku,* or *Cariniana pyriformis,* in the Lecythidaceae) were the most important and abundant trees in the forest.

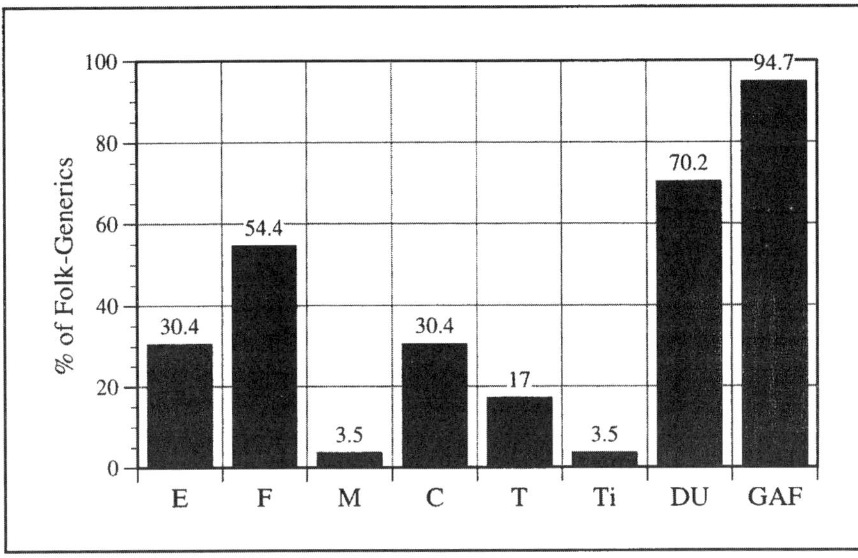

Figure 4.6. Percentages of folk-generics of trees plotted against type of Barí use ($n = 171$ of the 212 recorded). *Key:* E = food; F = fuel; M = medicine; C = construction; T = technology; Ti = commercial timber; DU = direct use; GAF = game animal food.

In terms of house construction, the Barí have potential use for 52 folk-generic taxa, or 30.4% of all folk-generic taxa registered in the plots. For artifact production (such as for baskets, bows, arrows, fishing spears, looms, etc.), 29 folk-generic trees have potential uses, representing 17% of the trees plotted.

The exploitation of commercial timber is a new use that the Barí have adopted in recent decades; that is, since they were contacted by the outside world. The Barí in my sample have 7 (3.5%) species of trees that are potentially useful as commercial timber. The most important are tropical cedars—namely, *Cedrela odorata* (*daiba,* in the Meliaceae)—as well as other tropical trees such as *Swietenia odorata* (*buyógbaa,* in the Meliaceae) and *Cariniana pyriformis* (*bahku,* in the Lecythidaceae). The Barí do not use 15 other commercially useful species of trees found in the plots since they are not as profitable as the three listed above. Among the species not exploited are *Copaifera langsdorffii* (*báahdã,* in the Caesalpinioideae), *Hymeneae courbaril* L. (*bwai bojkbá,* in the Caesalpinioideae), *Peltogyne purpurea* Pittier (*bokshi,* in the Caesalpinioideae), and *Parkia pendula* (*kõgdaig,* in the Mimosoideae).

The seventh type of use of trees I computed is one that is often neglected in ethnobotanical studies: trees that provide food for the animals the Barí hunt (referred to here as "game animal food trees"; see Lizarralde 1997). Even though this use is not as apparent, all of these trees are quite important and well known to the Barí. When the Barí walk in the forest in groups, they are constantly making remarks about which animals visit these trees or which one was hunted on a specific tree. Women, girls, and boys constantly hear the men's observations, such as where they are going to ambush a paca (*Agouti paca*) or agouti (*Dasyprocta punctata*) that evening or that the peccaries are regularly eating fruits of the big *abogboo* tree on the Karikogbay trail (each forest trail is identified by name, just as we name city streets). Trees like *logsorologsoro* (*Gustavia* sp., in the Lecythidaceae) are well known to every Barí even though they do not have any other particular cultural use, such as for food, firewood, house construction, or material culture. These two trees consistently attract pacas and agouties. For the Barí, 162 (94.7%) folk-generic taxa are potential food for the game they hunt. Almost all of the trees provide some food in the form of flowers or fruit for three species of the *Amazona* genus of parrots (*A. farinosa, A. ochrocephala,* and *A. frentirojo*) and two species of peccaries (*Tayassu pecari* and *T. tajacu*). Most monkeys, large rodents, tapir, and large frugivorous birds eat approximately 60–80% of the kinds of fruits produced by forest trees, according to what some Barí informants told me in the field.

Barí Use of Individual Trees in Plotted Forest

The previous analysis of Barí uses of trees shows a high potential use of their forest. However, I would like to point out that computing the number of folk-generic taxa may not provide the fullest account of the proportion of the forest that has potential uses. Therefore, I decided to also look at the number of individual trees used, since these data might provide a different perspective on the proportion of potential uses of the forest (Figure 4.7). The data do indeed reveal a higher potential use of the forest for the Barí.

Moreover, it is important to indicate at this point that the Barí tend to make greater use of those folk-generic taxa of trees that are represented by higher numbers of individuals in the forest. Trees such as *Sagotia racemosa* (*ahka,* in the Euphorbiaceae), represented by 209 individual plotted trees (6.6% of all plotted trees), and *Oenocarpus mapora* (*keki,* in the Arecaceae), represented by 276 individual plotted trees (8.7% of all plotted trees), are more likely to have a use than are species that are represented by few individuals, which tend to be rarely encountered and therefore are not well known to the people. This was shown in the forest plots, where only 18 folk-generic taxa represented by fewer than 5 individuals in the plots were found to have a direct use versus 41 such folk-generic taxa without any use. Thus, fewer than one-third (31%) of all folk-generic taxa represented by fewer than 5 individuals have some type of direct use. If we compare this with the data for folk-generic taxa with more than 10 individuals in the sample plots, the proportion more than doubles: 79% (54 of 68) of the folk-generic taxa have direct uses. Because many folk-generic taxa have large numbers of individuals and because the folk-generic taxa with numerous individuals are more likely to be used, using the number of individual trees instead of the number of species shows a higher proportion of the forest with some kind of potential use.

Of the 3162 trees plotted, the Barí designate 1350 (42.7%) as trees that provide food. Palms (Arecaceae) represent the largest number of individual trees used as a source of food (653, or 20.7% of all trees plotted), especially *Oenocarpus bataua* (*arikogbaa*), represented by 101 (3.2%) individuals, *O. mapora* (*keki*), represented by 276 (8.7%) individual trees, and *Attalea butyracea* (*araktogbaa*), represented by 83 (2.6%)

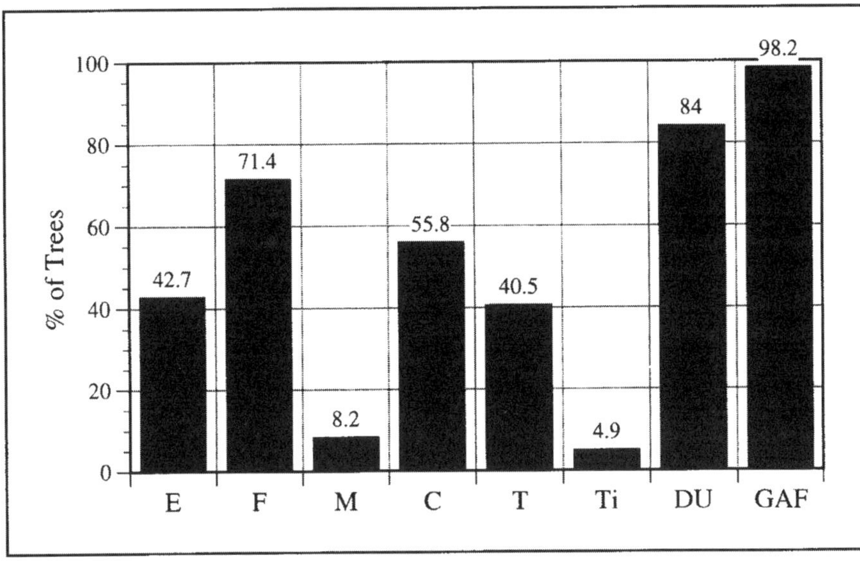

Figure 4.7. Percentages of all trees plotted against type of Barí use. *Key:* E = food; F = fuel; M = medicine; C = construction; T = technology; Ti = commercial timber; DU = direct use; GAF = game animal food.

individual trees. For firewood, the Barí have potential use for 2257 (71.4%) of the trees. Most important among these trees is *ahka* (*Sagotia racemosa,* in the Euphorbiaceae), the third most abundant tree in the region (209 of the 3162 trees plotted in 4.83 ha of forest). For medicinal use, 250 (7.9%) trees have potential uses. Of these trees, two species (*ahka,* or *Sagotia racemosa,* in the Euphorbiaceae, and *bahku,* or *Cariniana pyriformis,* in the Lecythidaceae) are the most important and abundant trees in the forest. In fact, these two species are represented by 241 individuals (209 *ahka* and 32 *bahku*), or 7.6% of all the trees plotted. Thus 96% of the medicinal trees in the tract are also in these two species (241 of the 250 medicinal trees). As previously mentioned, this percentage of medicinal trees is low due to the fact that most medicinal plants are herbs, shrubs, or vines.

For house construction, the Barí have potential use for 1764 (55.8%) of the trees. It is interesting to note that different species of trees are used for specific structural parts of the house. For example, the main pillar of the house is usually made with *Tabebuia chrysea* (*karika,* in the Bignoniaceae), *Bulnesia arborea* (*burubaakaa,* in the Zygophyllaceae), or *Dialium guianense* (*ogchiri,* in the Caesalpinioideae). In precontact times (before 1960) the proportion and densities of useful trees for house construction may have been higher since the Barí had a lower population density and were less sedentary, so their impact on the abundance of house-construction trees in any particular location was lower. In many of the studied forest plots, the impact of larger sedentary villages can easily be seen. Many of the house-construction trees are cut down faster than they can regenerate, and their stumps are visible in the nearby forest.

In the forest plots are also 1280 (40.5%) trees that are potentially useful for material culture. A very small proportion of the trees, 155 (4.9%), have potential use for commercial timber for the Barí, considering that not all commercial species are used. In total, 84% of all of the individual trees plotted in the 4.83 ha have direct uses for the Barí community.

In terms of indirect use, the proportion of game animal trees is 98.2% (3105), based on all the forest trees that were analyzed and on what the Barí place in this category. Thus the Barí's indirect use of the forest is comparable to what pasture would represent for cattle ranchers. The forest is a natural ranch providing food for the animals that the Barí harvest in their hunting activities.

Changes in Indigenous Peoples' Way of Life

The Barí are experiencing significant changes in their ways of using the forest. Their "traditional" way of life is currently different from the one they led 30–40 years ago. As with many other indigenous peoples worldwide, four basic factors are responsible for the changes the Barí are undergoing. First, their subsistence pattern has been shifting to a cash economy that demands higher resource extraction from their forest. Second, most of their territory was taken away by cattle ranchers, reducing it to 12% of what they had in 1900. Third, their population has increased from 850 in 1964, to 1560 in 1983, and to 1953 in 1992, while their territory has not increased in size (Lizarralde & Beckerman 1982; OCEI 1985, 1993; Lizarralde 1991; R. Lizarralde pers. comm.). This population increase has put a stress on the carrying capacity of their environment above what it can sustain. Finally, coal, oil, and lumber concessions have been granted to companies to exploit the 165,000-ha part of the Barí land that lies outside the national park.

Since the Barí were contacted in 1960, they have undergone rapid Westernization and engagement in the national market economy. Before 1960, their level of consumption of natural resources was in line with self-sufficiency in procurement of materials needed for subsistence (cf. Lizarralde 1991; Behrens et al. 1994). Thereafter, they began to acquire Western goods, such as shotguns, flashlights, knives, Western clothing, medicine, processed food, blankets, and, in the 1990s, television sets. At present, many of them are acquiring gas stoves, refrigerators, and even one washing machine (observed in July 2000 in the village of Saimadodyi). The increase in the "need" for Western goods has also increased their demand for cash, which they satisfy by selling their labor to cattle ranchers, working on cash crops, or processing lumber. Cutting down of trees either for lumber or to clear areas for cattle and crops is easily observed, with the larger gaps in the forest found around their permanent villages (cf. Behrens et al. 1994).

As indicated above, the Barí traditionally had a semisedentary settlement pattern. They would usually relocate the longhouse after 10 years, when surrounding resources were exhausted. The territory for a local group before the 1940s and 1950s was 1200–1600 km^2, which included a population of 200–400 Barí, or 0.11 people per km^2 (Lizarralde & Beckerman 1982: 49; Lizarralde & Lizarralde 1991: 453). This settlement pattern and low population density minimized degradation of the forest, because lower density means less use of resources to begin with and because frequent relocation tends to allow the forest to regenerate. Since 1900 the Barí have lost 88% of their territory, shrinking from 16,000 km^2 to today's 1900 km^2 (Lizarralde & Beckerman 1982: 49). By 1983 the Barí population was 1560, and its density was 0.82 people per km^2 (p. 49). Nine years later the population was 1953 (1503 in Venezuela and an estimated 450 in Colombia), and its density was 1.03 people per km^2 (R. Lizarralde pers. comm.). Currently, the Barí live in permanent settlements with a much higher population density. This high population density and traditional extracting methods are causing degradation of the forest and reduction of forest resources, as can be clearly observed in satellite images taken in 1991 (figure 3 in Behrens et al. 1994: 300). This is especially true of house-construction materials, which require the cutting down of many more trees than in precontact times and make extraction unsustainable.

With a current density of 1.03 people per km^2, or 10 times that in precontact times (cf. Beckerman 1978; Lizarralde & Beckerman 1982), the Barí are facing the need to change their relationship with the forest. Their awareness of this issue was raised in a Barí pan-village meeting in July 1994 to discuss strategies to protect and conserve their forest. Like the Borá indigenous people in Peru (Denevan et al. 1984; Denevan & Padoch 1987), the Barí are starting to manage their secondary-growth forest by selecting certain trees. There is no other solution, since their population seems to continue to increase while their territory remains the same.

Another problem the Barí are currently facing is exploitation of their territory for coal mining and lumbering. As mentioned above, large concessions were offered to private companies to exploit the approximately 165,000 ha of forest in the part of the Barí territory that lies outside the national park. Most outsiders and local non-Barí wrongly believe that the Barí have little knowledge of and no particular use for their forest. Thus many outsiders see them as "not using their resources" and as preventing companies from giving them "a better use." On the contrary, as I have shown, the Barí possess a very detailed knowledge of the local flora and fauna and of the wealth of useful natural resources the forest provides. This Barí ecological knowledge is proof of their long-term use of the forest. Since the Barí do not have official title recognized by the Venezuelan government for most of their land, this knowledge could be used as evidence to help them gain title to unclaimed land.

Use of Ethnobotanical Knowledge for Conservation

Barí knowledge of this forest is also a potential tool for its conservation, improvement, and reforestation. The Barí clearly understand the biological cycle and biogeographical distribution of most species. This knowledge has been essential to enable them to plan their resource use and locate the different tree resources at the right time of year. They are aware of their dependence on the forest for resources essential for their survival. They have noticed that certain tree species are decreasing in overall numbers due to overharvesting (e.g., *swai*, or *Geonoma stricta* Mart. var. *stricta*, and *araktá*, or *Attalea butyracea*, in the Arecaceae, have been overharvested to the point that there are none left near the larger villages). Since their population density is at its highest, in the past five years some Barí have been talking about alternatives and about protecting their flora and fauna. Understanding the ecology of the trees, they have the know-how to plant them if necessary, such as in many areas that have been deforested (Figure 4.8). They have the incipient idea of planting trees, although they have not fully utilized its potential. The problem is that they have not yet agreed on how to implement a plan to regenerate the forest tree species. This is a new situation for them, for in the past they would move to another forested area when resources became scarce (cf. Lizarralde & Beckerman 1982; Lizarralde 1991). Since they currently inhabit all of the best available lands, moving to another area is out of the question, and replanting trees seems to be the best choice. Some of them have recently begun to plant specific forest trees in order to improve their flora.

The forest can be improved for certain uses (i.e., increasing the number of fruit, house-construction, or lumber trees). As we have seen, many forest areas near villages have been harvested in an unsustainable way, resulting in depletion of certain tree species, especially those used for firewood and house construction, which I clearly observed in the forest near the large villages of Bachichida, Saimadodyi, and Bokshí. Replanting the most culturally important tree species offers potential for forest regeneration, since many of these tree species attract seed dispersers such as rodents and tapirs. Trees that attract these seed dispersers include *Attalea butyracea* (*araktaa,* in the Arecaceae), *Oenocarpus mapora* (*keki,* in the Arecaceae), and *O. bataua* (*arúu,* in the Arecaceae). The Barí have in fact started to plant trees like *A. butyracea* as a way to "add food" to the forest. Some wild fruit trees, like *Spondias mombin,* which are important as a food source for spider monkeys, tapirs, peccaries, and turtles, have also been planted in villages. Species of palms, lumber trees, and fruit trees can be easily planted or replanted. Their seedlings are quite abundant in the forest understory under the mature specimens of their species. I have observed dozens of immature palms such as *O. bataua* (*arúu*), *O. mapora* (*keki*), and *A. butyracea* (*araktaa*).

In the early stages of my research, back in 1988, I assumed that the Barí had some form of management of their forest. Later, however, I concluded that this assumption was wrong, because I could not find any substantial evidence of such practices. When I asked my Barí informants whether they managed the forest by planting trees and managing the growth of some, they stated that they did not see any reason for managing the forest, since all trees were abundant and had to be constantly cut down to create sunny patches for their gardens and villages. In those early stages of my work, I observed only one case of what could resemble forest management. In 1991 a Barí woman told me that, some 50 years earlier, her great-grandfather had planted many avocado trees (*kokwachi,* or *Persea americana,* in the Lauraceae) a 15-minute walk north of the village of Ohbadya (5 km northwest of Saimadodyi). However, since avocado trees are a domesticated, not wild, species, this cannot quite be considered a type of forest management. During a survey of Barí territorial boundaries in 1999 I was informed that two Barí men had already started to plant some *arúu* palms (*Oenocarpus bataua*) in the forest as a way of demarcating their territory and that they had edible plants on hand. In July 2000, in the village of Karñakaig, I also observed several individuals of *arúu* palms planted near houses and in the mule pasture.

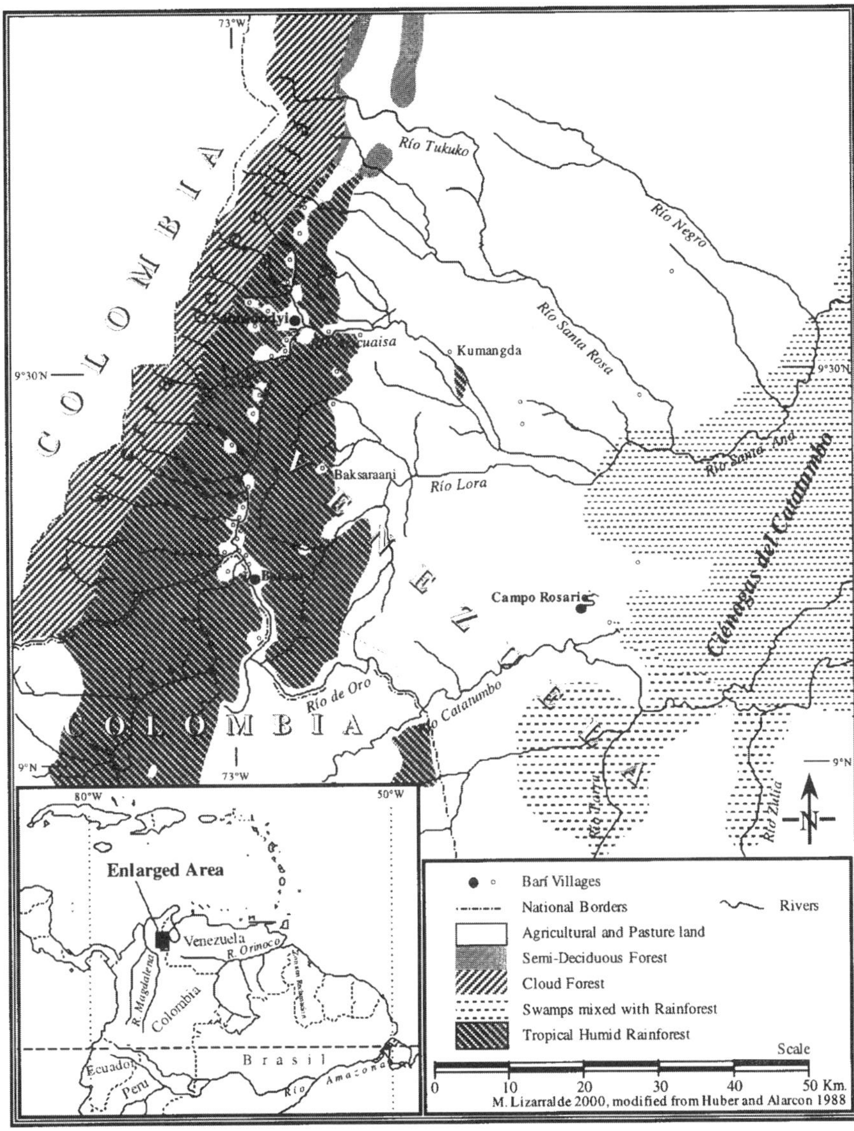

Figure 4.8. The Barí territory and locations of major types of forests.

Other than the planting of avocado or *arúu* palms, there is no evidence of traditional forest management.

In relation to tree management, another significant event was that in Saimadodyi, in July 1999, for the first time people were clearing a secondary forest to plant pasture for cows but left 40–50 individuals of *Tabebuia chrysea* (*karikã*, in the Bignoniaceae) in an area that covered nearly 1 ha. *Tabebuia chrysea* is an excellent tree for lumber and house construction. All of these trees were 10–15 years old. The same was observed in July 2000 in the village of Ishirakbadyiro (3 km south of Saimadodyi), where about two dozen trees of *Cedrela odorata* (*daiba*, in the Meliaceae) were not cut down in clearing for cattle pastures. However, this is clearly not an example of forest management, since the forest was being removed.

Until recently the Barí use of forest resources was in the guise of simply gathering them according to need. However, the new combination of planting *arúu* palms and the protection of *Tabebuia chrysea* and *Cedrela odorata,* in addition to planting avocado trees, are indeed incipient forms of tree management. These are clear examples that the Barí are beginning to shift from just gathering resources to managing them. Their knowledge of forest trees is a potential tool for conservation and sustainable use, the implications of which are just now beginning to come to light.

Conclusion

Like many ethnobotanists, I find myself caught between the traditional professional imperative of recording information required for scientific and academic purposes and a newer perceived obligation to address the urgent needs of indigenous peoples with whom I work by applying ethnobotanical information to current concerns. The ethnobotanical work of Brent Berlin and his colleagues with the Tzeltal Maya since the late 1960s provides a significant example of a shift from recording the Tzeltal's complete ethnobotanical knowledge to actually applying it to the alleviation of local health problems (Berlin et al. 1974; Berlin & Berlin 1996). Berlin and Berlin (1996) produced field manuals for the Tzeltal on the use of their traditional medicinal plants, written in both Spanish and the Tzeltal dialects. The Barí people, as well as Venezuelan officials, constantly raised questions about my research for solid reasons: They repeatedly asked how this information could be used to solve their own current problems.

My future goal is to integrate the information I have been gathering into action aimed at conserving the forest. If the forest has some use, it can be protected. Venezuelan officials who sympathize with the Barí's problems (e.g., those at the Instituto de Tierras, formerly the Instituto Agrario Nacional) observed that the information I collected was strong evidence in support of stopping coal concessions and helping the Barí obtain a communal title for the 150,000 ha still available, since the Barí have potential direct and indirect uses for 98% of the forest tree species (84% have direct use). This implies that the forest is not an unproductive mass of vegetation, as it has been seen by cattle ranchers and government administrators who granted permission to multinational companies (MAICCA, Corpozulia, and Carboca) to explore the coal potential of the region in dispute. The actions of government administrators clearly are geared toward opening paths to forest destruction, without understanding and valuing the potential sustainable uses of the forest for both Barí and non-Barí. Like cattle ranching, coal exploitation in the region will require removal of the forest and topsoil.

As another example, the government of the state of Zulia (Maracaibo) demonstrates a shortsighted approach by favoring the monetary value that coal exploitation can bring to the state while ignoring the ecological potential that Barí knowledge can offer for forest protection as well as for potential long-term extraction of forest products. Furthermore, it is ironic that the main problem of the state of Zulia in the past 25 years has been scarce supply of fresh water to the capital, 200 km from the Barí territory. Because they want to develop and destroy a vast portion of the Barí forest, the Zulia authorities clearly do not measure the ecological importance of the forest for the stability of the hydrological cycle in the region (cf. Molion 1989; Sponsel et al. 1996). On the other hand, it is not surprising that the region known as Sierra de Perijá, which includes a mountain range and has the largest block of rain forest in the state, is the state's wettest region.

Traditional ecological knowledge could be applied toward achieving sustainable development by helping the Barí themselves establish a reforestation program as well as find economic uses for the forest that will benefit them and the whole region. Clearly, the Barí need to have at their disposal ways of using their forest resources that will provide enough income to support their population. At the same time, bridging economic development and forest conservation requires identifying methods of sustainable use of the flora in order to avoid forest degradation. One possible course of action might consist of identifying all of the trees that have uses in other parts of South America and sharing this knowledge with the Barí. Applying other indigenous peoples' knowledge of uses of trees that are found in the Barí territory may increase the potential sustainable uses of this forest. My own goal in the near future is to provide the Barí with new data on the application of ethnobotanical knowledge in forest conservation and thus help them find sustainable uses for the forest in which they live and on which they depend.

Acknowledgments

Special thanks to Luisa Maffi and Thomas Carlson for inviting me to write a chapter for this volume. I would like to thank all of the Barí people for allowing me to reside and conduct my research in their cultural and natural environment, especially to Andrés Achirabu, Jaime, Atilio, Edgar, Raquelita, Nola, Orlando, Adrián Okboo, Akirida, and David Aleobadda. Funding for research to collect data in the field was provided by the Wenner-Gren Foundation (grant no. 5519), PREBELAC funds from the Institute of Economic Botany of the New York Botanical Garden, and Shaman Pharmaceuticals, Inc. I am grateful to Anne-Marie Lott for her meticulous editorial assistance with this chapter. I am further indebted to Luisa Maffi and Thomas Carlson for very careful reviewing and editing of the manuscript and offering very helpful suggestions. Any remaining errors in the text are my sole responsibility.

Literature Cited

Alcorn, J. B. 1984. Huastec Mayan Ethnobotany. Austin: University of Texas Press.

Balée, W. L. 1989. Nomenclatural patterns in Ka'apor ethnobotany. Journal of Ethnobiology 9: 1–24.

Balick, M. J. & P. A. Cox. 1996. Plants, People, and Culture: The Science of Ethnobotany. New York: Scientific American Library.

Beckerman, S. 1978. Datos etnohistóricos acerca de los Barí (Motilones). Vol. 8. Colección de Lenguas Indígenas, Serie Menor, 7. Caracas: Montalbán.

———. 1980. Fishing and Hunting by the Barí of Colombia. Working Papers on South American Indians 2: 67–109.

———. 1983. Optimal foraging group size for a human population: The case of Barí fishing. American Zoologist 23: 283–290.

———. 1994. Barí. Pp. 82–85 *in* J. Wilbert (ed.), South America. Vol. 7 of Encyclopedia of World Cultures. Boston: G. K. Hall.

——— & W. Sussenbach. 1983. A quantitative assessment of the dietary contribution of game species to the subsistence of South American tropical forest peoples. Pp. 337–350 *in* J. Clutton-Brock & C. Grigson (eds.), Animals and Archaeology, I: Hunters and Their Prey. Oxford: British Archaeological Report.

Behrens, C. A., M. G. Baksh & M. Mothes. 1994. A regional analysis of Barí land use intensification and its impact on landscape heterogeneity. Human Ecology 22: 279–316.

Berlin, B. 1981. The concept of rank in ethnobiological classification: Some evidence from Aguaruna folk botany. Pp. 92–113 *in* R. Carson (ed.), Language, Culture and Cognition: Anthropological Perspectives. New York: Macmillan.

———. 1992. Ethnobiological Classification: Principles of Categorization of Plants and Animals in Traditional Societies. Princeton, N.J.: Princeton University Press.

———, D. E. Breedlove & P. H. Raven. 1974. Principles of Tzeltal Plant Classification: An Introduction to the Botanical Ethnography of a Mayan Speaking Community in Highland Chiapas. New York: Academic Press.

Berlin, E. A. & B. Berlin. 1996. Medical Ethnobiology of the Highland Maya of Chiapas, Mexico: The Gastrointestinal Diseases. Princeton, N.J.: Princeton University Press.

Boom, B. M. 1987. Ethnobotany of the Chácobo Indians, Beni, Bolivia. Advances in Economic Botany, 4. Bronx: New York Botanical Garden.

Boster, J. S. 1985. Requiem for the omniscient informant: There's life in the old girl yet. Pp. 177–198 *in* J. Dougherty (ed.), Directions in Cognitive Anthropology. Urbana: University of Illinois Press.

Bulmer, R. 1974. Folk biology in the New Guinea Highlands. Social Science Information 13: 9–28.

Davis, W. 1999. Vanishing cultures. National Geographic Magazine, August, 62–89.

Denevan, W. M. & C. Padoch (eds.). 1987. Swidden-Fallow Agroforestry in the Peruvian Amazon. Advances in Economic Botany, 5. Bronx: New York Botanical Garden.

———, J. M. Treacy, J. B. Alcorn, C. Padoch, J. Denslow & S. F. Paitan. 1984. Indigenous agroforestry in the Peruvian Amazon: Bora Indian management of swidden fallows. Interciencia 9: 346–357.

Ellen, R. F. 1979. Omniscience and ignorance: Variation in Nuala knowledge, identification and classification of animals. Language and Society 8: 337–364.

Hays, T. E. 1974. Mauna: Explorations in Ndumba ethnobotany. Ph.D. diss., University of Washington.

Holmes, R. & J. Scorza. 1993. Estado de salud indígena: Los Barí. Maracaibo, Venezuela: Fundación Zumaque (MARAVEN).

Johnston, B. R. (ed.). 1994. Who Pays the Price? The Sociocultural Context of Environmental Crisis. Washington, D.C.: Island Press.

Lizarralde, M. 1997. Perception, knowledge and use of the rainforest: Ethnobotany of the Barí of Venezuela. Ph.D. diss., University of California, Berkeley.

———. 2001. Biodiversity and loss of indigenous languages and knowledge in South America. Pp. 265–281 *in* L. Maffi (ed.), On Biocultural Diversity: Linking Language, Knowledge, and the Environment. Washington, D.C.: Smithsonian Institution Press.

——— & R. Lizarralde. 1991. Barí exogamy among their territorial groups: Choice and/or necessity. Human Ecology 19: 453–467.

Lizarralde, R. 1991. Barí settlement patterns. Human Ecology 19: 428–452.

——— & S. Beckerman. 1982. Historia contemporánea de los Barí. Antropológica 58: 3–51.

Maffi, L. (ed.). 2001. On Biocultural Diversity: Linking Language, Knowledge, and the Environment. Washington, D.C.: Smithsonian Institution Press.

Majnep, I. S. & R. N. H. Bulmer. 1977. Birds of my Kalam country. Auckland: Auckland University Press.

Milliken, W., R. P. Miller, S. R. Pollard & E. V. Wandelli. 1992. The Ethnobotany of the Waimiri Atroari Indians of Brazil. Kew: Royal Botanic Gardens.

———, B. Albert & G. G. Gomez. 1999. Yanomami: A Forest People. Kew: Royal Botanic Gardens.

Molion, L. C. 1989. The Amazon forest and climatic stability. Ecologist 19: 211–213.

OCEI. 1985. Censo indígena de Venezuela 1982. Caracas: Oficina Central de Estadística e Informática.

———. 1993. Censo indígena de Venezuela 1992. Caracas: Oficina Central de Estadística e Informática.

Prance, G. T., W. Balée, B. M. Boom & R. L. Carneiro. 1987. Quantitative ethnobotany and the case for conservation in Amazonia. Conservation Biology 1(4): 296–310.

Reaka-Kudla, M. L., D. E. Wilson & E. O. Wilson (eds.). 1997. Biodiversity II: Understanding and Protecting Our Biological Resources. Washington, D.C.: Joseph Henry Press.

Sponsel, L. E., R. C. Bailey & T. N. Headland. 1996. Anthropological perspectives on the causes, consequences, and solutions of deforestation. Pp. 3–52 *in* L. E. Sponsel, T. N. Headland & R. C. Bailey (eds.), Tropical Deforestation: The Human Dimension. New York: Columbia University Press.

Weller, S. C. 1987. Share knowledge, intracultural variation, and knowledge aggregation. American Behavioral Science 31: 178–193.

Zent, S. & E. L. Zent. 2004. Ethnobotanical convergence, divergence, and change among the Hotï of the Venezuelan Guayana. Pp. 36–77 *in* T. J. S. Carlson & L. Maffi (eds.), Ethnobotany and Conservation of Biocultural Diversity. Advances in Economic Botany, 15. Bronx: New York Botanical Garden Press.

5

Ethnobotanical Ground-Truthing and Forest Diversity in the Western Amazon

Glenn H. Shepard Jr., Douglas W. Yu, and Bruce W. Nelson

The Paradox of Diversity

The numbers of species living at close quarters in an Amazonian forest can be staggering: 319 species of birds were counted in a census of 1 square kilometer of habitat on the Manu River (Terborgh et al. 1990); 1 hectare may contain up to 300 species of trees (Gentry 1988b); a single tree may contain more ant species than are present in all of Great Britain (Wilson 1986). A central goal of tropical ecology is to explain the origin and maintenance of high species diversity (Pianka 1966; Givinish 1999). The problem is more perplexing than just high species richness, however. Assemblages in the Tropics are made up of a few common species and large numbers of rare species. For instance, a systematic sample of 19,252 trees in the Peruvian Amazon turned up 829 species, of which 253 (31%) were represented by a single tree (Pitman et al. 1999). The problem of explaining high species diversity in the Tropics is the problem of explaining the paradoxical prevalence of rare species (Condit 1996).

Several non–mutually exclusive hypotheses have been proposed for the presence of rare species and overall species richness in tropical trees. First, some species found to be rare might simply be ones that are becoming extinct (Hubbell & Foster 1986b). Second, species-specific, density-dependent pathogens can maintain populations of many species at low levels (Janzen 1970; Wills & Condit 1999; Harms et al. 2000). Third, rare species might require specific but rarely attainable conditions to achieve regeneration (Ashton 1977; Grubb 1977; Chesson & Warner 1981; Terborgh 1985). The third case is included in what is known as the "habitat-mosaic" hypothesis of tropical diversity (see Terborgh 1992: 73–104). Different soils, topographies, disturbance histories, and drainage levels might combine to pose different challenges to plant regeneration and, when combined with the appropriate tradeoffs across plant species, will lead to each habitat's being dominated by a distinctive subset of species (Ashton & Hall 1992; Terborgh et al. 1996; Clark et al. 1998; Pitman et al. 1999). A mosaic of habitats, each supporting distinct assemblages of tree species, could "leak" tree species across habitat borders; most habitats would therefore host many locally rare species that are maintained by dispersal from abundant populations elsewhere.

Despite its conceptual simplicity, testing the model runs against several profound logistical obstacles. It is very difficult to define a habitat a priori, or even the likely size and shape of a typical habitat. Thus the only definitive way to test the habitat-mosaic hypothesis would be to carry out a thorough census of trees over tens of thousands of hectares and to correlate the local species assemblages with a number of environmental variables, such as soil characteristics, parent bedrock, topography, and drainage. An exhaustive (and exhausting!) survey of this scale would be expensive and time consuming to the point of impossibility, given current technologies, representing a level of effort analogous to the sequencing of an entire genome by hand. To date, perhaps only a few hundred hectares of Amazonia's 5 million square kilometers have been surveyed, mostly around cities, along major rivers and highways, and at a handful of well-studied research stations (Nelson et al. 1990; Tuomisto 1998).

Given the practical limitations, researchers studying habitat diversity in the Tropics must resort to smaller-scale sampling methods that can be divided into two types: intensive and extensive. In the intensive approach, a small area is densely sampled for tree species, which are correlated against local environmental variables such as soil type, light levels, and topography (for example, Hubbell & Foster 1986a; Tuomisto 1998; Clark et al. 1999; Webb & Peart 2000). The extensive method involves predefining habitats within a large area and then sampling trees within multiple examples of those habitats (Gentry 1988a; Ashton & Hall 1992; Duivenvoorden 1996; Terborgh et al. 1996). Yet there is no way in either method to know whether important source habitats have been left out, whether multiple habitats have been combined in a single habitat class, whether a single habitat has been artificially split, or, most important, whether a majority of the available habitats has been sampled. Both intensive and extensive approaches have provided some support for the habitat-mosaic hypothesis, but neither approach has been able to provide a quantitative assessment of the extent to which habitat mosaics contribute to high tropical tree species diversity and rarity. What is needed is some way to define a large number of candidate habitats over a broad area before sampling within each one, preferably erring on the side of overdifferentiation.

To this end, investigators have suggested that satellite remote sensing might provide a shortcut to such large-scale documentation of habitat diversity in tropical forests (Tuomisto et al. 1994), and early work in the western Amazon supported this contention. Analysis of satellite images of the southeastern Peruvian Amazon by Finnish scientists resolved 10–15 color/shade combinations (Salo et al. 1986), which corresponded to the habitats distinguished by scientists on the ground: e.g., mature floodplain forests, upland terra firma, swamps, dwarf forests on acidic white sand, and various swamps and riverine successional zones (Erwin 1984; Foster 1990a; Encarnación 1993). Many of these habitats are created in part by large-scale successional dynamics following river meandering (Foster 1990b), leaving approximately a quarter of the forest in various stages of succession at any given time (Salo et al. 1986). More recently, the same group of Finnish scientists reanalyzed satellite images to identify more than 100 candidate habitats in the Peruvian Amazon (Tuomisto et al. 1995), suggesting that habitat diversity might have been underestimated on the ground by as much as an order of magnitude. However, there is still little empirical evidence to support these conclusions (Condit 1996; Tuomisto 1998).

What we have attempted to do in our interdisciplinary research project is to take advantage of an already existing database of forest habitat diversity that covers tens of thousands of hectares: the forest classification system of the Matsigenka, an indigenous population of the southeastern Peruvian Amazon. In essence, we treat the Matsigenka system as a largely independent source of information regarding the spatial organization of tropical tree diversity, and we ask whether the high estimate of habitat diversity given by recent satellite studies is plausible and whether the different color/tone combinations seen in satellite images correspond to the habitats observed by local people on the ground. In short, we present a Matsigenka map of the world, and we take the first steps toward matching it against a satellite map of the world, a process we have dubbed "ethnobotanical ground-truthing."

The Contributions of Folk Science in Amazonia

Local Amazonian populations, especially indigenous societies, accumulate ecological observations throughout individual lifetimes and across generations, which provide them with a distinct practical advantage over professional biologists who make only occasional trips to the field. Historically, scientists in the Western tradition have tended to view indigenous and folk knowledge as suspect, the product of superstition, mysticism, and "primitive thinking" (see Frake 1964). In recent years, however, some biologists working in Amazonia have arrived at important insights by paying attention to the ecological knowledge of indigenous and other local populations. Pires and Prance's (1985) widely accepted forest classification for the Brazilian Amazon draws heavily upon the folk terminology of Brazil's Caboclos, riverine dwellers of mixed indigenous, European, and African descent whose ecological vocabulary is clearly indigenous (Tupi-Guarani) in origin. Encarnación (1993) likewise combines regional vernacular with scientific vocabulary in a description of lowland forest habitats in Peru. Goulding's (1980) pioneering study of Amazonian fish ecology is essentially an ethnoichthyology of the Brazilian Caboclos, verified empirically and translated into scientific terminology. The late Al Gentry learned from local woodsmen how to identify rain-forest trees based on trunk slash and other vegetative characters, revolutionizing Amazonian botany and creating the most useful field guide to the western Amazonian flora yet published (Gentry 1993). Marc van Roosmalen (pers. comm.; see van Roosmalen et al. 1998) has identified some 15 new taxa of monkeys from the Brazilian Amazon, gleaning most of his initial clues from interviews with indigenous and Caboclo informants. In short, indigenous informants and collectors have made significant contributions to the botanical and ethnobotanical sciences in the New World Tropics (Berlin 1984). With such a record of scientific contributions on behalf of indigenous and other local populations, we are confident that the Matsigenka may help us resolve the enigma of habitat diversity in tropical forests.

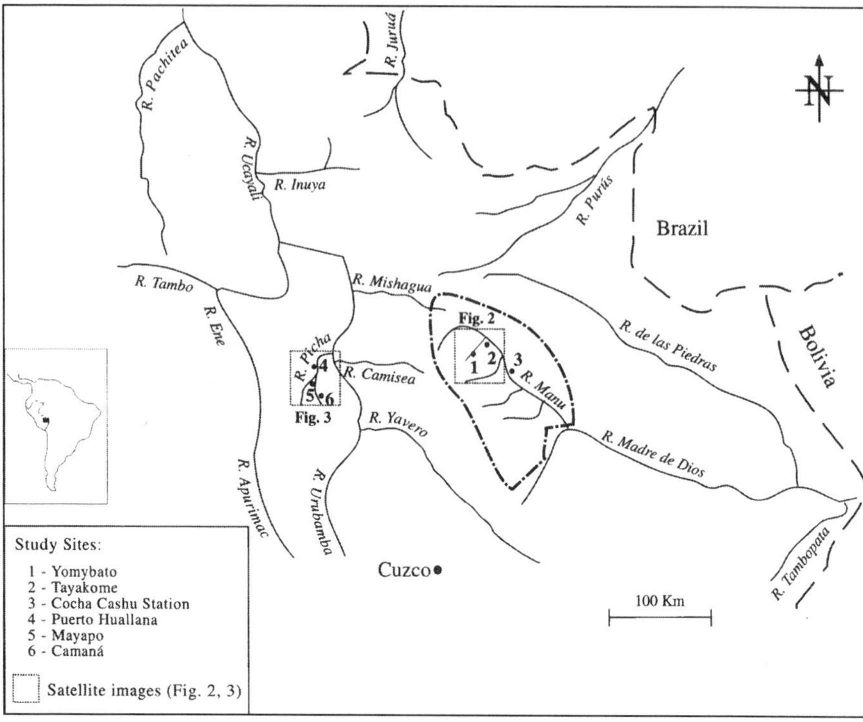

Figure 5.1. The study area, in southeastern Peru.

The Study Region

The Matsigenka are people of the *montaña,* the hilly rain forests that fringe the eastern slope of the Andes. They belong to the Arawakan cultural/linguistic family and have a current population of about 13,000, distributed in extended-family settlements and small communities along various tributaries of the Urubamba, Madre de Dios, and Manu Rivers. The Matsigenka cultivate manioc, maize, plantains, sweet potatoes, and other crops in small swiddens that are abandoned to forest regeneration after a few years of active cultivation (Johnson 1983). The Matsigenka also hunt, fish, and gather a wide range of forest products. Near mission towns and other trading centers, some Matsigenka engage in small-scale commercial cultivation of coffee, cacao, or annatto (Baksh 1995).

Our principal research sites are in the Matsigenka communities of Yomybato and Tayakome within the Manu Biosphere Reserve, a 1.6-million-ha area of protected tropical forest located in the department of Madre de Dios in southeastern Peru (Figure 5.1). The Manu Biosphere Reserve embraces the entire watershed of the Manu River, from alpine grasslands on Andean crests over 4000 m in elevation to the great expanses of lowland rain forest to the north and east. The Manu is a muddy, "white-water" river meandering in a wide floodplain that cuts through the sandy uplands of the alluvial horizon approximately 30 m above the river level. At the Cocha Cashu Biological Station, an average annual rainfall of just over 2000 mm was recorded, mostly during the rainy season (November–May) (Terborgh 1990). The marked dry season (June–October) receives less than 100 mm of precipitation. The Cocha Cashu Station, located in a 6-km-wide meander belt along the middle course of the Manu at an elevation of approximately 400 m above sea level, has been an important site for ecological research over the past 25 years (Terborgh 1983; Gentry 1990). The Native Community of Tayakome is located some 30 km upstream from the Cocha Cashu Station, close to the mouth of Quebrada Fierro. The Native Community of Yomybato is located along the upper course of Quebrada Fierro (known as *Yomuvaato* in Matsigenka), at approximately 450 m elevation.

Additional research was carried out in the Matsigenka communities of Mayapo, Puerto Huallana, and Camaná on the Picha River, some 150 km west of the Manu study site (Figure 5.1) and with an approximate elevation of 600 m above sea level. A tributary of the Lower Urubamba River, the Picha is a clear-water river in a rocky basin that descends from the Cordillera de Vilcabamba. The Vilcabamba mountain range is somewhat separated from the rest of the Andes and surrounded by lowland tropical rain forests to the north (Tambo River), east (Picha/Urubamba), and west (Ene and Apurimac Rivers). Since Inca times indigenous populations have migrated back and forth across the Cordillera de Vilcabamba to engage in trade or to escape political violence. Because of recent and past migrations across the cordillera, the Picha River Matsigenka are aware of high-elevation (montane) vegetation types of the Andes and the Andean foothills.

Personnel and Methods

Glenn Shepard has carried out ethnobotanical research in Yomybato, Tayakome, and other indigenous communities of the region since 1986 and is fluent in the Matsigenka language. In 1996 tropical ecologist Douglas Yu and ethnobotanist Manuel Lizarralde joined Shepard in the field for three months to carry out a preliminary study of Matsigenka forest classification in Yomybato (Shepard et al. 2001). In 1997 Shepard collaborated in Conservation International's Rapid Biodiversity Assessment of the Cordillera de Vilcabamba, carrying out a brief study of ethnoecology, forest classification, and resource-use patterns in the above-mentioned Matsigenka communities of the Picha River (Shepard & Chicchon 2001). In 1999 Shepard and Yu returned to the Manu for three months, armed with LANDSAT satellite images of the region that had been obtained and interpreted with assistance from ecologist and remote-sensing specialist Bruce Nelson. Mateo Italiano was our principal indigenous collaborator in the field during all three field seasons, though many other Matsigenka also contributed invaluable assistance to this study.

During the three research visits (two in Manu, one in the Picha River), we accompanied informants along forest transects and asked them to name and describe in their language the forest types we encountered (Matsigenka terms were transcribed using the practical orthography described in Snell 1998). When available, we used satellite images to plan, visualize, and interpret our transect walks. LANDSAT images were purchased from the Brazilian Space Research Institute and were displayed, enhanced, and analyzed using Idrisi (version 32, Clark University Labs) and Photoshop (versions 4.0/5.5, Adobe Systems) software. We took geographical coordinates of many localities using a hand-held global positioning satellite (GPS) unit, Garmin GPS models 45 and 12XL. Using the software package MacGPS Pro (version 2.6.7, James Associates), we are able to locate coordinates registered in our GPS database on the actual satellite image, within a margin of error of less than 50 m.

We selected a number of the most common forest types in the Manu study area and carried out botanical inventories in standardized forest plots and transects. The results of the botanical surveys are as yet incomplete, for voucher specimens are still being identified. Nonetheless, observations made by Matsigenka informants about the relative abundance of indicator species in different forest types appear to be confirmed empirically. In this chapter we explore Matsigenka forest classification in general terms and apply the method of ethnobotanical ground-truthing in a preliminary way to the available satellite images.

An Introduction to Satellite Imagery: Uses and Limitations

In recent years satellite imagery, GPS devices, and Geographic Information Systems (GIS) software have become progressively more available, affordable, and user friendly. First developed for military purposes, these technologies have countless applications in many fields and have proved especially important in recent studies of habitat diversity in Amazonia (Salo et al. 1986; Tuomisto et al. 1994; Tuomisto 1998). It is important that ethnobiologists and economic botanists also become aware of the potential applications of GIS and satellite technology in their own research. Many resources are available on the Internet, including Web-based tutorials, commercial and freeware GIS and image-processing software, bibliographies, map databases, and actual remote-sensing imagery, including satellite and radar images (see Appendix 5.1).

It must be remembered that satellites do not "see" species or species communities, they only record radiant energy reflected by the Earth. Each pixel in the satellite images employed in this study represents a 30-by-30-meter area (or ground-resolution unit) of the Earth's surface. For each ground-resolution unit, the Thematic Mapper (TM) sensor on board the LANDSAT satellite records six separate measures of reflected brightness: three in the visible portion (TM bands 1–3) and three in the infrared portion (bands 4, 5, and 7) of the electromagnetic spectrum. Scanning back and forth across the region or "scene," the satellite generates a digital matrix containing six values for each pixel. When decoded, the data reveal six superimposed grayscale images of the scene, one for each band of wavelength. Light tones represent high values of reflected brightness for a given band (the highest values being white), and dark tones represent low values (the lowest being black).

Researchers can use information from any or all of the six recorded bands in a variety of ways. To render the information into a visually assessable format, it is customary to create color composites. Three bands are chosen, the information from each band is arbitrarily assigned a different color (red, green, or blue), reflected brightness values are displayed as hue brightness values for the assigned color, and the three bands are viewed simultaneously as superimposed color channels. Thus reflected radiance values from multiple bands are mathematically combined into a single color for each pixel. The colors on a satellite image have nothing to do with normal color vision: information from visible as well as invisible (infrared) bands may be combined, and the assignment of bands to color channels is purely arbitrary. This is why such images are referred to as "false" color composites.

Visible light is not strongly reflected by healthy leaves, because plants make their living by absorbing most frequencies for photosynthesis. Over an ocean of healthy leaves such as the intact, moist, tropical forests of the southwestern Amazon, visible bands show little or no capacity to discriminate habitat types. For our field studies, we have made false-color composites[1] using the three infrared bands (bands 4, 5, and 7) of LANDSAT TM. Healthy leaves reflect bands 4 and 5; band 5 is reflected even more strongly by nonphotosynthetic material such as dead leaves, exposed branches, and bare soil. Dead leaves and exposed branches reflect band 7 most strongly. Even though it contains little additional information, band 7 appears to be marginally more useful in our images than does band 3 (visible), although the latter is more frequently used in LANDSAT TM imaging of Amazonia (e.g., Salo et al. 1986; Tuomisto et al. 1994, 1995). Assigning band 4 to the green channel, band 5 to the red channel, and band 7 to the blue channel, we generated visually appealing images in which (after contrast enhancement) secondary vegetation is white to light green, mature forest is dark green, bare soil is bright pink, semideciduous forest (combining the leafy response of band 4 with the exposed branch response of bands 5 and 7) is rosy to purple, and open water (absorbing all infrared bands) and shadows generated by topography are black.

For this publication, digital information from LANDSAT TM bands 3–7 was transformed into a single grayscale image (or derived band) called the Kauth-Thomas green vegetation index, available in the Idrisi software package (Eastman 1999: 119). The procedure, originally developed to detect agricultural crops throughout the growing cycle (Richards 1986: 142–145), consists of an image-independent linear transformation with positive coefficients for bands reflected by healthy, green vegetation and negative coefficients for bands absorbed by green vegetation. We found the resulting "greenness-band" image to be useful in rendering the information visible in our false-color composites accurately into publishable, black-and-white format with minimum loss of information.[2]

To understand the grayscale images presented here, it is useful to think of light tones as representing high "greenness" values (i.e., healthy leaves) and dark tones as representing nonphotosynthetic material (dead leaves, exposed branches, bare soil). White to light gray indicates a dense layer of healthy leaves on a uniform surface or canopy; for example, grass and living bamboo (white) and secondary forest (light gray). Black represents bare soil; for example, beaches or recently cleared gardens, as well as open water and topographic shadows. Darker grays (sometimes approaching black) appear to represent forest in a semideciduous state where exposed branches reduce the greenness response; for example, dead bamboo, living bamboo with sparse leaves, and water-adapted gallery forest during the peak of the dry season. Medium grays tend to represent mature, nondeciduous forest. Although the canopy is healthy and leafy, mature forest is darker and more textured compared with secondary forest, due to the shadows cast by different-sized trees.

Image enhancement (texture filters, contrast stretch) and digital classifications are tools used by remote-sensing scientists to maximize the information gleaned from satellite images. These tools must be used with caution, however, because noise generated by topography, canopy texture, detector defects, and pixel averaging can become amplified, generating a misleading classification of the forest. We applied a contrast stretch manually to each band in the false-color composite as well as to the grayscale image to help highlight differences between adjacent vegetation types, somewhat similar to adjusting the contrast on a television set. As we come to understand better the relationship between reflected radiance as measured by the satellite and actual ground features, we may test texture filters and digital classifications to assess their usefulness.

Each full LANDSAT scene shows an area of approximately 185 by 185 km. The satellite does not capture information for the entire scene in a single snapshot, but rather scans the scene pixel by pixel, back and forth across a matrix of columns and rows. During the process, which takes 24 seconds, the Earth rotates below the satellite's polar orbital path. The apparently rectangular image generated by the satellite represents, in fact, a skewed, rhombus-shaped area of the Earth's surface. To make satellite images compatible with GPS information, they must be corrected to make up for orbital skewing and other distortions and then assigned accurate geographical coordinates. Satellite images can be purchased with a standardized geometrical correction, or researchers can perform their own manual corrections using topographic maps or hand-held GPS units. Most GIS software packages include routines for assisting in the manual correction procedure. We performed a preliminary manual correction with topographic maps of the study sites, then fine-tuned the correction using actual GPS readings taken at landmarks that are both visible on satellite images and relatively stable through time; for example, sharp bends and mouths of entrenched streams (i.e., adjacent to terra firma), which less likely to shift position from year to year than are landmarks in more dynamic river floodplains.

The ground resolution for LANDSAT TM is limited to 30 by 30 meters. Each pixel in the image represents an average value over the entire ground-resolution unit. Any topographic or vegetational feature smaller than this size (e.g., a small stream or lake, treefall gap, or individual tree crown) is not possible to resolve. Features less than 60 by 60 meters tend to be fairly indistinct, because they are often split among a number of adjacent pixels. Like grainy photographs, LANDSAT images cannot be usefully magnified beyond a certain point. Furthermore, errors inherent in GPS readings and the geometric correction process place a limit on the precision of satellite images, especially for locating small features.

LANDSAT images are useful in distinguishing between deforested and forested areas, between forests with smooth and irregular canopy surfaces, between leafy and deciduous vegetation, and between forests with low and high leaf-area indexes. More subtle changes in species composition, however, may produce little or no alteration in the reflected optical properties. Moreover, even for those vegetation types that show up clearly on images, there is rarely a one-to-one correspondence between color or tone and vegetation. The same vegetation may show up as two different colors or tonalities on the satellite image, especially when topography changes. Forest on the sunny side of a slope appears lighter than does the same forest type in a flat area, and forest on the shady side of a slope is essentially black. This can be thought of as statistical Type 1 error, or overdifferentiation. On the other hand, different vegetation types may show the same color or tone if they happen to produce similar patterns of reflected radiance. For example, bamboo stands, grassy swamps, and low-canopy forest in seasonally flooded or eroded areas all have dense, uniform, leafy surfaces that reflect multiple bands strongly, generating a white tone in our images. Successional forests along the river, secondary forests in large windthrows, and some palm swamps all show similar light tones because of the even, relatively homogenous canopy. This corresponds to statistical Type 2 error, or underdifferentiation.

The prevalence of Type 1 and 2 errors does not render satellite images useless. When tone or shade changes between adjacent areas of the same satellite image, it is likely that something on the ground has changed. Exactly what has changed, however, may not always be apparent from the image alone. For this reason, ground-truthing is a crucial aspect of any remote-sensing research. At this point in the discussion, we turn to Matsigenka habitat knowledge as a source of information for ground-truthing of the available satellite images.

Matsigenka Habitat Classification

The Matsigenka use a sophisticated vocabulary for designating biotic as well as abiotic features of forest habitats (Shepard et al. 2001). Various categories of biotic and abiotic indicators constitute numerous subsystems of habitat classification, each treated somewhat independently from the others (Tables 5.1–5.4). Depending on the situation, a Matsigenka observer describing a given habitat may pay particular attention to soils, topography, drainage, vegetation, or disturbance regime, or to some combination of features. Though not organized into a single, unified hierarchy, the multiple systems of habitat description intersect to define forest types. We have attempted to represent the intersecting classificatory principles in the tables and figures. Note that habitat types in the tables are assigned reference numbers (hereafter, ref.) for ease of referral in the text and figures.

Abiotic features noted by the Matsigenka include topographic and hydrological features, disturbance regimes, and soil types (Tables 5.1–5.2). Topographic and hydrological criteria are used to distinguish two broad categories of habitats: lowlands or floodplain (*ovogeshi*), and upland terra firma or interfluvium (*nigankipatsa*), the same division currently used by tropical scientists (e.g., Terborgh et al. 1996). Montane forests of the Andean foothills (*otishi*) are treated as a separate category due to their distinctive topography, climate, and vegetation (Table 5.4). Disturbed or secondary (*tovaseri*, or weedy) vegetation (Table 5.4) also forms a salient category, contrasting

Table 5.1
Matsigenka habitats defined by topography and hydrology

Ref.	Name	Description
A1	*Ovogeshi*	Lowlands (floodplain), general term; also any lowland forest not included in a specific biotic/abiotic habitat type
T1	*otsegoa*	Seasonally flooded island, branch of river
T2	*imparage*	Open beach or wide streambed with sparse vegetation
T3	*oaaku, otapiku*	Semi-aquatic vegetation on or near river/stream bank
T4	*osateni*	Seasonal canal, depression in floodplain
T5	*otonkoatera ovogeshiku*	Levee island, small rise in floodplain forming island when river floods
T6	*nigankivoge*	Central floodplain, mature (late successional) lowland forest
T7	*choeni ovogeshi, choeni otishi*	Transitional zone from floodplain to uplands
T8	*ovogeshi niateni*	Large stream gallery forest
T9	*niateniku*	Small stream gallery forest
T10	*inkaare*	Swamp, lake; can occur in lowlands (e.g., oxbow lake) or uplands (palm swamps, rainwater lakes)
A2	*Nigankipatsa, Otishinapatsa*	Uplands (terra firma), hilly terrain, general terms; also any uplands not included in a specific biotic/abiotic habitat type
T11	*osateni niateni*	Ephemeral stream headwaters
T12	*pampa*	Flat area, especially flat uplands, alluvial terrace
T13	*agiringira, otonkoatera*	Downslope, upslope
T14	*otishi*	Hillcrest, ridge, mountain
T15	*imperita*	Cliff, rocky outcrop; also uplands on cliff adjacent to river
T16	*okonteaatira*	Spring, waterfall
T17	*oyashiaku*	River and stream headwaters; foothills, transition to montane habitat
T18	*otishipaketira*	Andean foothills
T19	*chovivanteni otishi*	High mountains

with primary forest (*inchatoshi;* Table 5.2). Soil classification is particularly important in choosing sites for swidden agriculture (Johnson 1983; Shepard et al. 2001).

Within the broad habitat categories defined by abiotic variables, the Matsigenka use biotic criteria to achieve a finer level of differentiation. Biotically defined habitats are distinguished according to dominant or indicator species, overall vegetative aspect ("phytoarchitecture"), and faunal indicators (Tables 5.3–5.4). We have divided indicator species into a number of subgroups inferred from Matsigenka botanical and ecological classification: palms, bamboos, ferns and herbs, and other indicator species (Table 5.3). It seems significant that palms, bamboos, grasses, and other monocots are so prominent as indicator species in Matsigenka forest classification. Palms are especially

Table 5.2
Matsigenka disturbance regimes and soil types

Ref.	Name	Description
		Disturbance regime
D1	*apamankera nia*	Area subject to flooding during rainy season
D2	*otarankira*	Landslide, cliff, eroded area
D3	*oterongera inchato*	Treefall gap
D4	*potagarine*	Burned hillside; natural fire landscape in Andean foothills with moss and ferns
D5	*tsamairentsi*	Recently felled swidden garden
D6	*magashipogo*	Productive swidden from prior years with some degree of secondary growth
D7	*pairani magashipogo*	Old swidden fallow
D8	*inchatoshi*	Primary forest, forest with large-diameter trees that has largely recovered from past disturbance
		Soils
S1	*jenkivane*	Sandy loam; flat uplands
S2	*potsitapatsa*	Black soil, river sediments; river floodplain
S3	*kiteri kipatsi*	Yellow (ferralitic?) soil; hills, uplands
S4	*impaneki*	Sand, beach
S5	*sokopane*	White sand soils in uplands
S6	*kiraapatsa*	Red clay or clay/sand loam; hilly uplands, stream floodplains, mineral licks
S7	*kusomiriakipatsa*	Contractile clay soil (forms hard lumps when dry)
S8	*inkaarepatsa, jampovatsa*	Mud, seasonally or permanently waterlogged soil; swamps, lakes
S9	*mapuseku*	Rocky soil
S10	*omakaramangaitira*	Mossy soil in cloud forest

important economic species for the Matsigenka and other indigenous groups of Amazonia (Balick 1984). Some species of palms, bamboos, grasses, and other monocots are specific to certain soils and drainage properties (Gentry 1988a; Encarnación 1993; Clark et al. 1999); furthermore, monocots are often colonial, abundant, or highly apparent in the understory, making them useful and salient habitat indicators.

Because habitat definitions overlap to some extent, and because our list may not yet be exhaustive, it is difficult to count the exact, total number of forest types recognized by the Matsigenka. In all, informants from five study communities named 77 different (though not necessarily exclusive) biotically defined habitats, including 51 primary or natural forest types[3] defined by indicator species, 7 kinds of secondary vegetation, 6 montane-only vegetation types, 6 forest types defined by overall vegetative aspect, and 7 habitats defined by faunal associations (Tables 5.3–5.4). The Matsigenka also distinguish 19 habitat zones defined by topography and hydrology, 8 degrees of forest disturbance, and 10 soil types affecting vegetation (Tables 5.1–5.2). Some habitat types are discussed in greater detail in the satellite ground-truthing exercise below, as well as in Shepard et al. (2001).

Table 5.3

Matsigenka forest and other natural habitats defined by indicator species. Dialect variants are separated by a slash (Manu/Picha). Habitat size, in hectares per patch: XS = <0.1; S = up to 1; M = up to 10; L = up to 100; XL = >100. TM = apparency on LANDSAT TM.

Ref.	Name	Description	Size	TM
		Palm forests		
1	*tirotishi*	*Astrocaryum murumuru* forest; floodplain only	M	–
2	*shevoshi*	*Attalea butyracea* forest; transition from floodplain to uplands	S	+/–
3	*tsigaroshi*	*Attalea macrocarpa* forest; floodplain only	M	–
4	*kontashi*	*Attalea tesmannii* forest; rare, along stream banks	S	+/–
5	*shiantishi*	*Bactris concinna* understory palm forest, wet areas, especially in floodplain	S	+
6	*tsirerishi*	*Euterpe precatoria* forest; poorly drained uplands, swamp margins	M	–
7	*tyonkintoshi / choginashi*	*Geonoma* sp. in understory; flat uplands	M	–
8	*tsikeroshi*	*Geonoma* sp. in understory; flat uplands	M	–
9	*memerishi / metakishi*	*Geonoma* sp. in understory; flat uplands	M	–
10	*kapashi*	*Hyospathe* (?) sp. in understory; hill crests at higher elevations	M	–
11	*kamonashi*	*Iriartea deltoidea* forest; ubiquitous, especially in floodplain	M	–
12	*koshishi / toturokishi*	*Mauritia flexuosa* palm swamp	M/L	+
13	*kinirishi*	*Mauritiella* sp. palm swamp	M/L	+
14	*segashi*	*Oenocarpus bataua* forest; poorly drained uplands, swamp margins	M	+/–
15	*kompiroshi*	*Phytelephas macrocarpa* forest; hillcrests and floodplain	M	+/–
16	*vakirintsishi / kontirishi*	*Socratea exorrhiza* forest; stream gallery forest	M	+/–
17	*kompaparishi / konkaparishi*	*Socratea salazarii* forest; common in flat uplands	M/L	+/–
18	*kepitoshi*	*Wettinia augusta* forest; common in flat uplands at higher elevations	M/L	+/–
		Bamboo forests		
19	*samatsishi*	*Aulonemia* sp. forest; uplands and slopes	S/M	+
20	*songarentsishi*	*Chusquea* spp., *Olyra* spp.; grassy bamboos on slopes, also montane	S/M	+
21	*manipishi*	*Guadua angustifolia* ; small stands near river margin	XS	–
22	*kapiroshi*	*Guadua sarcocarpa/weberbaueri* forest; floodplain and uplands	L/XL	+
23	*yaiveroshi*	*Guadua* sp. forest; floodplain only on clay soils	L/XL	+
24	*shinkeroshi*	*Guadua* sp. forest; uplands only	M	NA

Table 5.3, *continued*

Ref.	Name	Description	Size	TM
		Fern and herb habitats		
25	*tinkanarishi*	*Cyathea* spp. (tree fern) stand; streambeds, also montane	S	–
26	*tsirompishi*	Fern stands (Pteridophyta, miscellaneous species); moist or rocky areas, also montane	XS	–
27	*itsirianeshi matsontsori*	*Aechmea* sp. stand ("jaguar's pineapple plantation"); uplands	XS	–
28	*imereshi*	*Diplasia* sp. stand; moist areas	XS	–
29	*savoroshi*	*Gynerium saccharoides* cane thickets; beaches	S	+
30	*chakopishi*	*Gynerium sagittatum* thicket; arrow cane	S	NA
31	*sagontoshi*	*Heliconia metallica* in understory; wet areas in floodplain	XS	–
32	*tsipanashi*	Marantaceae spp. in understory	XS	–
33	*shimpenashi, tiposhi*	Poaceae spp. (terrestrial grasses); slopes, dry seasonal lakes, also montane grasslands	S/M	+
34	*sampetashi / kentakorishi*	Poaceae spp. (aquatic grasses); swamps, lakes	M/L	+
35	*porenkishi*	*Renealmia* sp.; swamp margins	S	–
36	*impomerishi*	*Tessaria integrifolia*; beach vegetation	S	+
		Other indicator species		
37	*matsityananashi*	*Alibertia pilosa* forest; floodplain with open understory	S	+/–
38	*toaroshi*	*Apuleia leiocarpa* forest; near streams	M	+/–
39	*kovantishi*	*Calliandra amazonica*; along stream banks	XS	–
40a	*setikoshi / inkonashi*	*Cecropia membranacea* forest; branch islands, floodplain near river	L	+
40b	*tonkoshi*	*Cecropia polystachya* forest; branch islands, floodplain, eroded riverbanks; much like *C. membranacea*, sometimes used interchangeably with *setikoshi* (40a)	M	+
41	*santarishi*	*Cedrela odorata*		
42	*pariashi*	*Cedrelinga caeteniformis* forest; flat uplands	M	–
43	*kachopitokishi*	*Chrysochlamys ulei* forest; on slopes, with *Tovomita weddeliana* and *Wettinia augusta*	M	–
44	*pakitsashi / piamentsishi*	*Clavija longifolia* forest; on slopes, with *Aulonemia*	M	–
45	*taiirishi, songaareshi*	*Erythrina* spp. forest; riverbank	S	–
46	*potogoshi*	*Ficus* spp. forest (especially *F. insipida*); floodplain near river	M	+/–
47	*tiirokishi*	*Ficus trigona* swamp	M	(+)
48	*koñorishi / konorishi*	*Hevea brasiliensis* forest; flat uplands	M/L	–
49	*intsipashi*	*Inga* spp.; forest; border of swamps and lakes	S	–
50	*kovuvapishi*	*Sagotia* sp.; on slopes	S	–

Table 5.4

Matsigenka secondary and montane vegetation, habitats defined by overall vegetative aspect, and faunal indicators. Dialect variants are separated by a slash (Manu/Picha). Habitat size, in hectares per patch: XS = <0.1; S = up to 1; M = up to 10; L = up to 100; XL = >100. TM = apparency on LANDSAT TM.

Ref.	Name	Description	Size	TM
		Weedy, secondary vegetation		
51	*tovaseri*	Weeds, weedy secondary growth	S/M	+
52	*narongashi*	Dense, tangled secondary growth of herbs, creeping vines	M	+/–
53	*shivitsasemai*	Liana forest	M	+/–
54	*yaaroshi*	*Cecropia sciadophylla* and other pioneer species in old fallows, windthrows	L/XL	+
55	*shintishi*	*Guazuma crinita* and other weedy species in fallows	S	+
56	*kogi oshivokera*	*Tephrosia* sp.; hardy cultigen (fish poison) belying old swiddens in apparently "primary" forest	S	–
57	*pugoroshi*	*Vernonia* spp.; weedy growth in recent swidden fallows	S/M	+
		Montane-only vegetation		
58	*katarompanaki*	*Clusia* sp. forest	?M	NA
59	*kasankari koka*	*Erythroxylum coca* ("fragrant coca"); ancient Inca plantations?	?S/M	NA
60	*kashikarishi*	*Polylepis* sp. forest	?M	NA
61	*yaviroshi*	*Puya* sp. grasslands	?M/L	NA
62	*tsiriantiniroshi*	*Tillandsia* sp. (Spanish moss); cloud-forest epiphtye	?M	NA
63	*tipeshi*	*Spaghnum* sp. (moss), moist and rocky areas	?M	NA
		Overall vegetative aspect		
64	*kusokiri inchato*	Hardwood forest (notably Chrysobalanaceae); uplands	M/L	–
65	*kurayongashi / karororoempeshi*	High-canopy forest; mature lowland and stream gallery forest	M/L	+/–
66	*okametira*	"Good place"; i.e., for walking; forest with open understory	M	–
67	*oshavishitira*	Low-canopy forest	S/M	+
68	*otyomiaigeni inchato*	Dwarf forest, cloud forest; montane	?M	NA
69	*terira ontime inchato*	"No trees," Andean grasslands above tree line	?M/L	NA
(D8)	*inchatoshi*	Primary forest (Table 5.2)		
(51)	*tovaseri*	Weedy vegetation (above)		
(52)	*narongashi*	Tangled secondary growth (above)		
(53)	*shivitsasemai*	Liana forest (above)		

Table 5.4, *continued*

Ref.	Name	Description	Size	TM
		Faunal indicators		
F1	*irapitari kimaro*	Macaw clay lick; riverbank	XS	–
F2	*itsimini*	Animal mineral lick; stream bank or stream headwaters	XS/S	+/–
F3	*vuimpuyoseku / kovutatsirira*	Singing grounds of the screaming piha (*Lipaugus vociferans*); primary upland forest	M	–
F4	*matyaniroshi*	Fire-ant forest; secondary, liana forest	S/M	–
F5	*matyagirokishi*	*Cordia* ant-shrub forest; uplands	S/M	–
F6	*sakaroshi*	Ant-garden forest; swamps	S	–
F7	*kepage*	Animal den in overturned tree roots; high-turnover uplands	XS	–

Ethnobotanical Ground-Truthing

Habitat classifications and remote-sensing analysis can be applied at different spatial scales: global, continental, biome-level (for instance, Amazon rain forest), national (Peru), regional (Manu National Park), and local or community-level. Indigenous habitat classification, and therefore the procedure of ethnobotanical ground-truthing, is most useful at the local or community level. Indigenous people have extremely detailed knowledge of the habitats within a few days' walk or canoe trip from their communities. By working with several indigenous communities within the same or different regions, it may become possible to extrapolate indigenous habitat knowledge over a broader regional scale.

Here we use the Matsigenka habitat classification system to interpret two LANDSAT TM images encompassing the five communities visited, one image of the Upper Manu River basin (Figure 5.2) and the other, of the Picha River basin (Figure 5.3). All images presented here are grayscale, "greenness-band" renderings of LANDSAT TM bands 3–7 with a manually applied contrast enhancement, as explained above.

The spatial scale represented in Figures 5.2 and 5.3 lies intermediate between the community and regional levels. At this scale, the habitat types most evident are those associated with topography and hydrology. Rivers and major tributaries are clearly etched in black across the images; hilly and mountainous areas are obvious because of the distinctive shadows they cast on the terrain. Also evident are certain kinds of large-scale disturbance; for example, long-term river dynamics and large windthrows. The vegetation type most evident at this scale is bamboo-dominated forest (*kapiroshi*, ref. 22), which is responsible for most of the white and light gray tones in both images and which contrasts with the medium to dark gray of dense or primary forest.

In Figure 5.2, the Manu River flows southeastward across the upper right quadrant of the image, and three major tributaries flow eastward and northward. Andean foothill ridges (*otishipaketira*, ref. T18) are visible in the southwestern corner of the image as

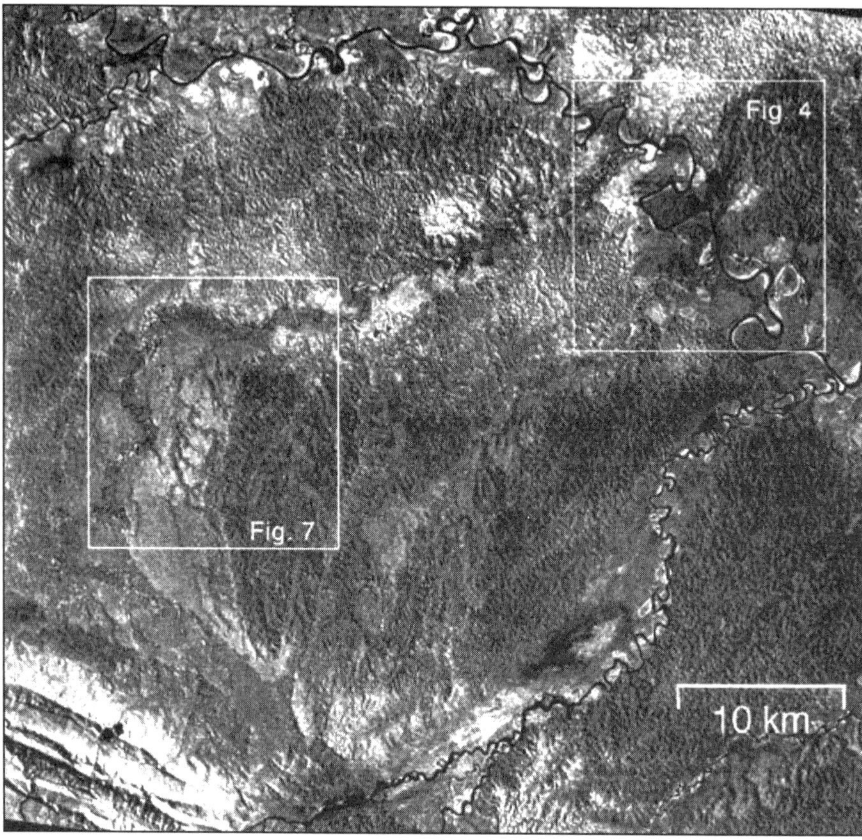

Figure 5.2. Satellite image of the study region on the upper Manu River. The image is a contrast-enhanced, geometrically corrected portion of LANDSAT TM scene 04/68, recorded on September 19, 1997. Enlarged views of the same image showing the communities of Tayakome and Yomybato are found in Figures 5.4 and 5.7. In Figures 5.2, 5.3, 5.4, and 5.7, multiple spectral bands were converted into a derived band, grayscale image using the Kauth-Thomas green vegetation index.

parallel lines of light and shadow. This mountainous region, home to a number of isolated Matsigenka settlements, contains distinctive vegetation and fauna; for example, the high-elevation thatch palm *kapashi* (*Hyospathe* sp., ref. 10), spectacled bear (*maeni*), cock-of-the-rock (*oe*), forest snails (*sankero*), and other species of the cloud-forest margin.

The meander belt of the Manu River is clearly seen in the northeastern quadrant of Figure 5.2 and includes a number of large oxbow lakes, remnants of former river meanders isolated from the main waterway when the river cuts across a narrow loop[4] to seek a new course (see Foster 1990b). Most oxbow lakes are dominated by aquatic grasses (*sampetashi,* ref. 34), visible as light-colored fan or crescent shapes, sometimes bordered by black areas representing exposed mudflats. Light areas along the inside of

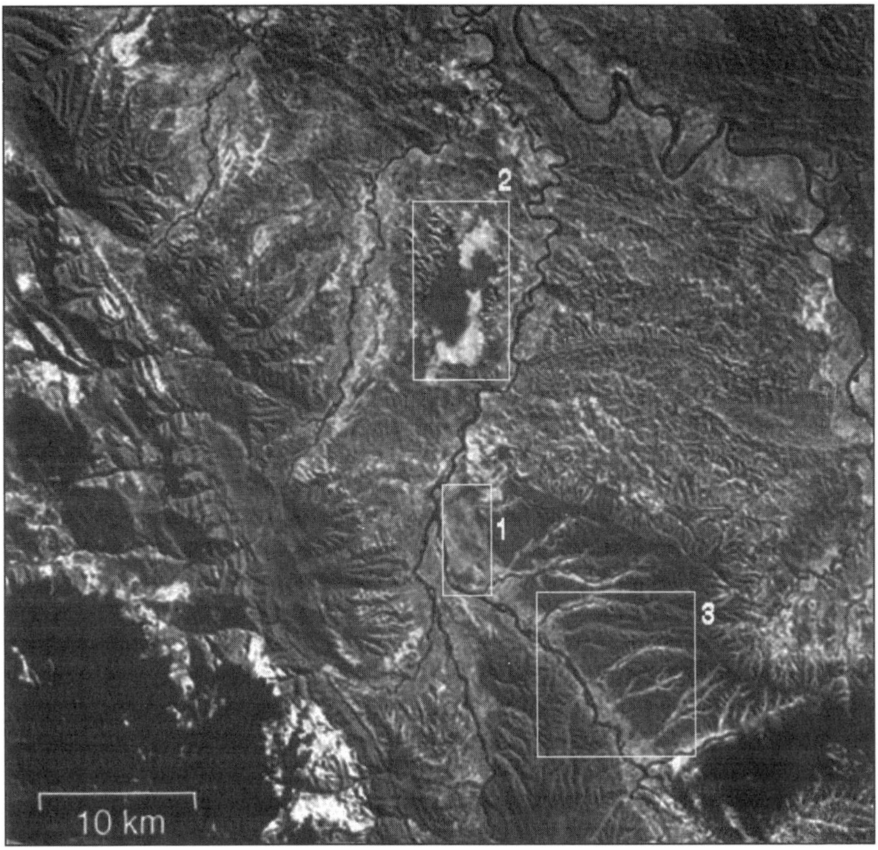

Figure 5.3. Satellite image of the study region on the Picha River. The image is a contrast-enhanced enlargement (not geometrically corrected) of LANDSAT TM scene 05/68, recorded July 5, 1996. See the text for discussions of Details 1–3.

expanding river bends are zones of primary succession from bare sand, including beach vegetation (*imparage, savoroshi, impomerishi;* refs. T2, 29, 36) and *Cecropia* stands (*setikoshi, tonkoshi,* refs. 40a, 40b). Dark colors in the center of large meander loops on the Manu River represent mature floodplain forest (*nigankivoge,* ref. T6), having a high, textured canopy with many shadows cast by giant emergent trees (*kurayongashi,* ref. 65). Dark areas along smaller tributaries, especially in the Figure 5.7 detail, represent stream gallery forests (*ovogeshi niateni, niateniku;* refs. T8, T9). Adapted to moister soils, some tree species near streams appear to enter into a semideciduous state during the peak of the dry season, when the image was recorded. Dead leaves and exposed branches contribute to a decreased greenness response and hence darker coloration.

Large white patches in roughly circular to elliptic forms, particularly visible in the northeastern corner of Figure 5.2, are mostly dense stands of the bamboo *Guadua weberbaueri*[5] (*kapiroshi,* ref. 22) in flat areas of the floodplain and uplands. Diffuse, black-and-white checkerboard patterns, also in the northeastern corner of the image, represent eroded, hilly terrain with bamboo (white) and hill shadows (black). Near the center of the Figure 5.7 detail (see also Figures 5.8, 5.9), light gray patches amid the dark, dendritic patterns of stream gallery forest appear, at first glance, to be bamboo-dominated forests. On the ground, however, they turned out to be large expanses of secondary forest dominated by *Cecropia sciadophylla* (*yaaroshi*, ref. 54), apparently the aftermath of large windstorms (Nelson & Amaral 1994). In addition to the white of *kapiroshi* bamboo forests and the light gray of *yaaroshi* secondary forests, we note a number of variations of medium to dark gray forests in the interfluvial areas, discussed in greater detail below (see Figures 5.4–5.9).

Comparing Figures 5.2 and 5.3, we notice tremendous differences in the geomorphology and vegetation of the Manu and Picha River regions, separated by less than 150 km. In Figure 5.3 the Picha River and its tributaries flow from south to north, meeting the Urubamba, which flows north and west in this image. The foothills (*otishipaketira*, ref. T18) and high mountains (*chovivanteni otishi*, ref. T19) of the Cordillera de Vilcabamba dominate the western third of the image, representing a much more abrupt transition from lowlands to mountains than that in the Manu region. The cordillera contains distinctive montane vegetation that the people of the Picha described in extraordinary detail (see Table 5.4, refs. 58–63), drawing on collective memories and oral histories of past migrations throughout the region (Shepard & Chicchon 2001).

What little mature floodplain forest (*nigankivoge*, ref. T6) is present on the Picha River is restricted to a few small, dark gray patches adjacent to the river (see Detail 2, below). Instead, the floodplains (*ovogeshi*, ref. A1) of the Urubamba, especially of the Picha and its tributaries, are dominated by dense stands of bamboo (*kapiroshi*, ref. 22), light gray to pure white on the image. Stream gallery forests (*niateniku*, ref. T9), which show distinctive semideciduous forest vegetation in the Manu study site, are mostly dominated by *kapiroshi* bamboo (ref. 22) in the Picha basin. Especially in the southeastern part of the image, the white, bamboo-dominated river and stream margins take on a dendritic appearance, unlike the circular patterns of bamboo forest in the Manu image. The former are clearly limited by changes in substrate, whereas the latter resemble the "biological borders" (i.e., substrate independent) described for Acre by Vidalenc (2000). Both the Picha and the Urubamba have rocky basins and are fairly stable in their courses, compared with the meandering and constantly shifting Manu. Most lakes in the Picha basin are small and not visible at this scale.

Just south of the center of Figure 5.3 we note an elongated patch of medium gray surrounded by light gray bamboo (Detail 1). This is a large swampy area that, according to Matsigenka lore, was formerly an enormous lake inhabited by dangerous spirits. The lake is said to have been drained and the dangerous spirits banished by a powerful Matsigenka shaman, perhaps in the late 1960s. A few areas of open water are still present, visible as tiny black circles and ellipses. These contain aquatic grasses (*kentakorishi*, ref. 34) and herbs such as *Renealmia* (*porenkishi*, ref. 35). Swampy areas are visible as larger, gray ellipses in the upper two-thirds of Detail 1 and are dominated by the palm *Mauritiella* (*kinirishi*, ref. 13), not present in the Manu region. (The related *Mauritia flexuosa,* ref. 12, ubiquitous in swamps of the Manu, is not present in the Picha except for a few individuals in old village sites, grown from seedlings brought from other regions.) Also present in swampy areas are stands of the palms *Oenocarpus* (*segashi,* ref. 14), *Euterpe* (*tsirerishi,* ref. 6), and *Socratea exorrhiza* (*vakirintsishi/kontirishi,* ref. 16), mostly along the margin between the *Mauritiella* stands and the surrounding bamboo forest. South of the swamps is one of the few fragments of mature floodplain forest (*nigankivoge,* ref. T6) present on the Picha. Containing canopy giants such as *Ceiba pentandra* (*shirigari*) and *Sloanea* (*terorivanteki/tarovenki*), it is visible as a dark gray patch adjacent to the river near the southern edge of Detail 1.

Most of the interfluvial area between the Urubamba and the Picha basin in Figure 5.3 is a hilly checkerboard of white bamboo-dominated forest (*kapiroshi,* ref. 22) and black hill shadows, especially in the eastern third of the image. Only a few large areas of dense, primary forest (*inchatoshi,* ref. D8) are found on the entire image, visible as dark gray wedges in interfluvial areas. Just north of the center of the image is an irregularly shaped patch of dark gray primary forest (Detail 2) flanked by crescents of pure white to the north and south. The area is a terraced upland plateau (*pampa,* ref. T12) on a high cliff (*imperita,* ref. T15) adjacent to the river. The white crescents show a response pattern similar to that of bamboo forest (*kapiroshi*) but in fact represent treeless grasslands (*shimpenashi, songarentsishi;* refs. 33, 20) on slopes along the edge of the plateau. Flat, poorly drained areas at the center of the plateau have stands of *Cedrelinga cataeniformis* (*pariashi,* ref. 42) and show an even, dark gray tone. Slopes between terraces are visible as a few fine, dark shadows running mostly north–south and are dominated by mixed stands of the small tree *Sagotia* sp. (*kovuvapishi,* ref. 50) and the palm *Wettinia augusta* (*kepitoshi,* ref. 18).

The largest areas of continuous upland forest in the image are found in the southeastern quadrant of the image, on opposite sides of the river (Detail 3). (The black area southeast of Detail 3 appears to be a cloud shadow.) River and stream margins are covered in dense *kapiroshi* bamboo stands (ref. 22), forming a distinct dendritic pattern. Moving from the river floodplain inland, the land rises in a series of broad, forested terraces (dark gray) separated by abrupt slopes, visible as fine black creases roughly parallel to the river. Flat, well-drained areas have stands of *Hevea brasiliensis* (*konorishi,* ref. 48) and *Cedrelinga* (*pariashi,* ref. 42), with *Wettinia augusta* (*kepitoshi,* ref. 18) in the understory. Slopes between terraces have *Sagotia* stands (*kovuvapishi,* ref. 50). In small patches of poorly drained and white sand soils toward the center of the terraces (not visible at this scale) we found a low-canopy forest (*oshavishitira,* ref. 67) with stands of the palm *Oenocarpus* (*segashi,* ref. 14) and lianas (*shivitsasemai,* ref. 53).

Moving farther inland on either side of the river in Detail 3, the landscape begins to crease and fold, crisscrossed by a series of ridges (*otishi,* ref. T14) and small streams. High-canopy forest (*karororoempeshi;* ref. 65) and *Phytelephas macrocarpa* palm forest (*kompiroshi,* ref. 15) are found along ridges. On slopes, there occurs a low canopy forest containing a mixture of *Wettinia* (*kepitoshi,* ref. 18) and two small Clusiaceae tree species, *Chrysochlamys ulei* (*kachopitokishi,* ref. 43) and *Tovomita weddeliana* (*tegarintsipini;* see ref. 43). This forest shows a very distinctive silhouette looking up from the forest floor because of the contrasting shapes of the three species' large leaves. Closer to the stream headwaters are stands of the bamboo *Aulonemia* (*samatsishi,* ref. 19) on steep slopes leading down to the streambed, visible as a slightly darker tone of gray in dendritic patterns along the upper courses of stream tributaries on the western part of Detail 3. Along ridges at the stream headwaters (upper right corner of Detail 3) occur stands of the palm *kapashi* (*Hyospathe,* ref. 10), harvested for roof thatch.

Unfortunately, our Picha River expedition was brief, and a GPS unit was not available, limiting our ability to carry out more detailed ground-truthing. Future fieldwork in the region, including GPS mapping and botanical inventories of Matsi–genka-defined forest types, is required to gain a better understanding of the distinctive habitats occurring in this landscape of abrupt topographic transitions and sharp contrasts in vegetation between adjacent areas.

The Manu Study Sites: A Closer Look

Because of more extensive research in the Manu region, including botanical inventories and a large database of GPS readings, we are able to present a more detailed study of habitat diversity at the local level for the communities of Tayakome and Yomybato. At this spatial scale, ethnobotanical ground-truthing proves to be a powerful tool for locating and characterizing diverse habitat types. Figures 5.4 and 5.7 are magnifications of the satellite image shown in Figure 5.2, corresponding to the environs of the native communities of Tayakome and Yomybato, respectively. Figures 5.4–5.6 and Figures 5.7–5.9 represent multiple views of the same areas around the two communities: an actual satellite image (Figures 5.4 and 5.7), a schematic map showing the Matsigenka topographic/hydrological classification of the landscape[6] (Figures 5.5 and 5.8), and a map showing Matsigenka-recognized vegetation types (Figures 5.6 and 5.9), using GPS data projected onto the satellite imagery. Though not possible in this publication format, the ideal way to view the multiple maps of each community would be as a series of superimposed transparencies. Throughout the following discussion, readers may find it helpful to first locate specific habitats by their respective reference numbers on the vegetation maps (Figures 5.6 and 5.9) and then refer back to the satellite images (Figures 5.4 and 5.7).

Tayakome

The Native Community of Tayakome is located on a forested cliff overlooking the Manu floodplain, just downstream from the mouth of Quebrada Fierro or *Yomuvaato* (Figures 5.4–5.6). House and recent garden sites (ref. D5) are visible on the satellite image as circles of dark gray or black, mostly west of a large, diamond-shaped meander loop on the Manu River. East of the village site and within the diamond-shaped meander loop is mature floodplain forest (*nigankivoge,* ref. T6), with an uneven, dark gray coloration due to the textured canopy. Near the center of the meander is a fine, white line. This is a seasonal watercourse (*osateni,* ref. T4) in the floodplain, dominated by lianas and dense stands of spiny *Bactris concinna* (*shiantishi,* ref. 5; see Figures 5.5 and 5.6).

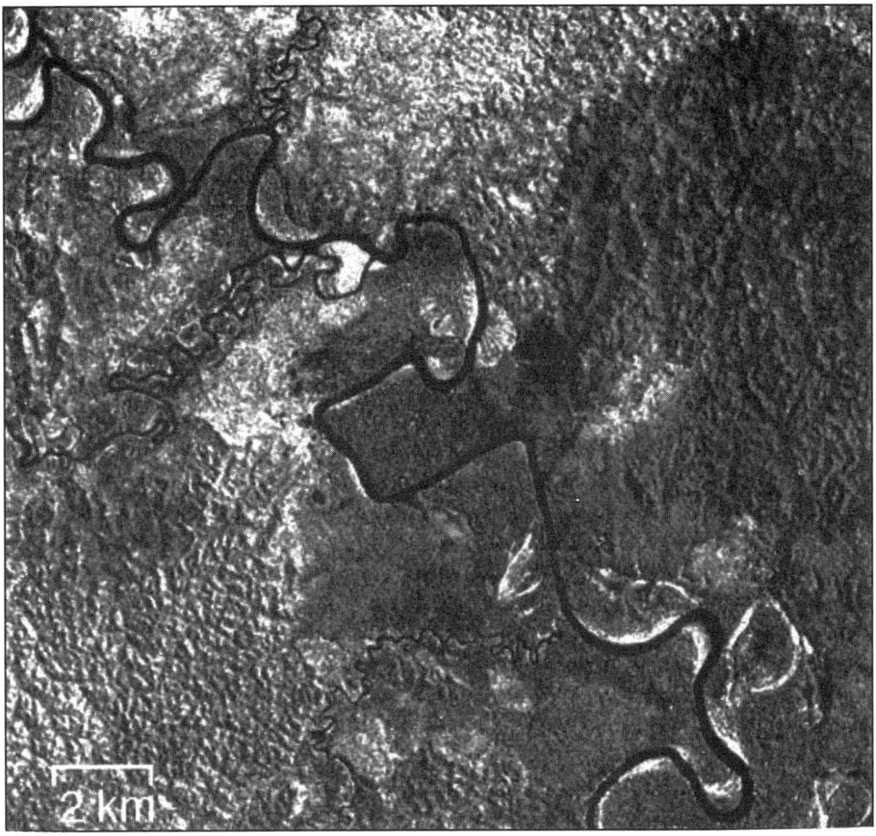

Figure 5.4. Satellite image of the Native Community of Tayakome and environs, Manu National Park.

Successional forests along the Manu are visible at several places on the image where white arcs run parallel to black river loops, especially in the southeastern quadrant. Moving inland from the beach, the vegetation sequences begin with *Tessaria integrifolia* stands (*impomerishi*, ref. 36) and cane thickets (*savoroshi*, ref. 29). This transition is especially clear on the inside curve of a tight meander loop at the extreme southeastern corner of the image. Dark tones indicating exposed sand give way to lighter grays as the vegetation cover becomes thicker away from the river. Distinct white belts and semicircles adjacent to the river, visible along most meander loops, represent *Cecropia membranacea* forest (*setikoshi*; ref. 40a). A similar species, *C. polystachya* (*tonkoshi*, ref. 40b), also occurs adjacent to the river, especially on branch islands and along eroded cliff bases. Early successional forests of *Cedrela odorata* (*santarishi*, ref. 41) and *Ficus insipida* (*potogoshi*, ref. 46) occur frequently in a belt adjacent to the river or inland from the *Cecropia* zone, sometimes visible where the white of *Cecropia* fades to medium gray (see Figure 5.6, ref. 40/41).

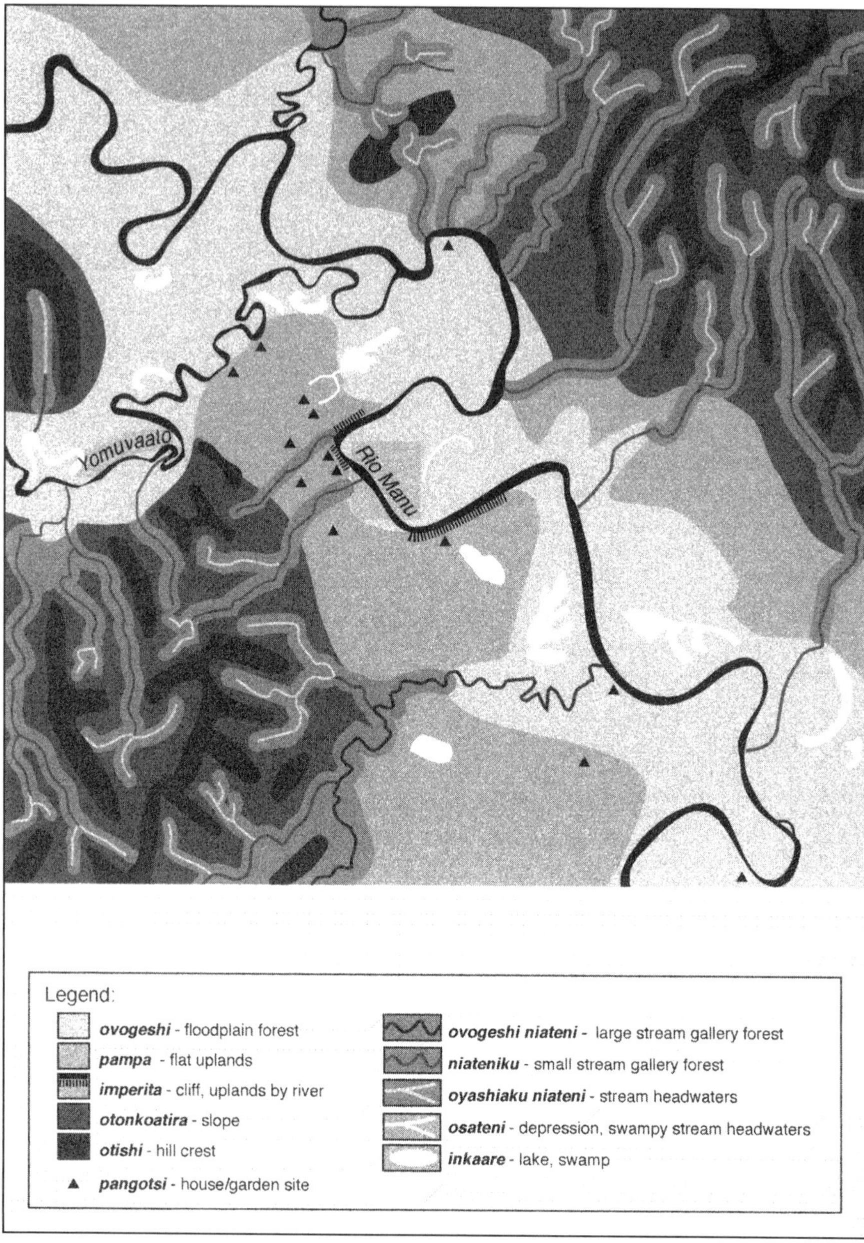

Figure 5.5. Schematic representation of habitats defined by topography/hydrology (see Table 5.1) around Tayakome, drawn to the same scale as Figure 5.4.

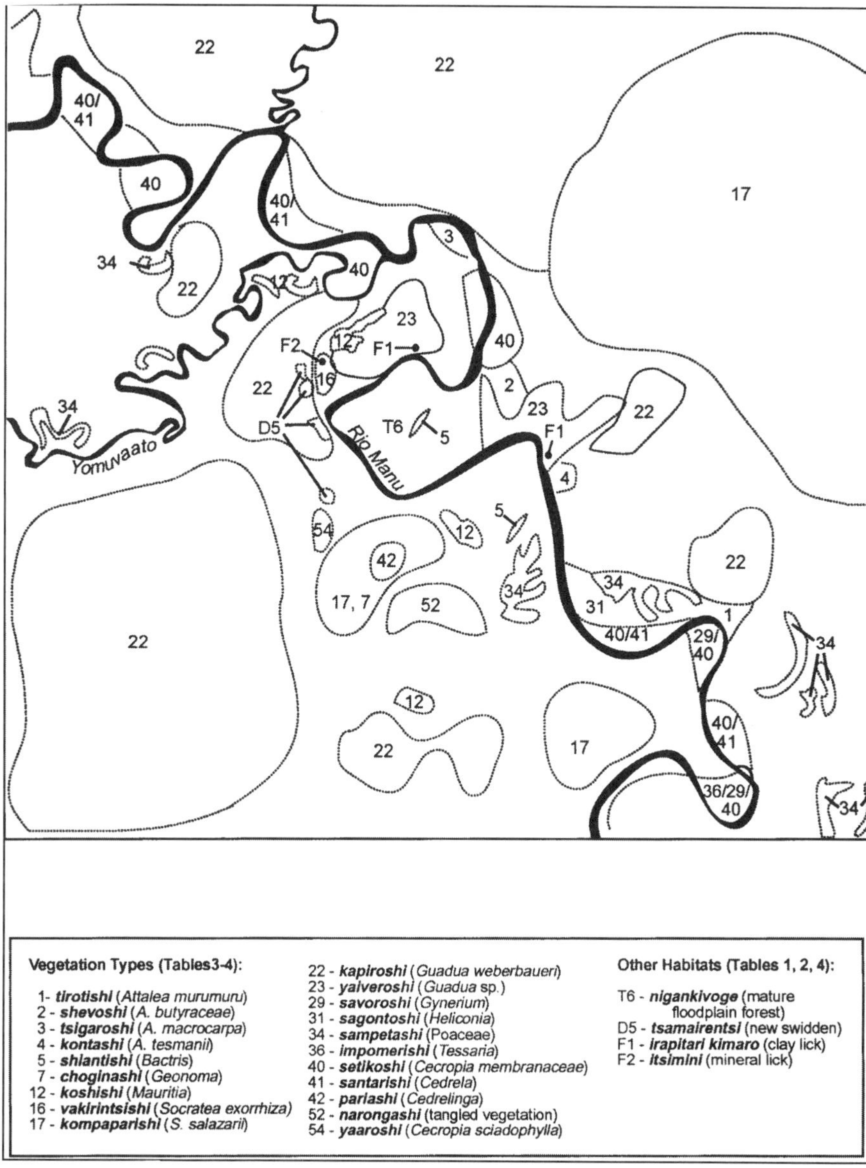

Figure 5.6. Matsigenka-defined vegetation types around Tayakome, identified through ethnobotanical ground-truthing, drawn to the same scale as Figure 5.4. Reference numbers follow Tables 5.1–5.4.

Oxbow lakes, dominated mostly by aquatic grasses (*sampetashi*, ref. 34), are visible on the image as white fan or crescent shapes not connected with the river. Large oxbow lakes are found along the Manu River in the southeastern part of the image; smaller ones are seen along the lower course of Yomuvaato stream in the northwestern quadrant. At least four *Mauritia flexuosa* palm swamps (*koshishi*) are visible on the image as elliptical areas of light to medium gray in poorly drained uplands and old oxbow lakes (see Figure 5.6, ref. 12). Several vegetation types occur within floodplain forests, including stands of *Astrocaryum murumuru* (*tirotishi*, ref. 1), *Attalea macrocarpa* (*tsigaroshi*, ref. 3), *Iriartea deltoidea* (*kamonashi*, ref. 11), and *Heliconia metallica* in depressions near the river (*sagontoshi*, ref. 31). Though distinctive on the ground, the various floodplain forest microhabitats are fairly difficult to distinguish on the satellite images employed, generally all a nondescript, mottled gray. *Attalea butyracea* stands (*shevoshi*, ref. 2) occur in transitional areas to uplands, visible as a lighter gray patch between the black *yaiveroshi* bamboo (ref. 23) and the white *Cecropia* forest (ref. 40) near the center of the Tayakome image (see Figure 5.6).

Much of the upland forest around Tayakome is dominated by *Guadua weberbaueri* bamboo (*kapiroshi*, ref. 22). *Kapiroshi* bamboo forests are visible as large patches of white or light gray. Almost unbroken patches of *kapiroshi* occur throughout the western half of Figure 5.4, especially on eroded, hilly terrain. The roughly circular border between dark, primary upland forest and white bamboo visible in the northeastern corner of both Figures 5.2 and 5.4 may represent an advancing wave of bamboo, invading from the north. Two smaller *kapiroshi* forests are visible as light ellipses surrounded by darker gray upland forest in the eastern third of Figure 5.4. These may represent areas of relatively recent dispersal of *kapiroshi* into the uplands.

Yaivero, an unidentified species of *Guadua* (with affinities to *G. glomerata*),[7] occurs in two fairly large stands in the vicinity of Tayakome (*yaiveroshi*, ref. 23). The *yaiveroshi* bamboo forests are visible near the center of the Tayakome image as two irregularly shaped, dark gray to black patches, one on either side of the Manu River (see Figure 5.6, ref. 23). In the case of *G. weberbaueri*, black indicates dead bamboo after an event of synchronous flowering and dieback (Nelson 1994; Vidalenc 2000). These forests do not appear to have suffered recent bamboo mortality, however. Instead, their black tone may be a result of this *Guadua* species' highly clustered and vertically angled leaves or of its slender stems, which tend to fall and lie almost horizontally, providing a strong "exposed branch" response. *Yaiveroshi* stands are much denser than *kapiroshi* bamboo forest, and very few trees grow in the densest areas. *Yaiveroshi* occurs on contractile clay soils, what the Matsigenka describe as "hard, lumpy earth" (*kusomiriakipatsa*, ref. S7) because of its tendency to form discrete lumps when dry. The Matsigenka observe that contractile clay soils and *yaiveroshi* are associated with macaw clay licks (*irapitari kimaro*, ref. F1), which form on the eroding side of riverbanks. Bearing out this observation, both *yaiveroshi* stands in this image are adjacent to clay licks (see Figure 5.6).

Upland forests without bamboo in the vicinity of Tayakome are characterized by the palm *Socratea salazarii* in the understory (*kompaparishi*, ref. 17). There are three principal *kompaparishi* forests near Tayakome, two patches south of the community on flat uplands, and one large area across the Manu from the community in hilly uplands traversed by small streams. The *kompaparishi* uplands closest to the community (due south) contain scattered stands of *Geonoma deversa* (*choginashi*, ref. 7) and a stand of *Cedrelinga cataeniformis* (*pariashi*, ref. 42; see Figure 5.6). The area shows a mottled, medium gray tone with irregular lighter patches throughout, perhaps due to the numerous treefall gaps observed. Just northwest of this *kompaparishi* (ref. 17), on the way to the main community, is a small, light gray patch on a ridge crest, representing secondary forest with *Cecropia sciadophylla* (*yaaroshi*, ref. 54) in 30-year-old garden fallows. Just southeast of this *kompaparishi* is an area of slightly darker, more uniform gray, representing stream gallery forests with dense, tangled vegetation (*narongashi*, ref. 52).

Yomybato

The Native Community of Yomybato[8] is located along the upper course of Yomuvaato, a shallow, rocky stream in a narrow basin with rusty, reddish brown floodwaters whose color inspired the Spanish name of the stream, Quebrada Fierro (Stream of Iron [Oxide]) (Figures 5.7–5.9). The central community of Yomybato is visible on the image as a dense cluster of black circles representing house compounds and recently cleared swiddens (ref. D5). Also visible are white circles representing weedy growth in older but still active swiddens (ref. D6) and diffuse, elliptical areas of light gray representing young secondary forest in old swiddens in various phases of regrowth (*shintishi, pugoroshi;* refs. 55, 57). Numerous house settlements and gardens in various phases of the swidden cycle are visible at a considerable distance from the central community area (see Figure 5.8).

The forests around Yomybato Community show several important differences when compared with those near Tayakome. Except near its mouth, Yomuvaato stream is relatively stable in its course, and the processes of river meandering, oxbow lake formation, and beach-to-forest succession evident along the Manu River are much less important. Inundated forests near Yomuvaato stream are confined to a much narrower floodplain flanked by hilly uplands and show a vegetation that mixes floodplain, stream gallery, and upland elements (*choeni ovogeshi, choeni otishi,* ref. T7). Floodplain and stream gallery forests (*ovogeshi niateni, niateniku,* refs. T8, T9) along the upper Yomuvaato show a dark coloration on the satellite image (Figures 5.2 and 5.7), apparently the result of dry-season semideciduousness. South of the main community area, in a stream gallery forest in a delta between Yomuvaato stream and a tributary, we were shown a forest containing a diffuse stand of *Apuleia leiocarpa* (*toaroshi,* ref. 38; see Figure 5.9), a tree previously considered rare for Manu.

Kapiroshi bamboo forests (ref. 22) are visible in Figures 5.2 and 5.7 as white patches north and northeast of the community. Small patches of *yaiveroshi* bamboo forest (ref. 23) are found along Yomuvaato stream near the community, but they are not very distinct on the satellite image. There are no oxbow lakes near Yomybato, though there are many small, seasonal lakes (*inkaare,* ref. T10) and swampy, ephemeral stream headwaters (*osateni niateni,* ref. T11) in flat uplands, only a few of which are visible at this

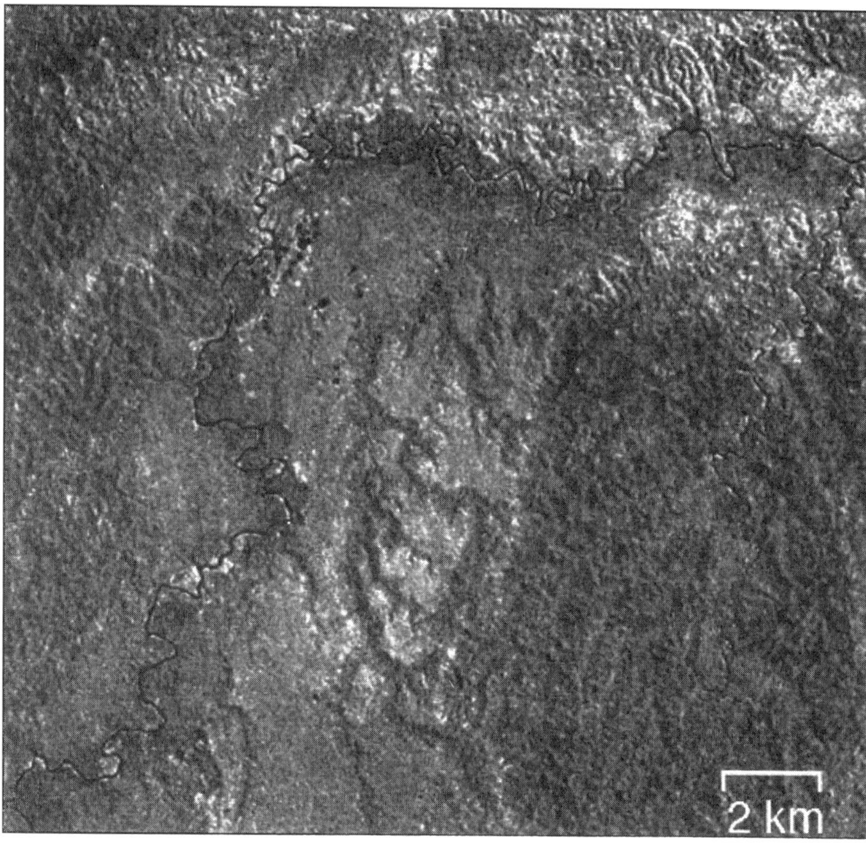

Figure 5.7. Satellite image of the Native Community of Yomybato and environs, Manu National Park.

spatial scale (see Figure 5.9). *Mauritia flexuosa* palm swamps (*koshishi,* ref. 12) occur in poorly drained areas of the uplands and are often found in mixed stands with *Oenocarpus bataua* (*segashi,* ref. 14), so they present a less homogenous canopy and a less distinctive signature on the satellite image than do those of Tayakome.

Upland forests around Yomybato are characterized by the subcanopy palm *Wettinia augusta* (*kepitoshi,* ref. 18). *Kepitoshi* is synonymous with flat, well-drained uplands for the people of Yomybato and is the preferred forest type to clear for swidden agriculture. Intact *kepitoshi* is found throughout the broad upland terrace (*pampa,* ref. T12) between Yomybato Community and the closest tributary stream, Potsitakigia (see Figures 5.8, 5.9). We found a few upland *kepitoshi* areas with rare individuals or patches of *Socratea salazarii* (*kompaparishi,* ref. 17), the inverse of the situation in Tayakome, where *kompaparishi* is common and *kepitoshi* is rare.

Kepitoshi forests are characterized by loose, sandy loam soils, sparse emergent trees (including *Parkia, Cedrelinga, Protium,* and *Hevea*), and a high degree of turnover due to treefalls and windthrow. Wind is an important cause of natural disturbance in the

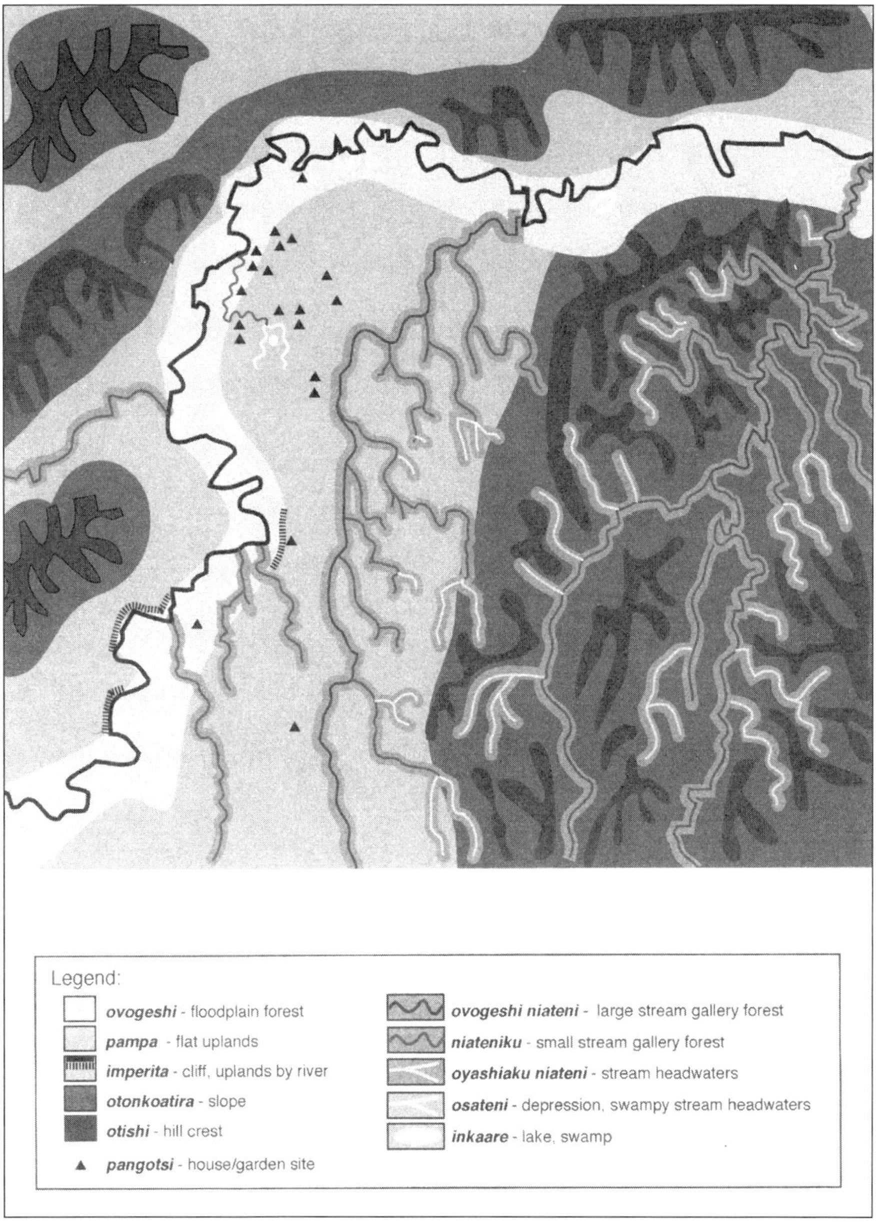

Figure 5.8. Schematic representation of habitats defined by topography/hydrology (see Table 5.1) around Yomybato, drawn to the same scale as Figure 5.7.

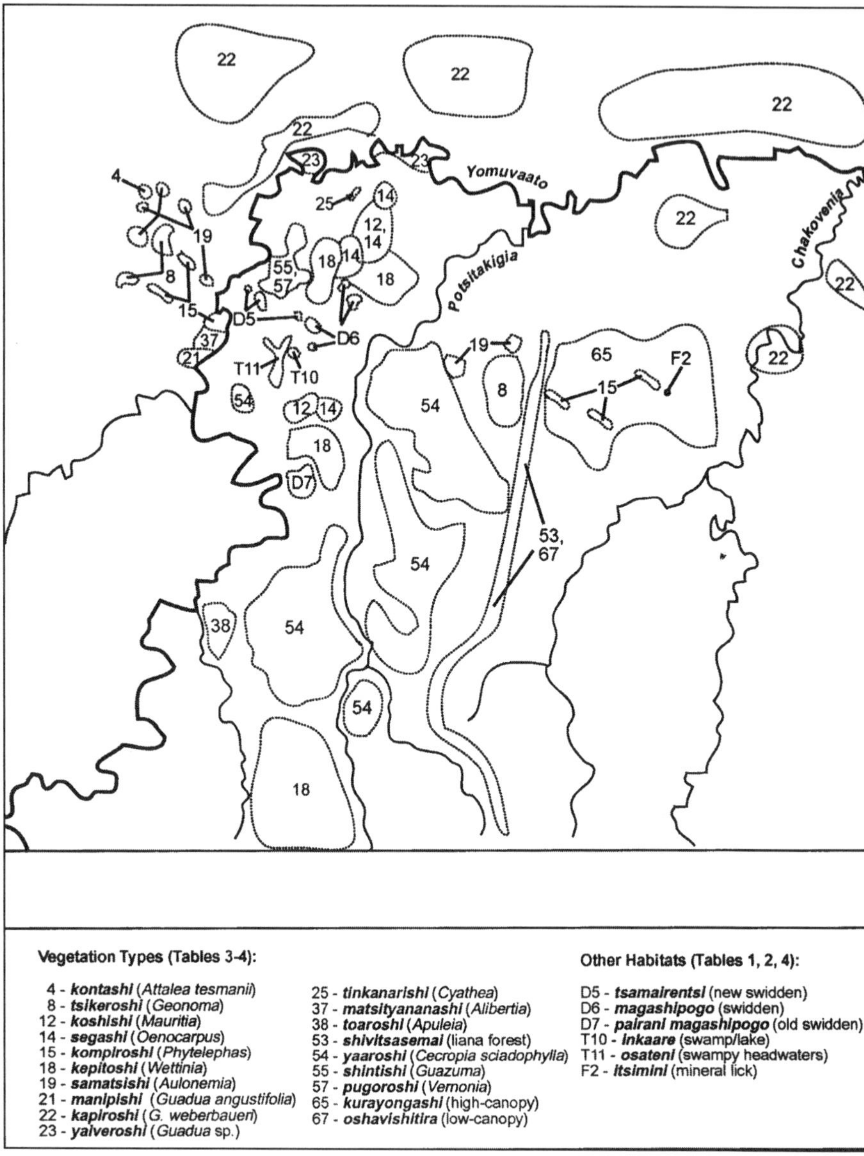

Figure 5.9. Matsigenka-defined vegetation types around Yomybato, identified through ethnobotanical ground-truthing, drawn to the same scale as Figure 5.7. Reference numbers follow Tables 5.1–5.4.

upland forests of Yomybato. Moderate winds fell single trees quite commonly. Strong windstorms are rare but can topple trees throughout tens and even hundreds of hectares. The Matsigenka pointed out several large areas south and east of the community that had been affected by windstorms in the past and are now characterized by secondary forests of pioneer species, notably *Cecropia sciadophylla* (*yaaroshi*, ref. 54).[9] Most of the central portion of Figure 5.7 is wind-generated *yaaroshi*, showing a white to light gray tone much like that of bamboo forest, contrasting with the dark, dendritic pattern of stream gallery forest. Both *C. sciadophylla* (*yaaroshi*, ref. 54) and *Wettinia augusta* (*kepitoshi*, ref. 18) are rarely found in the *Socratea salazarii*–dominated (*kompaparishi*, ref. 17) upland forests around Tayakome (Figures 5.4–5.6). We hypothesize that the Yomybato terrace is a mosaic of forests recovering from windthrow, in a process of succession beginning with *yaaroshi*, followed by *kepitoshi*, and ending at (though rarely attaining) *kompaparishi*.

Farther to the east on the Yomybato image we note an abrupt transition in topography. The flat terrain ends, delineated by a fine white line toward the middle of the satellite image running roughly north–south. On the ground, the white line consists of a low-canopy forest (*oshavishitira*, ref. 67) of shrubs and lianas (*shivitsasemai*, ref. 53) on an eroded slope. Continuing eastward, the terrain rises in a series of ridges traversed by many small streams. On some ridges and in stream galleries are high-canopy forests (*kurayongashi*, ref. 65) of giant emergent trees with high, spreading crowns. These have a dark tone on the image, due both to the ridge shadows and to an apparent semideciduous nature in the dry season. A large animal mineral lick (*itsimini*, ref. F2) is found at a cliff of eroded red clay in the center of the eastern third of the image, faintly visible as a black spot on the satellite image. This and other smaller mineral licks are important places for hunting many bird and mammal species (see Figures 5.8, 5.9).

On many ridge crests (usually with light tones on this image) in both the eastern and the northwestern portions of the image, the understory is dominated by the palm *Phytelephas macrocarpa* (*kompiroshi*, ref. 15). *Phytelephas* forest also occurs in at least one patch within the floodplain near the northwestern corner of the image, adjacent to a flooded forest with an open understory containing *Alibertia pilosa* (*matsityananashi*, ref. 37; see Figure 5.9). Adjacent to this is an unusually large patch of the bamboo *Guadua angustifolia* (*manipishi*, ref. 21). *Manipi* is the largest of the local bamboos, with tall, elegant stems that can exceed 10 cm in diameter and 12 m in height. It forms dense, circular, single-species stands surrounded by thorny branch shoots reminiscent of barbed wire. Absent around Tayakome, *manipishi* is common along the upper course of Yomuvaato stream. Though distinctive, in fact unmistakable, on the ground, *manipishi* stands rarely exceed 10 m in diameter and are hence usually not visible on LANDSAT TM images. However, this particular patch, actually a cluster of adjacent circular patches along the river margin, is large enough to show up as a small white circle on the image, contrasting with the light gray surrounding forest (see Figure 5.9, ref. 21).

Throughout the image, especially in the northwestern portion, are patches of the bamboo *Aulonemia* (*samatsishi*, ref. 19), apparently a specialist on slopes and disturbed uplands. We witnessed healthy *samatsishi* stands in this region in 1996, which are apparent as small, whitish elliptical patches on the image. In 1998–1999 all *samatsishi* patches in the northwestern part of the image underwent a natural, synchronous dieback and were in the early phases of regeneration during our 1999 field stay. Also in the northwestern portion of the image adjacent to a small stream is a single stand of the palm *Attalea tessmanii* (*kontashi,* Figure 5.9, ref. 4), containing a delicious, almond-like nut guarded within a hard mesocarp that is used to make tobacco pipes. A single stand of *A. tessmanii* forest likewise occurs near Tayakome, close to the mouth of a small stream south of the community (Figure 5.6, ref. 4).

Matsigenka Habitat Types, Satellite Imagery, and Forest Diversity

Excluding 8 montane habitats not treated in this chapter, we have drawn on a total of 62 Matsigenka-defined vegetation types (refs. 1–57, 64–67) and 7 faunal habitats (F1–F7) to carry out ethnobotanical ground-truthing of satellite images. From the ground-truthing exercise, it becomes apparent that the Matsigenka see forests in much greater detail than do satellites. The spatial scale of many of the vegetation types recognized by the Matsigenka is small (see Tables 5.3–5.4). Ten of the Matsigenka-defined habitats belong to the scale category XS (<0.1 ha), corresponding roughly to a maximum size of a single pixel (30 × 30 m, or 0.09 ha), beyond LANDSAT TM's powers of resolution. Some 24 habitat types correspond to the scale category S (up to 1 ha), corresponding roughly to a maximum size of 10 pixels (0.9 ha), close to the limits of LANDSAT TM resolution. Thus 34 of 69, or half, of Matsigenka-recognized vegetation types are close to or beyond the limits of LANDSAT TM pixel resolution.

Vegetation types that cover large areas and directly affect canopy structure are among the easiest to locate on satellite images, especially bamboo-dominated forests (see Nelson 1994), swamps, and secondary forests. Some stream gallery forests also show up clearly with a dark tone on our images. However, some Matsigenka-defined habitats occurring at medium to large scales are difficult to differentiate using satellite imagery alone. For example, understory or subcanopy vegetation, important in Matsigenka habitat classification, may change without greatly altering reflectivity of the upper canopy. Such habitats can be differentiated only by subtle variations of tone, shade, and/or texture between adjacent regions, if at all. This is particularly true for habitats defined by subcanopy palms, such as *choginashi* (ref. 7), *kamonashi* (ref. 11), *kompaparishi* (ref. 17), and *kepitoshi* (ref. 18) and for habitats defined by the presence of salient but sparsely abundant emergent trees, such as *toaroshi* (ref. 38), *pariashi* (ref. 42), and *koñorishi* (ref. 48). In other cases, even habitats with markedly different canopy flora and structure show similar reflected radiance values in LANDSAT TM. Prime examples are *kapiroshi* bamboo forest (ref. 22), *yaaroshi* secondary forest (ref. 54), and *shivitsasemai* low-canopy liana forest (ref. 53), all three of which have light tones approaching pure white. Even full-color composite images can fail to separate these large-scale Matsigenka-defined habitats from satellite images.

The final column (TM) in Tables 5.3 and 5.4 provides a rough assessment of relative apparency or distinctiveness of each Matsigenka-defined habitat type on the

LANDSAT TM images employed in this study. Nineteen habitats are included in the category of greatest apparency (+). Few of these would surprise tropical ecologists or remote-sensing specialists, however, for they include salient and well-documented vegetation types such as bamboo forests, palm swamps, terrestrial and aquatic grasses, and various kinds of succession zones and secondary forest. Perhaps of greatest scientific interest in this category are a number of bamboo forests other than *Guadua weberbaueri*, some of which appear to have received little or no attention by ecologists working in the Manu and other parts of Peru: *yaiveroshi* (ref. 23), *samatsishi* (ref. 19), *shinkeroshi*, (ref. 24), and *manipishi* (ref. 21). Even where native-defined forest types are not novel to science, the ability of local people to recognize, describe, and locate them within their territory is of invaluable assistance when carrying out field studies of habitat diversity.

Fourteen Matsigenka-defined habitats are included in the category of marginally apparent (+/–), hardly obvious but nonetheless sometimes visible upon closer study as a subtle shift in color/tone or texture from adjacent areas on the image. Some well-documented habitats are included in this group; for example, *Cedrela-Ficus* forests (refs. 41, 46; see Foster 1990b). However, other important Matsigenka-defined habitats in this group may provide novel information to ecologists, notably upland palm forest types *kompaparishi* (ref. 17), common near Tayakome but rare near Yomybato, and *kepitoshi* (ref. 18), common near Yomybato but rare near Tayakome. The ecological, soil, or disturbance conditions that lead to this reversal in prevalence of these palm species are worthy of further study. The stream gallery forest *toaroshi* (ref. 38), containing stands of the elsewhere rare tree *Apuleia leiocarpa*, and *kontashi* (ref. 4), containing dense stands of the rare palm *Attalea tesmanni*, also provide support for the habitat-mosaic hypothesis on a larger scale than would be possible without the Matsigenkas' detailed knowledge about the local environment.

Thirty-four Matsigenka-defined vegetation types and faunal habitats were not apparent (–) on the satellite images employed, either because they were too small or because they produced no noticeable change in the canopy reflectance. The remaining two, plus eight montane vegetation types, were mentioned by Matsigenka informants but not observed at the study sites (NA). Thus nearly half (34 of 69 types) of the forest diversity recognized by the Matsigenka is simply beyond LANDSAT TM's powers of resolution. Perhaps recent improvements in satellite technology—for example, LANDSAT 7 and IKONOS high-resolution satellite imagery or more sophisticated digital analysis techniques—will allow additional habitats to be resolved. What is needed now is systematic sampling of Matsigenka-defined habitats, both apparent and non-apparent on satellite images, in order to establish the extent and nature of habitat differences in terms of their flora and vegetative structure.

Results and Applications of Ethnobotanical Ground-Truthing

Returning to the questions posed at the beginning of this chapter, we ask whether the estimate of 100 or more habitats for the Peruvian Amazon by Tuomisto et al. (1995) is plausible and whether the different color/tone combinations seen in satellite images correspond to the habitats observed by local people on the ground. Our answer to the first question is, Yes—although not for quite the same reasons Tuomisto and colleagues propose. Their high estimate of habitat types ("biotopes") was based exclusively on naked-eye visual assessment of color patterns (including topographic shadows) detected on LANDSAT false-color composites printed at a scale of 1:250,000. In such an assessment, Type 1 and Type 2 errors (overdifferentiation and underdifferentiation), as well as noise generated by canopy texture, topography, atmospheric effects, and pixel averaging, may confound the recognition of actual vegetation types. Tuomisto et al. complement their satellite data with ground studies of floristic composition at different sites and along straight-line transects, demonstrating a high level of variation at both medium scales (from one site to the next) and small scales (along the same transect). However, the floristic inventories were not tied to the 100+ biotopes identified from satellite data, since these were mostly inaccessible and in any case too numerous to conveniently study. Furthermore, the floristic comparison was based on a limited number of easily identified indicator species (ferns and Melastomataceae only), quite narrow compared with the large range of indicator species found in indigenous habitat classification. Based on our study, however, and despite the noted limitations, the high estimate indeed appears to be plausible. Matsigenka informants from five communities identified a total of 62 primary and secondary forest vegetation types (excluding 8 montane types and 7 faunal habitats) from two regions a mere 150 km apart. Fleck (1997) likewise describes 47 indigenous-defined habitats among the Matses Indians of the Galvez River, some of which overlap the Matsigenka categories and several of which are unique. Given the high degree of habitat diversity at quite local scales, an estimate of 100 habitats for the entire Peruvian Amazon seems plausible, if not conservative.

The answer to the second question is, Sometimes yes, sometimes no. Half of Matsigenka-defined habitats were either apparent or marginally apparent on LANDSAT TM images of the study sites; the other half were not apparent at all. Even with highly apparent habitat types, there is rarely a one-to-one correspondence between color/tone and vegetation type: different habitats may look the same (underdifferentiation), and similar habitats may look different (overdifferentiation[10]), depending on how canopy structure, topography, and the atmosphere interact in reflecting the sun's energy toward the satellite sensor. Satellites provide a convenient tool for arriving at a preliminary assessment of habitat diversity over large areas, but the prevalence of Type 1 and Type 2 errors renders blind interpretation (i.e., without ground-truthing) of satellite imagery quite problematic. Although they appeal to our Star Wars–era imagination as all-seeing, robotic "eyes in the sky," satellites are in fact far from omniscient when it comes to habitat recognition. The essential, and most difficult, part of satellite ground-truthing lies, of course, on the ground. In this regard, native habitat knowledge combined with GPS technology proves to be a powerful set of tools for interpreting satellite images and carrying out studies of local habitat diversity.

Because of multiple threats to tropical forest, rapid ecological assessment techniques are a keystone in conservation efforts throughout Latin America (Sobrevilla & Bath 1992). Much of Amazonia remains virtually unstudied in terms of basic floristic and faunal composition (Emmons & Feer 1990; Nelson et al. 1990), not to mention more complex aspects of ecology and habitat structure. Indigenous and folk knowledge about the environment represents a vast database about habitat diversity, species distributions, ecological interactions among organisms, economically important species, and sustainable management practices (Posey 1983). Furthermore, local involvement in resource management is emerging as an important element of conservation strategies for Amazonia (see Bodmer 1994). In any project concerned with evaluating local biodiversity, habitat diversity, resource distribution, conservation, or management strategies in indigenous territories, ethnobotanical ground-truthing may prove to be a practical method for integrating folk and scientific knowledge.

Several important scientific classifications of regional Amazonian habitats use many local terms, mostly drawn from mestizo or Caboclo dialects of Spanish or Portuguese (Pires & Prance 1985; Encarnación 1993). Habitat knowledge among indigenous people appears to be equally if not more detailed in describing local forest features. Comparing our study with Fleck's (1997; Fleck & Harder 2000) independent study of Matses Indian habitat classification in northern Peru, an extraordinary concordance is found in terms of the organizational and nomenclature principles, salient biotic and abiotic variables, and certain specific habitat types recognized (Shepard et al. 2001). Considering that the Matsigenka and Matses belong to different language families (Arawakan and Panoan) and live some 600 km apart, one senses the potential power of ethnobotanical ground-truthing as a method for studying habitat diversity at larger, regional scales. Other studies of indigenous forest classification (Carneiro 1978; Parker et al. 1983; Posey 1983, 1985; Balée 1994) could be compared and synthesized, and on this basis, common aspects of organization and nomenclature might be identified (cf. Berlin 1992) and additional studies conducted. Eventually, one might hope to generate a mosaic of indigenous forest classifications covering substantial portions of the Amazon basin, many of which have been poorly studied, if at all, by scientists. As Fleck and Harder (2000) observe, however, indigenous habitat knowledge is maintained through active involvement in traditional hunting and resource-use practices. As indigenous societies, economies, and languages are transformed by contact with global culture, traditional ecological knowledge may diminish or be lost altogether. For this reason, both documentation and practical applications of indigenous knowledge are necessary and urgent.

While encouraging other researchers to explore ground-truthing studies, we note that a single researcher alone is not likely to succeed in carrying out a complete ground-truthing project. Collaboration among multiple researchers, disciplines, and cultures is essential. Commercially available GIS and image-processing software can be complex as well as expensive, and proper use of it requires extensive training. Interpretation of satellite imagery and other remote-sensing data is an area of specialization unto itself. Ground surveys and botanical collection and identification require special skills, methods, and equipment, whereas work in local communities necessitates ethnobiological, anthropological, and linguistic training. Local people have the specific knowledge of habitats necessary to carry out research on the ground and should be included as an integral part of the collaborative team. Finally, scientists, policymakers, and activists in host countries should be involved in the research and have access to its results.

In this chapter we have demonstrated the potential usefulness of combining indigenous habitat knowledge with satellite technology. Like any method, ethnobotanical ground-truthing has many limitations. Specific ethnoecological knowledge may not necessarily be applicable at broader scales but is certainly relevant for assessing the range of habitat variation within a local landscape (1–100 km^2), the so-called mesoscale (Clark et al. 1998). By comparing multiple systems of indigenous habitat classification, it may be possible to extend the relevance of local knowledge to wider, regional levels. Given the tremendous diversity of Amazonian forests, the widely recognized shortcomings of scientific knowledge about them, the clear relevance of indigenous knowledge to science, and the accelerating rates of destruction of both the forests and their native peoples, ethnoecological studies in Amazonia should be high on the research agendas of anthropologists, ecologists, and conservationists alike.

Acknowledgments

We dedicate this chapter to the memory of Darrell A. Posey, in honor of his pioneering work in ethnoecology, including habitat classification, and his passionate advocacy of indigenous people's territorial, cultural, and resource rights. The A. L. Green Fund of Harvard University provided major funding for this study. We thank Edward O. Wilson and Kathleen Horton for their generous support of our project with the fund. Additional funding for preliminary phases of the project was provided by the Wenner-Gren Foundation (1995, 1997), the National Science Foundation (1996), Conservation International (1997), and the Centro de Orientamento Educacional (1998). Institutional support has been provided by Asociación Peruana para la Conservación de la Naturaleza and by the herbarium of the Universidad Nacional San Marcos, Lima. We thank our Peruvian colleagues Joquina Albán and Hamilton Beltrán of the Universidad Nacional San Marcos, Lima, for their collaboration on the project. We would also like to thank Ada Castillo of Manu National Park, Cesar Uchima of the Instituto Nacional de Recursos Naturales, and Silvia Sánchez and Alejandro Smith, both of the Asociación Peruana para la Conservación de la Naturaleza for their kind assistance in many matters. Robin Foster and Tyana Wachter of the Field Museum have assisted us in the storage and identification of botanical vouchers from this project. We thank Luisa Maffi and Thomas Carlson for their helpful comments on an earlier draft of this chapter and for their hard work editing this volume. To David Fleck we extend our appreciation for a productive exchange of ideas and for sharing unpublished data. We thank Patrick Klemawesch and "la comisión de visita conyugal," Maria Nazareth F. da Silva, and Sophie Adams for their hard work, patience, and excellent company in crossing the Y2K threshold without mishap. On this note, we also thank the elusive Togyeri for holding their fire. We thank Manuel Lizarralde for his assistance during the 1996 field season and David Sidaris for comic relief during an arduous week of portage. Mateo Italiano of the Native Community of Shipetiaari was of invaluable assistance during all field trips and will, we hope, continue to assist us in the future. Finally, thanks to Alejandra Araos, Carlos

Avanti, Cesar Avanti, Alejo Matsipanko, Merino Matsipanko, Manuel Maviti, Casiano Nontyogitso, Agosto Oyeyoyeyo, Rani Rios, Ismael Vicente, Mariano Vicente, Maura Vicente, Lucho Yoveni, Celsor, Manuel, Lucio, and many other Matsigenka companions, *Noshaninkaige Matsigenkaige.*

Notes

1. Selected false-color composites of satellite data from our research project will be made available for viewing on the Internet. For more information, contact the authors.

2. Several other derived bands available in the Idrisi software were evaluated, including the commonly used normalized difference vegetation index (NDVI) and brightness and hue ratios. The Kauth-Thomas "greenness band" was preferred because it adequately reflects the distinctions observed on the false-color composite, does not remove topographic detail (as NDVI does), and does not suffer from nearly the degree of "striping." NDVI images of our study site show a series of alternating, rectangular dark and light stripes in forested areas that are clearly artifacts of background detector noise. Because visible light is mostly absorbed by the forest canopy, TM bands 1–3 have a weak signal and thus a low signal-to-noise ratio. Due to the particularities of the NDVI transformation, such detector noise becomes amplified and results in striping.

3. There are 51 types if refs. 40a (*setikoshi*) and 40b (*tonkoshi*) are counted separately; however, sometimes the terms are used interchangeably, referring to two similar *Cecropia* species.

4. The tight river loop visible near the northwestern corner of Figure 5.4 was pinched off several months after the image was captured. During the 1997–1998 rainy season, the Manu River cut across a narrow neck of land at the base of the loop, isolating it and thus forming a new oxbow lake, now known as "Cocha Nueva."

5. Some identify the species as *Guadua sarcocarpa,* a close relative of *G. weberbaueri* (Londoño & Petersen 1991). The two are difficult to distinguish in the normal sterile condition. Flowering and fruiting events occur at great intervals, as much as 15–30 years, so the taxonomy of the genus is poorly understood.

6. With the kind consent of the authors, we have employed some mapping conventions used by Fleck and Harder (2000), both for clarity and for purposes of comparison between the two studies.

7. According to the Matsigenka, "true" *yaivero* (*yaivero-sano*), an unidentified *Guadua* species, is closely related to *yaivero-tavanakota* ("*tavana*-blade *yaivero*"), identified as *G. glomerata*. The two species are similar but appear to be distinctive.

8. The officially recognized name of the community, Yomybato, is an alternate transcription of the Matsigenka name for the stream, *Yomuvaato*.

9. A smaller, manmade *yaaroshi* of secondary growth in an old garden site is visible as a white circular patch near Yomuvaato stream, south of the central community area; see Figure 5.9.

10. Overdifferentiation is especially a problem with bamboo forests, which show different color patterns at different life stages (Nelson 1994).

Literature Cited

Ashton, P. S. 1977. A contribution of rain-forest research to evolutionary theory. Annals of the Missouri Botanical Garden 64: 694–705.

——— & P. Hall. 1992. Comparisons of structure among mixed dipterocarp forest of northwestern Borneo. Ecology 80: 459–481.

Baksh, M. G. 1995. Changes in Machiguenga quality of life. Pp. 187–205 *in* L. E. Sponsel (ed.), Indigenous Peoples and the Future of Amazonia: An Ecological Anthropology of an Endangered World. Tucson: University of Arizona Press.

Balée, W. 1994. Footprints of the Forest: Ka'apor Ethnobotany—The Historical Ecology of Plant Utilization by an Amazonian People. New York: Columbia University Press.

Balick, M. J. 1984. Ethnobotany of palms in the Neotropics. Pp. 9–23 *in* G. T. Prance & J. A. Kallunki (eds.), Ethnobotany in the Neotropics. Advances in Economic Botany, 1. Bronx: New York Botanical Garden.

Berlin, B. 1984. Contributions of Native American collectors to the ethnobotany of the Neotropics. Pp. 24–33 *in* G. T. Prance & J. A. Kallunki (eds.), Ethnobotany in the Neotropics. Advances in Economic Botany, 1. Bronx: New York Botanical Garden.

———. 1992. Ethnobiological Classification: Principles of Categorization of Plants and Animals in Traditional Societies. Princeton, N.J.: Princeton University Press.

Bodmer, R. E. 1994. Managing wildlife with local communities: The case of the Reserva Comunal Tamshiyacu-Tahuayo. Pp. 113–134 *in* D. Western, M. Wright & S. Strum (eds.), Natural Connections: Perspectives in Community-Based Conservation. Washington D.C.: Island Press.

Carneiro, R. 1978. The knowledge and use of rainforest trees by the Kuikuru Indians of central Brazil. Pp. 201–216 *in* R. I. Ford (ed.), The Nature and Status of Ethnobotany. Anthropological Papers, 67. Ann Arbor: University of Michigan, Museum of Anthropology.

Chesson, P. L. & R. R. Warner. 1981. Environmental variability promotes coexistence in lottery competitive systems. American Naturalist 117: 923–943.

Clark, D. B., D. A. Clark & J. M. Read. 1998. Edaphic variation and the mesoscale distribution of tree species in a neotropical rain forest. Journal of Ecology 86: 101–112.

———, M. W. Palmer & D. A. Clark. 1999. Edaphic factors and the landscape-scale distributions of tropical rain forest trees. Ecology 80(8): 2662–2675.

Condit, R. 1996. Defining and mapping vegetation types in mega-diverse tropical forests. Trends in Ecology and Evolution 11: 4–5.

Duivenvoorden, J. F. 1996. Patterns of tree species richness in rain forests of the middle Caqueta area, Colombia, NW Amazon. Biotropica 28: 142–158.

Eastman, J. R. 1999. Idrisi32 Guide to GIS and Image Processing Vol. 2. Worcester, Mass.: Clark University, Clark Labs.

Emmons, L. H. & F. Feer. 1990. Neotropical Rainforest Mammals. Chicago: University of Chicago Press.

Encarnación, F. 1993. El bosque y las formaciones vegetales en la llanura amazónica del Perú. Alma Máter [Universidad Nacional de San Marcos, Lima] 6: 95–114.

Erwin, T. 1984. Tambopata Reserved Zone, Madre de Dios, Perú: History and description of the reserve. Revista Peruana de Entomología 27: 1–8.

Fleck, D. W. 1997. Mammalian diversity in rainforest habitats as recognized by Matses Indians in the Peruvian Amazon. M.S. thesis, Ohio State University.

——— & J. D. Harder. 2000. Matses Indian rainforest habitat classification and mammalian diversity in Amazonian Peru. Journal of Ethnobiology 20(1): 1–36.

Foster, R. B. 1990a. The floristic composition of the Río Manu floodplain forest. Pp. 99–111 *in* A. H. Gentry (ed.), Four Neotropical Rainforests. New Haven, Conn.: Yale University Press.

———. 1990b. Long-term change in the successional forest community of the Río Manu floodplain. Pp. 565–572 *in* A. H. Gentry (ed.), Four Neotropical Rainforests. New Haven, Conn.: Yale University Press.

Frake, C. O. 1964. The ethnographic study of cognitive systems. Pp. 28–41 *in* S. Tyler (ed.), Cognitive Anthropology. Prospect Heights, Ill.: Waveland Press.

Gentry, A. H. 1988a. Changes in plant community diversity and floristic composition on environmental geographical gradients. Annals of the Missouri Botanical Garden 75: 1–34.

———. 1988b. Tree species richness of upper Amazonian forests. Proceedings of the National Academy of Science 85: 156–159.

——— (ed.). 1990. Four Neotropical Rainforests. New Haven, Conn.: Yale University Press.

———. 1993. A Field Guide to the Families and Genera of Woody Plants of Northwest South America (Colombia, Ecuador, Peru). Washington, D.C.: Conservation International.

Givinish, T. J. 1999. On the causes of gradients in tropical tree diversity. Journal of Ecology 87: 193–210.

Goulding, M. 1980. The Fishes and the Forest: Explorations in Amazonian Natural History. Berkeley and Los Angeles: University of California Press.

Grubb, P. J. 1977. The maintenance of species richness in plant communities: The importance of the regeneration niche. Biological Review 52: 107–145.

Harms, K. E., S. J. Wright, O. Calderon, A. Hernandez & E. A. Herre. 2000. Pervasive density-dependent recruitment enhances seedling diversity in a tropical forest. Nature 404: 493–495.

Hubbell, S. P. & R. B. Foster. 1986a. Biology, chance and history and the structure of tropical rainforest tree communities. Pp. 314–332 *in* J. M. Diamond & T. J. Case (eds.), Community Ecology. New York: Harper & Row.

——— & ———. 1986b. Commonness and rarity in a neotropical forest: Implications for tropical tree conservation. Pp. 205–231 *in* M. E. Soulé (ed.), Conservation Biology: The Science of Scarcity and Diversity. Sunderland, Mass.: Sinauer Associates.

Janzen, D. H. 1970. Herbivores and the number of tree species in tropical forests. American Naturalist 104: 501–528.

Johnson, A. W. 1983. Machiguenga gardens. Pp. 29–63 *in* R. B. Hames & W. Vickers (eds.), Adaptive Responses of Native Amazonians. New York: Academic Press.

Londoño, X. & P. M. Petersen. 1991. *Guadua sarcocarpa* (Poaceae, Bambuseae), a new species of Amazonian bamboo with fleshy fruits. Systematic Botany 16(4): 630–638.

Nelson, B. W. 1994. Natural forest disturbance and change in the Brazilian Amazon. Remote Sensing Review 10: 105–125.

——— & I. L. do Amaral. 1994. Destructive wind effects detected in TM images of the Amazon basin. International Archives of Photogrammetry and Remote Sensing 30(7a): 339–343.

———, C. A. C. Ferreira, M. F. da Silva & M. L. Kawalski. 1990. Endemism centers, refugia and botanical collection density in Brazilian Amazonia. Nature 345: 714–716.

Parker, E., D. A. Posey, J. Frechione & L. F. da Silva. 1983. Resource exploitation in Amazonia: Ethnoecological examples from four populations. Annals of the Carnegie Museum of Natural History 52(8): 163–203.

Pianka, E. R. 1966. Latitudinal gradients in species diversity: A review of concepts. American Naturalist 100: 33–46.

Pires, J. M. & G. T. Prance. 1985. The vegetation types of the Brazilian Amazon. Pp. 109–145 *in* G. T. Prance & T. E. Lovejoy (eds.), Amazonia. Key Environments Series. Oxford and New York: Pergamon Press.

Pitman, N. C. A., J. Terborgh, M. R. Silman & P. Nuñez. 1999. Tree species distributions in an upper Amazonian forest. Ecology 80: 2651–2661.

Posey, D. A. 1983. Indigenous ecological knowledge and development of the Amazon. Pp. 225–258 *in* E. F. Moran (ed.), The Dilemma of Amazonian Development. Boulder, Colo.: Westview Press.

———. 1985. Indigenous management of tropical forest ecosystems: The case of the Kayapo Indians of the Brazilian Amazon. Agroforestry Systems 3: 139–158.

Richards, J. A. 1986. Remote Sensing Digital Image Analysis: An Introduction. Berlin and New York: Springer-Verlag.

Salo, J., R. Kalliola, I. Häkkinen, Y. Mäkinen, P. Niemelä, M. Puhakka & P. D. Coley. 1986. River dynamics and the diversity of Amazon lowland forest. Nature 322: 254–258.

Shepard, G. H., Jr. & A. Chicchon. 2001. Resource use and ecology of the Matsigenka of the eastern slopes of the Cordillera Vilcabamba. Pp. 164–174 *in* A. E. Alonso, A. Alonso, T. S. Schulenberg & F. Dallmeier (eds.), Biological and Social Assessments of the Cordillera de Vilcabamba, Peru. RAP Working Papers, 12. Washington, D.C.: Conservation International.

———, D. W. Yu, M. Lizarralde & M. Italiano. 2001. Rainforest habitat classification among the Matsigenka of the Peruvian Amazon. Journal of Ethnobiology 21(1): 1–38.

Snell, B. E. 1998. Pequeño diccionario Machiguenga-Castellano. Documento de Trabajo No. 32. Lima: Instituto Lingüístico de Verano / Summer Institute of Linguistics.

Sobrevilla, C. & P. Bath. 1992. Evaluación ecológica rápida: Un manual para usuarios de América Latina y el Caribe. Preliminary ed. Arlington, Va.: The Nature Conservancy, Latin American Science Program.

Terborgh, J. 1983. Five New World Primates: A Study in Comparative Ecology. Princeton, N.J.: Princeton University Press.

———. 1985. The vertical component of plant species diversity in tropical and temperate forests. American Naturalist 120: 760–776.

———. 1990. An overview of research at Cocha Cashu Biological Station. Pp. 48–59 *in* A. H. Gentry (ed.), Four Neotropical Rainforests. New Haven, Conn.: Yale University Press.

———. 1992. Diversity and the Tropical Rain Forest. New York: Scientific American Press.

———, S. Robinson, T. A. Parker III, C. A. Munn & N. Pierpont. 1990. Structure and organization of an Amazonian forest bird community. Ecological Monographs 60(2): 213–238.

———, R. B. Foster & P. Nuñez. 1996. Tropical tree communities: A test of the nonequilibrium hypothesis. Ecology 77: 561–567.

Tuomisto, H. 1998. What satellite imagery and large scale field studies can tell about biodiversity patterns in Amazonian forests. Annals of the Missouri Botanical Gardens 85: 48–62.

———, A. Linna & R. Kalliola. 1994. Use of digitally processed satellite images in studies of tropical rain forest vegetation. International Journal of Remote Sensing 15(8): 1595–1610.

———, K. Ruokolainen, R. Kalliola, A. Linna, W. Danjoy & Z. Rodriguez. 1995. Dissecting Amazonian biodiversity. Science 269: 63–66.

Van Roosmalen, M. G. M., T. van Roosmalen, R. A. Mittermeier & G. A. B. da Fonseca. 1998. A new and distinctive species of marmoset (Callitrichidae, Primates) from the lower Río Aripuanã, state of Amazonas, central Brazilian Amazon. Goeldiana Zoologia 22: 1–27.

Vidalenc, D. 2000. Distribuição das Florestas Dominadas pelo Bambu *Guadua weberbaueri* em Escala de Paisagem no Sudoeste da Amazônia e Fatores Edáficos que Afetam sua Densidade. M.S. thesis, University of Amazonas.

Webb, C. O. & D. R. Peart. 2000. Habitat associations of trees and seedlings in a Bornean rain forest. Journal of Ecology 88(3): 464–478.

Wills, C. & R. Condit. 1999. Similar non-random processes maintain diversity in two tropical rainforests. Proceedings of the Royal Society of London, Ser. B, 266: 1445–1452.

Wilson, E. O. 1986. The arboreal ant fauna of Peruvian Amazon forests: A first assessment. Biotropica 19(3): 245–251.

Appendix 5.1: Selected GIS Web Resources

General

Fundamentals of Remote Sensing. Canada Centre for Remote Sensing. Free, Web-based tutorial. <http://www.ccrs.nrcan.gc.ca/ccrs/learn/tutorials/fundam/fundam_e .html>.

Guide to GIS Resources on the Internet. Berkeley Digital Library. Links to software, tutorials, directories, publications, and other web resources. <http://sunsite .berkeley.edu:80/GIS/gisnet.html>.

Satellite Imagery

Global Rain Forest Mapping Project. NASA. JERS–1 radar images of Amazonia. <http: //southport.jpl.nasa.gov/GRFM/>

LANDSAT Images Amazon (TRFIC). Tropical Rain Forest Information Center. Low-priced satellite imagery of Amazonia, mostly Brazil; low-resolution preview images download at no cost. <http://www.bsrsi.msu.edu/trfic/index.html>.

LANDSAT 7. NASA. Continually updated, global database of recent satellite images; low-resolution preview images download at no cost. <http://landsat7.usgs.gov/ index.php>.

Space Imaging: IKONOS Satellite Images. Entopos. Commercial, high-resolution satellite images. <http://www.entopos.co.id/ikonos/ikonos.html>.

Software

ER Mapper. Commercial GIS software, free trial and viewing software. <http://www .ermapper.com/online/software/index.htm>.

GIS.com. ESRI. Commercial GIS software, free trial and viewing software, introductory information and Internet links about GIS. <http://www.gis.com>.

Idrisi Project Home Page. Clark University Labs. Commercial GIS software, free trial software, GIS research bibliography, and Internet links. <http://www.clarklabs.org /home.asp>.

MacGPS/MacGPS Pro. James Associates. Freeware and low-priced commercial software for transferring and displaying data from Garmin GPS units to database files and satellite images (Macintosh only). <http://www.macgpspro.com/>.

NIH Image. National Institutes of Health. Image-processing freeware. <ftp://codon .nih.gov/pub/nih-image/>.

SPRING. Instituto Nacional de Pesquisas Espaciais (INPE), Brazil. Full-feature GIS freeware (PC and Linux only). <http://www.dpi.inpe.br/spring/>.

TNT Lite. Microimages. Freeware version of TNT-mips GIS software. <http://tnt .microimages.com/tntlite/>.

Part 2

Knowledge and Sustainable Use of Traditional Plant Resources

6

Indigenous Nomenclature, Classification, and Utilization of African Rattans

Terry C. H. Sunderland

Introduction

The recording of vernacular names as part of ethnobotanical research has become routine, so that contemporary works on ethnobotany provide detailed methodologies on how to collect local nomenclatural information (Given & Harris 1994; Martin 1995; Alexiades 1996). While the limitations of studying vernacular names in isolation have been recognized (Wilkie & Saridan 1999), it is known that the careful study of plant names reveals a wealth of knowledge on how indigenous people perceive and utilize their plant resources (Berlin 1973, 1977, 1992; Berlin et al. 1973; Brown 1977; Hays 1983; Balée 1989). It has also been proved that studies of nomenclature can provide deep insights into the ethnobotanical taxonomic structures elaborated by indigenous societies (Berlin 1973, 1977, 1992; Berlin et al. 1973; Balée 1989).

This cognitive approach to the study of ethnobotany has been widely applied, and, despite considerable regional, ethnic, and social differences in the communities studied, many researchers have identified similar systems of folk classification in different places around the world (Ekandem 1955; Berlin 1973, 1977, 1992; Berlin et al. 1973; Friedberg 1974; Brown 1977; Balée 1989). The recurrent features of these systems have led to the formulation of basic principles of ethnobiological classification utilized by many folk societies (Berlin 1973, 1977, 1992; Berlin et al. 1973), principles that provide useful models for the study of how indigenous societies view their plant resources.

The majority of folk biological classifications, it is claimed, are based on the universal significance of the morphological discontinuities among organisms, and rarely is classification based on functional considerations such as cultural utility (Berlin 1973, 1992). However, utilitarian approaches to classification are also considered significant in understanding how indigenous peoples relate to species that are of functional importance and to those that are not. In addition, comparison of such "special classifications" with the general-purpose classification system utilized by indigenous societies provides significant insight into how useful plants and animals are classified within, and relate to, the latter classifications (Hunn 1976, 1982; Atran 1983; Randall & Hunn 1984; Gianno 1986; Balée & Daly 1990).

The rattans of Africa are taxa of immense functional value, and the classification systems of which they are part are based as much on recognition of this functional utility as they are on morphological discontinuities. In this chapter I present the findings of a recent study of a number of rattan and general-purpose classification systems in selected languages of the Bantu language subgroup of Central Africa in particular, as well as a cursory review of those of the non-Bantu languages within the greater Niger-Congo language family of West Africa. Based on extensive fieldwork, as well as a broad review of herbarium collections and available literature, a comprehensive analysis of rattan nomenclature and systems of classification is presented for the first time. This review is based on the vernacular names listed by species in Appendix 6.1. The analysis is preceded by a botanical description of rattans, with special reference to African species, and a discussion of indigenous utilization of African rattans, as well as by an outline of the history and linguistics of sub-Saharan Africa and a summary of Berlin's model of ethnobiological classification that provides the framework for the classification of rattans.

A Brief Introduction to Rattans

Rattans are climbing palms belonging to the Palm family (Palmae or Arecaceae). There are around 650 known different species of rattan belonging to 13 genera (Uhl & Dransfield 1987). These taxa are concentrated solely in the Old World Tropics; there are no true rattans in the New World.[1] Rattans belong to the Calamoideae, a large subfamily of palms whose members are characterized by overlapping, reflexed scales on the fruit. These climbing palms are always spiny, a necessary preadaptation to the climbing habit (Dransfield 1992).

The flexible stems of rattan provide the source of cane for the well-established rattan industry, currently worth some U.S.$6.5 billion per year (ITTO 1997). As well as forming the basis of a thriving export market, rattans contribute significantly to the economies of forest-based communities throughout their range. The harvest and sale of rattan and rattan products are an integral part of subsistence strategies for many rural communities and provide an opportunity for many forest-dwellers to enter the cash economy (Burkill 1935; Corner 1966; Godoy & Feaw 1989; Godoy 1990; Davies & Richards 1991; de Fretas 1992; Peluso 1992; Dransfield & Manokaran 1994; Siebert 1997).

The Rattans of Africa

Twenty species of rattan, representing four genera, are known to occur in the lowland tropical forests of Africa (Sunderland 1999a, 1999b). Three of these genera are endemic to the African continent: *Laccosperma* (syn. *Ancistrophyllum*), *Eremospatha*, and *Oncocalamus*. These taxa climb with the aid of a whiplike organ, a cirrus. Unlike the cirrate Asian members of the Calamoideae, the cirrus of the African taxa represents a marked extension of the rachis between the distal leaflets. These leaflets are present as reduced, reflexed, thornlike organs termed "acanthophylls." The fourth genus present in Africa, *Calamus,* is specifically of Asian distribution, with more than 370 species there, but is represented by only one highly variable species in Africa (Uhl & Dransfield 1987; Sunderland 1999a, 1999b). This species, *Calamus deërratus* G. Mann & H. Wendl, differs from the other African taxa in that it climbs with the aid of an armed flagellum, a long appendage arising directly from the stem, which is actually a modified sterile inflorescence.

Rattans in Africa occur in a wide range of ecological conditions. The majority of the species are most commonly encountered in closed forest systems, where their light-demanding nature makes them well adapted for early gap colonization. As such, they respond well to a small reduction in the forest canopy. Increases in forest disturbance, caused by selective logging, for example, encourage the regeneration of rattans, and these climbing palms are often a common feature along logging roads and skid trails. Even after initial germination, the seedlings of some rattan species can remain on the forest floor for a number of years, waiting for the optimum light conditions needed for the stems to begin their long journey to the canopy. Other species, in particular *Calamus deërratus,* grow in swamps or along riverbanks prone to flooding; still others, such as *Laccosperma opacum* and *L. laeve,* are highly shade tolerant and are more commonly found under the forest canopy.

Indigenous Utilization of African Rattans

As is the case in Southeast Asia, the rattans of Africa play an integral role in maintaining the economy of forest peoples and are a crucial component of subsistence-based systems (Abbiw 1990; Davies & Richards 1991; Falconer 1994; Morakinyo 1994; Townson 1995; Defo 1997, 1999; Defo & Sunderland 1999; Sunderland 1999a, 1999b; Minga in press; Tenati in press). The most important product of these rattan palms is cane; this is the stem stripped of its leaf sheaths and stem epidermis, although the epidermis is also often utilized for basic weaving. The inner stem is solid, strong, and uniform yet highly flexible. The canes are used either in whole or round form or are split, peeled, or cored. This latter raw material can then be used as simple rope or is woven into baskets or other products. The range of indigenous uses of rattan canes across Africa is vast. Aside from the extensive use of this inner stem, other plant parts of some species of rattan are utilized for a wider range of purposes, and some plant parts have alimentary, household, or medicinal purposes.

The extensive nature and wide range of uses for African rattans have led to the proliferation of a common misconception, particularly among the conservation and development community active in many areas in Africa, that all rattans are useful and, therefore, that all rattan species have potential commercial applications. This is not the case. Although there is a substantial use for many African rattan species and a wider, somewhat more spontaneous[2] use for many others, recent taxonomic and ethnobotanical studies indicate that a number of species are not utilized in any way or are used only in the absence of other, more desirable species. This lack of utilization is due to inflexibility or a tendency to break when being worked or a tendency of the species to produce aerial branches and hence provide cane of limited length. In this respect, the overview of nomenclature and classification of African rattans presented in this chapter highlights the importance of such studies in focusing conservation and development initiatives on the most appropriate taxa. Tables 6.1 and 6.2 summarize the taxonomy and uses of African rattans.

A Brief Introduction to Language and History in Sub-Saharan Africa

The Niger-Congo family of languages dominates the forested zone of sub-Saharan Africa. From Senegal to Lake Chad a number of distantly related languages within this language family occur (Oliver & Fage 1975; Iliffe 1995; Grimes 1996). These non-Bantu languages are differentiated from one another to such an extent that, through the study of glottochronology, most linguists postulate that they have been growing apart for around 8000 years, at least as long as their speakers have been sedentary agriculturists (Iliffe 1995; Oliver & Fage 1975).

In contrast, almost all of the peoples south of a line across from the Cross River area of southeastern Nigeria to southern Somalia speak more closely related languages (Vansina 1990; Iliffe 1995). These Bantu languages form a subgroup within the Niger-Congo family of languages that are relatively homogenous yet are distributed across a vast area of sub-Saharan Africa. From linguistic studies it has been determined that the center of origin of the Bantu languages is probably the Benue region of Nigeria, from which, around 5000 years ago, this Bantu family split into two branches: Eastern and Western (Vansina 1990; Iliffe 1995). The former group moved slowly eastward along the northern edge of the Congo Basin to the great lakes area of East Africa, whereas Western Bantu developed east of the Cross River in the highlands of western Cameroon.

Around 3000 years ago the Western Bantu speakers began to migrate slowly southward, reaching as far as what is now northern Namibia. During this migration, pioneer groups broke off, traveling eastward up the river valleys through the topical forest and settling as far inland as southern Sudan and the western Zambezi. The introduction of the banana and plantain, postulated to have occurred around A.D. 500 (Vansina 1990), made this rapid expansion through the tropical forest possible, because it provided the means to farm an otherwise inhospitable environment.

The study of the ancestral Bantu language indicates that, at the time of separation, Bantu speakers made pottery and had begun to farm yet had not begun to use metals. Both Bantu groups had words for "oil palm" and "yam" and their cultivation, although there existed no vocabulary for cereal cultivation in Western Bantu until this group's expansion reached the savanna areas of eastern and southern Africa. Most terms for root and tree cultivation are of Western Bantu origin, and it is clear that the Western Bantu vocabulary and system of classification of living things has developed to reflect the group's familiarity with their forest surroundings (Guthrie 1948, 1969–1971; Vansina 1990; Oliver 1999).

Table 6.1

The rattans of Africa: taxonomy and utilization

Genus, species	Description	Distribution	Cane use	Non-cane use
Calamus				
deërratus G. Mann & H. Wendl.	Clustering flagellate species; stems up to 20 m, up to 35 mm in diameter; leaves ecirrate	Senegal to Angola, west to Uganda	Yes, but only in the absence of other, more desirable species	Many
Eremospatha				
barendii Sunderland	Clustering; stems up to 30 m, up to 25 mm in diameter; conspicuous knee and bracts on inflorescence	Coastal forests of Cameroon	None recorded	None recorded
cabrae (De Wild. & Th. Dur.) De Wild.	Clustering; stems to 50 m, up to 25 mm in diameter; leaflets obovate; papillose inflorescence	Gabon & DR Congo to northern Angola	Yes	Few
cuspidata (G. Mann & H. Wendl.) H. Wendl.	Clustering; stems to 15 m, 25 cm in diameter; leaflets with conspicuous apiculum	Congo Basin	Few	None recorded
dransfieldii sp nov.	Clustering; stems to 30 m, up to 30 mm in diameter; knee conspicuous; lowermost leaflets clasping stem	Upper Guinea forests (Sierra Leone to western Nigeria)	Yes, particularly traded in Ghana	Few
haullevilleana De Wild.	Clustering; stems to 25 m, up to 25 mm in diameter; ocrea striate; leaflets spathulate–ovate	Congo Basin	Yes, highly prized and widely traded	Many
hookeri (G. Mann & H. Wendl.) H. Wendl.	Clustering; stems to 30 m, up to 30 mm in diameter; knee conspicuous, leaflets rhomboid to obovate	Eastern Nigeria to Gabon	No	Few

Table 6.1, *continued*

Genus, species	Description	Distribution	Cane use	Non-cane use
Eremospatha, continued				
laurentii De Wild.	Clustering; stems to 30 m, up to 30 mm in diameter; knee conspicuous; lowermost leaflets clasping stem	Congo Basin, with outliers in Upper Guinea forest	Few recorded	None recorded
macrocarpa (G. Mann & H. Wendl.) H. Wendl.	Clustering; stems to 50 m, 10–18 mm in diameter; juvenile leaves bifid, adult leaflets linear lanceolate	Senegal to DR Congo	Yes, juvenile form reputed to be the best small-diameter cane in Africa; widely traded	Many
quinquecostulata Becc.	Clustering; stems to 15 m, 10 mm in diameter	Southeastern Nigeria to southern Cameroon	Few	None recorded
tessmanniana Becc.	Clustering; stems to 100 m (although branching is common), up to 15 cm in diameter; glaucous gray-green leaflets	Southern Cameroon to eastern Guinea	None recorded	None recorded
wendlandiana Dammer ex Becc.	Clustering; stems to 60 m, up to 30 mm in diameter; conspicuous knee and rhomboid leaflets	Southeastern Nigeria to Gabon	Yes, but poor-quality cane	
Laccosperma				
acutiflorum (Becc.) J. Dransf.	Clustering; stems to 70 m, up to 60 mm in diameter; yellowish appearance; nonpendulous leaflets	Upper Guinea to DR Congo	None recorded	None recorded
korupensis sp. nov.	Clustering; stems often branching, to 10 m, up to 15 mm in diameter; acanthophylls absent	Coastal forests of Cameroon	None recorded	None recorded
laeve (G. Mann & H. Wendl.) H. Wendl.	Clustering; stems often branching, to 10 m, up to 15 mm in diameter; leaflet margins unarmed; seeds smooth	Upper Guinea to DR Congo	None	Few

opacum (G. Mann & H. Wendl.) Drude	Clustering; stems often branching, to 10 m, up to 15 mm in diameter; leaflet margins armed; seeds warty	Upper Guinea to DR Congo	Yes, but poor-quality cane	Few
robustum (Becc.) J. Dransf.	Clustering; stems to 45 m, 50 mm in diameter; leaflets conspicuously pendulous, glaucous blue-green	Southeastern Nigeria to DR Congo	Yes, highly prized cane; widely traded	Many
secundiflorum (de Beauv.) Kuntze	Clustering; stems to 30 m, up to 35 mm in diameter; leaflets sigmoid, dark green	Senegal to DR Congo	Yes, highly prized cane; widely traded	Many
Oncocalamus				
macrospathus Burr.	Clustering; stems to 35 m, up to 30 mm in diameter, sheaths well armed; rachillae bright yellow, seeds smooth	Southern Cameroon to northern Angola	No, poor-quality cane	None recorded
mannii (H. Wendl.) H. Wendl.	Clustering; stems to 30 m, 28 mm in diameter, sheaths well armed; rachillae bright crimson, seeds warty	Southern Cameroon to Gabon	No, poor-quality cane	None recorded
tuleyi Sunderland	Clustering; stems to 30 m, up to 45 mm in diameter, sheaths sparsely armed or unarmed; seeds smooth	Southeastern Nigeria and southwestern Cameroon	No, poor-quality cane	None recorded
wrightianus Hutch.	Clustering ?; stems to 10 m, up to 10 mm in diameter; leaflets sigmoid	Southern Nigeria	Yes, but for cane rope and twine only	Few

Table 6.2

Summary of the non-cane uses of African rattans.
(See Appendix 6.1 for further elaboration.)

Species	Use	Region
Calamus deërratus	Palm heart eaten	Ghana, Sierra Leone
	Young shoots roasted and eaten	Ghana
	Grilled leaves macerated and made into tea to promote weight loss and to treat edema caused by vitamin deficiencies	Senegal
	Ash from burned roots used as salt substitute	Guinea-Bissau
	Sheath twisted and used to clean cooking pans	Ghana
	Sheath twisted to make rope	Nigeria
Eremospatha cabrae	Base of leaf sheath used as a chew stick	DR Congo
E. haullevilleana	Palm heart eaten	Congo
	Fruits used for decoration	DR Congo
	Acanthophylls used as fishhooks	DR Congo
	Sap used as arbortifacient	DR Congo
E. macrocarpa	Powdered root used to treat syphilis	Ghana, Nigeria
E. wendlandiana	Palm heart eaten	Congo
	Base of leaf sheath used as a chew stick	Cameroon
Laccosperma laeve	Roasted roots eaten to improve virility	Central African Republic
L. opacum	Sap potable & drunk by forest workers	Gabon
	Palm heart eaten	Congo
L. robustum	Palm heart eaten	Cameroon to Gabon
	Young leaves eaten in stews	Equatorial Guinea
L. secundiflorum	Palm heart eaten	Throughout its range
	Young shoots eaten	Throughout its range
	Sap potable & drunk by forest workers	Senegal, Gabon
	Tea from young shoots used as vermifuge	Ghana, Gabon
	Sap, when mixed with other species, used to treat dysentery	DR Congo
Oncocalamus tuleyi	Base of leaf sheath used as a chew stick	Cameroon
O. wrightianus	Base of leaf sheath used as a chew stick	Nigeria

Berlin's Model of Ethnobiological Classification

It has been long recognized that the presence of hierarchically arranged folk taxonomies is probably universal and that, as such, this feature is shared by a wide range of unrelated human societies (Conklin 1962; Berlin 1973, 1977, 1992; Berlin et al. 1973). This hypothesis has been supported by comparison of the classifications of a number of indigenous groups and led to the development of a theoretical model by Berlin and his collaborators (Berlin 1973, 1977, 1992; Berlin et al. 1973). Berlin's model is based on hierarchy and recognizes five (sometimes six) mutually exclusive ranks that are logically comparable to those of contemporary scientific taxonomies (Figure 6.1). The folk classification systems of rattan encountered through our studies are discussed within this framework. A brief description of Berlin's model is thus in point.

The highest, or most inclusive, category is the unique beginner (or kingdom), which is implicitly recognized by most folk societies and distinguishes between the plant and animal kingdoms. However, this distinction is commonly of a covert nature for most indigenous societies, and explicit terms for "plant" or "animal" do not exist in most folk taxonomies (Berlin 1992). Indeed, Berlin (1973) argues that the terms for "plant" or "animal" are quite recent constructs, citing linguistic evidence that a term for plant is first found in the writings of Albertus Magnus in the thirteenth century and appeared only as recently as 300 years ago in French (Berlin 1973: 267). It is hence unsurprising that in many prescientific societies the designation for kingdom is covert.

Life-form classification has been the subject of much debate, particularly with regard to the relative influence of universality versus utility (Berlin et al. 1973; Berlin 1977, 1992; Brown 1977; Hunn 1982; Atran 1983; Ellen 1998). In general, life-forms are broad classes usually recognized by their distinctive morphology (Berlin 1973, 1992; Berlin et al. 1973). Berlin views the life-form categories of folk systematics as very similar to those of early classical botanical classification (see Atran 1983) involving the major categories, "tree," "vine," and "herb" (Berlin 1977). Indeed, so universal is the application of these life-form categories that Berlin suggests: "These three major groupings represent such distinct perceptual discontinuities that their recognition may constitute a substantive near-universal in pre-scientific man's view of the world" (Berlin 1977: 385). However, it has been noted that other life-forms may also be recognized according to their utility or ecological distribution, as well as morphological discontinuities or, indeed, a combination of these (Hunn 1982; Atran 1983; Randall & Hunn 1984). Hence, in most folk taxonomies there may be more than the three major life-form categories suggested by Berlin.

Intermediates are small groupings of generics (see below) that are similar to each other through shared morphological or functional characters and fall in between life-forms and generics. These are also often referred to as "covert categories," because they are often unnamed and not easily detected in folk systematics or may be known only by a small subsection of a community (Berlin 1973; Martin 1995).

Generics are the most salient categories in folk classification systems and as such are the most commonly encountered during initial ethnobotanical surveys. Generics are the smallest units in nature that are easily recognized on the basis of relatively large numbers

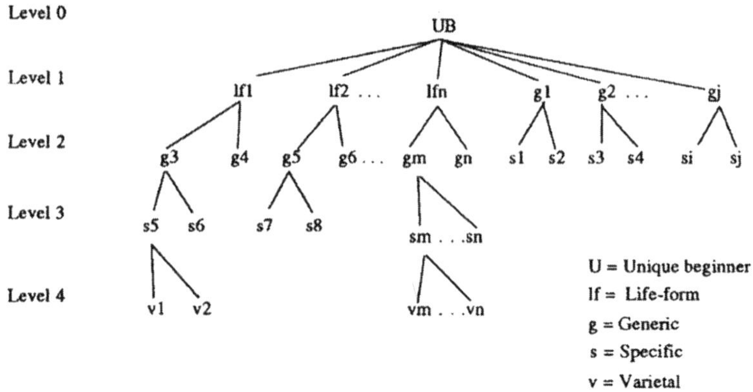

Figure 6.1. Berlin's theoretical model of ethnobiological classification, showing "the schematic relationship of the universal ethnobiological ranks and their relative hierarchical positions as shown in an idealised system" (Berlin 1992:16).

of gross morphological characters. The majority of generic taxa within folk taxonomies are included in (or affiliated with) one of the recognized life-form taxa. However, there are generic taxa that are sufficiently morphologically distinct or economically important to be classified as unaffiliated; that is, they are independent of life-forms.

Some generics are further divided into specific or even varietal categories. These differ from both life-form and generic categories in that they are conceptually distinguished from one another on the basis of very few morphological characters. The two levels of classification are more widely applied to plants of considerable cultural or utilitarian value.

Methods Used in This Research

Fieldwork for the study of the taxonomy, ecology, and utilization of African rattans was undertaken primarily in Ghana, Nigeria, Cameroon, and Equatorial Guinea. The field surveys concentrated both on the collection of herbarium voucher specimens to provide representative material for the completion of a taxonomic revision of the rattans of Africa and the recording of vernacular names and uses for each species. Where possible (and where appropriate), vernacular names were recorded in the field using a tape recorder and later transcribed and interpreted with the assistance of translators who were native speakers of the languages in question. English (both "grammar" and Pidgin English) and French were used as lingua francas. Further study of existing herbarium voucher material held in herbaria,[3] as well as an extensive literature search, provided a plethora of vernacular names that could be compared with the actual species to which the local name referred. This latter study confirmed the presence of significant consistencies within a number of classification systems first detected during our field research.[4]

For the purposes of this chapter, vernacular names are spelled phonetically, but the use of the phonetic alphabet has been avoided for ease of reading. Language and tribal names have been standardized by referring to the Summer Institute of Linguistics' *Ethnologue* (Grimes 1996).

African Rattan Folk Nomenclature with Reference to Berlin's Model

The Unique Beginner or Kingdom

In common with many folk classifications throughout the world, in the Niger-Congo language family there exist no words for "plant" (Guthrie 1948, 1969–1971; Westermann & Bryan 1952), and the classification of the plant kingdom is covert; that is, this taxonomic rank is not lexically recognized. Although in many languages in the Western Bantu subgroup a root word for animal, -*namma,* is commonly encountered, this term is polysemous with "meat" (Guthrie 1948, 1969–1971; Sharpe pers. comm. 2000). This observation conforms to Berlin's assertions that, in traditional societies, when a kingdom is named the word is often polysemous with some life-form class or reflects an overt recognition of function (Berlin 1973, 1992).

Life-Form Categories

In the Niger-Congo language family, explicit terms for "tree," "vine," and "herb" are widely recognized life-form categories (Guthrie 1948, 1969–1971; Westermann & Bryan 1952; Oliver & Fage 1975; Vansina 1990; Profizi & Makita-Madzou 1996).[5] However, both on the basis of morphological discontinuities and as a reflection of cultural utility, a number of additional life-form classifications within the plant kingdom are also recognized. In many Western Bantu languages, for example, at the life-form level of classification domesticated plants are often distinguished from wild plants despite shared morphological similarities (Vansina 1990). In addition to domesticated plants, other plants of extreme functional utility are accorded separate life-form categories. For example, "palm" is recognized as a separate life-form category within many folk taxonomies. Despite the fact that the Palmae are considered a "natural" plant family and have long been classified by Western science as such, many indigenous societies do not share this view and include only arborescent taxa (standing palms) within their palm life-form category (see Ellen 1998 for a discussion of this). Indeed, it is commonly the case that the life-form category "palm" of many folk taxonomies does not include climbing palms or other palms exhibiting significant morphological discontinuities, such as acaulescence (Ellen 1998).

For example, in his research in Peru Berlin (1977) recorded that the Aguaruna recognize "palm" as a life-form, distinct from "tree," "vine," or "herb." Yet they exclude from this category slender climbing palms (members of the genus *Chamaedorea*), as well as a number of acaulescent species (Berlin 1977). Ellen (1998) also excludes rattans from his discussion of palm life-form categories in a number of indigenous societies for the same reason: they are not included in the same category as standing palms. As such, rattans are commonly classified apart from other palms as a separate life-form (Ellen 1998). Indeed, the word "rattan" is derived from the Malay word *rotan,* the life-form category for all climbing palms.

In the Niger-Congo language family, "palm" is recognized as a separate life-form, albeit covertly.[6] Yet, as with many other indigenous societies (Berlin 1977; Ellen 1998), this life-form category encompasses the standing palms only (*Elaeis, Raphia,*[7] *Phoenix,* and *Borassus*) and excludes both the climbing palms and the small acaulescent genus *Sclerosperma,* as well as the stoloniferous *Podococcus barteri.*[8] In this Niger-Congo context, climbing palms are recognized as a distinct life-form category. This life-form category, when applied, is explicit and overt and, as such, is commonly named.

Intermediate and Generic Categories

Although rattan as a life-form category is distinct in many non-Bantu languages, the Western Bantu languages do not commonly recognize rattans as a distinct life-form, or, if they do, the recognition is covert. In such instances, these societies categorize rattans within intermediate categories by grouping closely related folk generics (see below). The separation of the climbing palms in this manner is based on utilization or, less commonly, on morphological discontinuities such as diameter or the presence or absence of spines on the leaf sheath. In these instances the biological genera *Eremospatha, Calamus,* and *Oncocalamus,* as well as some small-diameter members of *Laccosperma,* are often grouped together at the intermediate level.

When this intermediate category is recognized, it excludes the large-diameter members of the biological genus *Laccosperma.* These taxa, which represent a number of biological species, are highly prized as a source of good-quality cane and are categorized in many folk classifications only at the generic level. Table 6.3 presents the folk classification of rattan canes in selected Niger-Congo languages at the life-form, intermediate, and generic ranks. Interestingly, some generic terms are polysemous with the products that are derived from the plant itself. Some examples are included in Table 6.4.

Specific Categories

Although some rattan folk genera are monotypic, a considerable proportion of the remaining genera are polytypic and are further split into folk-specific categories. Within the polytypic genera of rattan, the recognition and natural grouping by indigenous societies of folk-specific taxa often reflect the intrinsic relationship of these specifics both to each other and to the ranks superordinate to them. Commonly, these are lexically recognized by means of terms derived from kinship and descent terminology. Such kinship metaphors have been identified in a number of folk taxonomies (Hays 1983; Berlin 1992) and have been widely encountered in rattan classification systems, particularly within the Western Bantu subgroup.

For example, the Denya of Cameroon, who speak a language belonging to the Western Bantu subgroup, use the intermediate category *e-chié* for all canes not included in the group of large-diameter, heavily armed species of *Laccosperma.* However, as is commonly the case with intermediate lexicons in other Bantu languages, this term is

Table 6.3
Life-form, intermediate, and generic folk classification of rattan canes in selected Niger-Congo languages

Folk name (-root)	Language (country)	Language subgroup[a]	Ethnobiological category
ailé (all spp. except large-diameter *Laccosperma* spp.)	Anyin (Côte d'Ivoire)	Non-Bantu	Intermediate
ahike (large-diameter *Laccosperma* spp.)			Generic
nwatia (all climbing palms)	Akan-Asanti (Ghana)	Non-Bantu	Life-form
dekun (all climbing palms)	Gun-Gbe (Benin)	Non-Bantu	Life-form
ikan (all climbing palms)	Edo (Nigeria)	Non-Bantu	Life-form
egbèé (all climbing palms)	Yoruba (Nigeria)	Non-Bantu	Life-form
kogiri (all climbing palms)	Fulfulde	Non-Bantu	Life-form
kwagiri (all climbing palms)	Hausa	Non-Bantu	Life-form
uga (all climbing palms)	Igbo (Nigeria)	Non-Bantu	Life-form
-chié (all spp. except large-diameter *Laccosperma* spp.)	Denya (Cameroon)	Bantu	Intermediate
gekwiya (large-diameter *Laccosperma* spp.)			Generic
edju (*Oncocalamus* spp.)	Oroko language group (Cameroon)	Bantu	Generic
ndongo (*Eremospatha* spp.)			Generic
mekah (large-diameter *Laccosperma* spp.)			Generic
nloun (all spp. except large-diameter *Laccosperma* spp.)	Bassa (Cameroon)	Bantu	Intermediate
? (large-diameter *Laccosperma* spp.)			Generic
mokolo (small-diameter canes)	Bakossi (Cameroon)	Bantu	Intermediate
mekah (large-diameter *Laccosperma* spp.)			Generic
nlon (all spp. except large-diameter *Laccosperma* spp.)	Bulu (Cameroon)	Bantu	Intermediate
nkan (all *Laccosperma* spp.)			Generic
nlong (all spp. except large-diameter *Laccosperma* spp.)	Fang (Equatorial Guinea & Gabon)	Bantu	Intermediate
nkan (all *Laccosperma* spp.)			Generic
mikaana (all climbing palms)	Téké (Congo)	Bantu	Life-form
kekelé (small-diameter canes)	Zande, Lingala, Swahili-Zaire (DR Congo)	Bantu	Intermediate
likaw (large-diameter *Laccosperma* spp.)			Generic

[a] Bantu languages are characterized by the possession of root terms that are distinguished as either singular or plural by independent prefixes. These root terms are commonly shared among related languages, and it is variation within the prefixes that is reflected in the variation in names for plants, for example (see Guthrie 1948, 1953).

Table 6.4
Selected cane product names

Name	Language (country)	Description
		Palm heart
mekah[a]	Balundu-Bima (Cameroon)	Apex of *L. robustum*
baa ndanga	Téké (Congo)	Apex of *E. haullevilleana*
mukaana a ngomu[a]	Téké (Congo)	Apex of *L. secundiflorum*
mukaana a buulu[a]	Téké (Congo)	Apex of *E. wendlandiana*
ngodji	Lomdo (DR Congo)	Apex of *L. robustum*
		Cane and cane rope
aka	Fang (Equatorial Guinea)	Cleaned stems of *L. robustum* / *L. secundiflorum*
ukpa	Ijo-Izon (Nigeria)	Split stems of *L. secundiflorum*
ukwen	Edo (Nigeria)	Split stems of *L. secundiflorum*
ekwe oya[a]	Igbo (Nigeria)	Split stems of *L. opacum* for tie-tie
ekwele / *akwala*	Igbo (Nigeria)	Split stems of *O. wrightianus* (coarse cordage)
udo	Igbo (Nigeria)	Split stems of *O. wrightianus* (fine twine)
elili	Igbo (Nigeria)	Split stems of *O. wrightianus* (string or thread)
apié[a]	Igbo (Nigeria)	Cane rope of *C. deërratus*
		Baskets
kenten	Akan-Asanti (Ghana)	Long baskets made from stems of *L. opacum*
penja	Bakossi (Cameroon)	All cane baskets
mbaka	Denya (Cameroon)	Farm baskets made from *E. macrocarpa*
bi-dong	Fang (Equatorial Guinea)	Fish baskets made from split stems of *L. robustum* & *E. macrocarpa*
be-koro	Fang (Equatorial Guinea)	Fish traps made from split stems of *L. robustum* & *E. macrocarpa*
nkeuiñ	Fang (Equatorial Guinea)	Farm baskets made from split stems of *L. robustum* & *E. macrocarpa*
maa kutu	Téké (Congo)	Baskets made from *E. haullevilleana* (*baana* = small; *mwana kutu* = medium; *kiana* = large)

[a] The product name is also the generic category or derives directly from it.

polysemous for "cane rope" and is used, in the folk-generic sense, to recognize one highly valuable species of small-diameter rattan in particular, *Eremospatha macrocarpa*. At the folk-specific level the Denya lexically recognize all other species of cane in the *e-chié* intermediate category, even though many are not actually valued as a source of cane, with the lexical designation often reflecting kinship metaphors. For example, two other species of *Eremospatha*, *E. tessmanniana* and *E. quinquecostulata*, are both referred to as *calumé-e-chié*, "the uncle of cane rope," pointing to the poor quality of the cane yielded

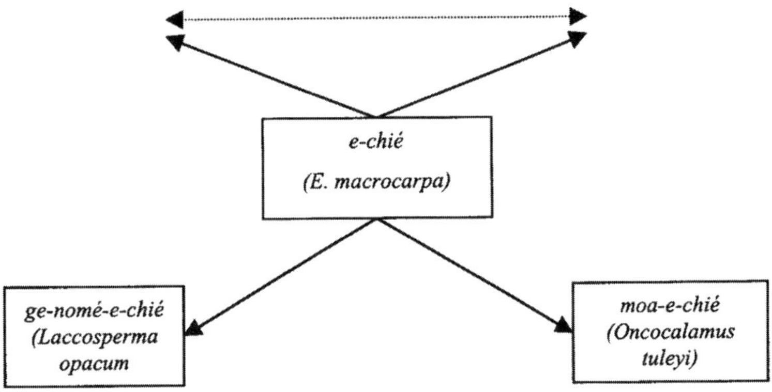

Figure 6.2. Selected nomenclatural relationships within the Denya language group, Cameroon.

by these biological species. In the same context, *E. wendlandiana* is referred to as *mua-e-chié* or "brother to cane rope." Two other unrelated taxa, despite being morphologically distinct, are also referred to in relation to the true cane rope: *Laccosperma opacum* as *ge-nomé-echié,* or "slave to cane rope," and *Oncocalamus tuleyi* as *moa-e-chié,* or an undefined "relative to cane rope." The nomenclatural relationships of these cane categories are depicted in Figure 6.2.

Further examples of such metaphors include the recognition of species that are related yet separate in biological terms. *Laccosperma opacum,* a small-diameter cane, is called *npue-nkan* by the Fang of Equatorial Guinea; it is considered the "child of" true *nkan* (generic term for *L. robustum* and *L. secundiflorum;* large-diameter canes and widely utilized species). Similarly, the same species, *L. opacum,* is also categorized at the specific level by the Mokpwe around Mount Cameroon as *liko ko'ko,* or "close to cane."

The application of the language of descent is commonly encountered within taxa that have manifestly different morphological characteristics in the juvenile and adult phases, particularly reflected in the leaf shape; flabellate when juvenile and pinnate when adult. Interestingly, these juvenile stages can be so morphologically distinct that they have sometimes been described by Western botanists as "new" species (Drude 1895). A number of these species—for example, *Eremospatha macrocarpa* and *E. haullevilleana*—are widely utilized in the juvenile stages but are too inflexible for use when mature. The recognition in some folk taxonomies of this within-species morphological variation and distinct utilization is effected by using what are different specific-level terms for the juvenile and adult growth phases. All of the prefixes listed in Table 6.5 are glossed as "child of."

Varietal Categories

Although varietal categories are well recorded for a few African palms (particularly the oil palm, reflecting its long history of farmer-led selection and breeding; see Burkill 1997), varietal categories for rattan have not been encountered.

Table 6.5
Folk-specific taxa of *Eremospatha macrocarpa*, based on morphological and utilization characteristics

Species name		
Juvenile phase	Adult phase	Language (country)
koto-mbalu	*mbalu*	Mende (Sierra Leone)
asa-nlong	*nlong* (semi-adult); *ongam* (adult)	Bulu (Cameroon)
bana-ndongo	*ndongo*	Balundu-Bima (Cameroon)

Table 6.6
Summary of rattan classification in the Niger-Congo language group, with reference to Berlin's model

		Language	
Rank	Type of name	Bantu	Non-Bantu
Kingdom	Unnamed	Covert	Covert
Life-form	Primary or unnamed	Rattan life-form category named	Not recognized or covert
Intermediate	Primary or unnamed	Commonly no members (but see Table 6.2)	Commonly two members (but see Table 6.2)
Generic	Primary	2–3 members	4–6 members
Specific	Usually secondary	No generics contain known specifics	Most generics contain specifics
Varietal	None	None recorded	None recorded

Table 6.6 presents a summary of rattan classification in Africa with reference to Berlin's conceptual model. For ease of reference, the theoretical structure of vernacular names is presented in Figure 6.3.

Discussion and Conclusions

It is known that the peoples within the Niger-Congo language family consider rattans as life-forms lying outside the standard life-form classifications of "tree," "vine," and "herb" (Berlin 1973, 1977, 1992; Berlin et al. 1973). As we have seen, this is undoubtedly due both to morphological discontinuities and to considerations of functional utility. As is the case with other indigenous classification systems that regard the climbing palms as separate from standing palms (Berlin 1977; Ellen 1998), throughout their range African rattans are accorded life-form status of their own, whether covertly or overtly recognized.

In the case of many of the non-Bantu language groups that predominate in the forests of Upper Guinea, the hierarchical classification of rattans is relatively straightforward and closely adheres to Berlin's model. The fact that the Upper Guinea forests

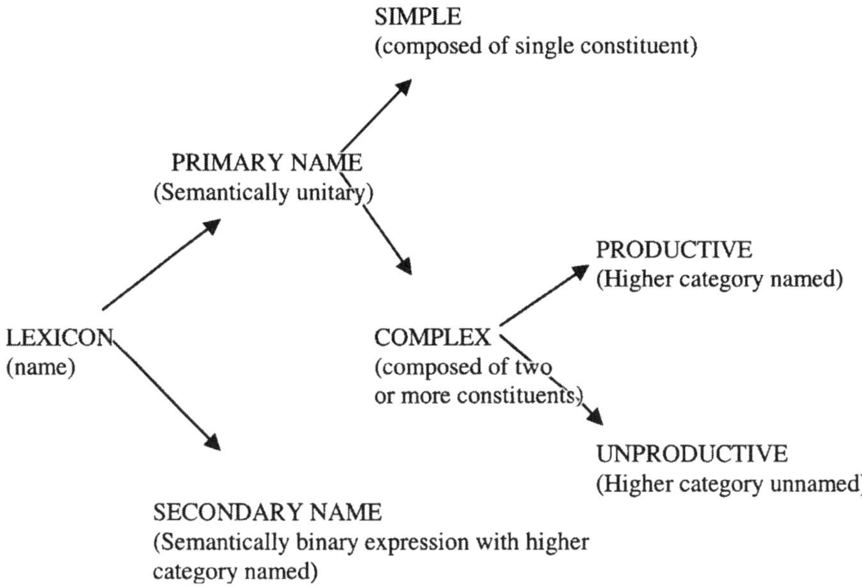

Figure 6.3. The structure of vernacular names. *Source:* Modified from Berlin et al. 1973.

Table 6.7

Selected examples of parity of generic categorization of African rattans between folk taxonomic systems and Western classification[a]

Language (country)	*Laccosperma*	*Eremospatha*	*Calamus*
Loko (Sierra Leone)	*kafo*	*mbalu*	*tambe*
Akan-Asanti (Ghana)	*ayie*	*mfia*	*demm̄eré*
Ijo-Izon (Nigeria)	*ukpa*	*boru*	*apie*

[a] The genus *Oncocalamus* does not occur in the Upper Guinea forest region.

are relatively species poor in terms of rattan means that somewhat parsimonious classification systems are encountered, with very few categories recognized below the generic. Life-form categories within these classifications are named, are clear and unambiguous, and refer to all palms with the climbing habit. At the generic level, in particular, it is also clear that there is considerable correspondence between indigenous classifications and Western taxonomy (Table 6.7). In this case it may be argued that morphological considerations take precedence in the non-Bantu folk taxonomies for rattan.[9]

In contrast, in the Bantu language subgroup, the use of intermediate categories is more widespread, and the life-form category for rattan is covert. As discussed, the climbing palms are essentially subdivided into two main categories, and one intermediate and one to several generic categories are commonly recognized. This may be a reflection of both increased rattan diversity and increased ethnic complexity (Brown

1977), particularly in the region from southeastern Nigeria to northern Gabon, the most biologically and linguistically diverse area of Africa.

Berlin (1973) suggests that intermediate categories are rare in folk taxonomies and that, because they often lack names, some doubt has been expressed as to whether they should be included as an ethnobiological rank at all. However, the explicit use of intermediate categories for African rattans, particularly in the Western Bantu language subgroup, is evidence to the contrary. It is likely that the widespread use of the intermediate category is indicative of the recognition of functional utility over morphology.

More specific examples can further illustrate this point. The use of an intermediate category for what are biologically separate groups is widespread within the Western Bantu subgroup, and it is important to note that the name given to this intermediate category is commonly polysemous with the generic root name for the most widely utilized rattan species, *Eremospatha macrocarpa*. Further splitting at the folk-specific level relates each folk taxon with this species, often through kinship metaphors or other relational terminology, regardless of often significant morphological differences among these folk specifics. The nomenclature used is based on relating each species to the main taxa that are useful (in this case *E. macrocarpa*) and those others that are not as widely used.

The most obvious example of the importance of functional utility in indigenous classification of rattan is with regard to the large-diameter species of *Laccosperma*. Although *Laccosperma* was previously considered a single species, it is now recognized that there are in fact three biologically distinct species of large-diameter *Laccosperma: L. secundiflorum, L. acutiflorum,* and *L. robustum* (Sunderland 1999a, 1999b). However, most folk taxonomies recognize only one, and sometimes two, generics for this group of biological species, despite significant and easily observable morphological discontinuities obvious in the field. This is because *L. secundiflorum* and *L. robustum* provide the best-quality cane in Africa. They are widely utilized at the subsistence level and also provide the basis for a thriving trade. Although traded in the same way, these species are obviously different when encountered in the field, and, because they are sympatric, their morphological differences are obvious. *Laccosperma secundiflorum* has broad, sigmoid leaflets that are held more or less horizontally and possesses a relatively long petiole; it is a species found in the forest understory and at forest margins. *L. robustum* has fine, linear leaflets that are almost completely pendulous on the rachis; it occurs in forest gaps and is extremely light demanding. Yet more than 100 informants have told me that the species "are the same" and as such are included in a single generic category in local classifications.

It is clear that within the two main classification systems discussed above both morphology and functional utility are taken into account in relation to African rattans. However, the differences between the two approaches can be considered cognitive. In the case of the non-Bantu classifications, the important criteria for classification are "climbing palm" and the morphological differences within that group; in the Western Bantu classifications, what takes precedence are the aspects of functional utility—i.e., "cane"—and how taxa relate to each other in that functional context.

Although correspondences have been found between folk classification and scientific taxonomy at the generic level in some classifications (Table 6.7), this equivalence is not commonly encountered in African rattans. As discussed, within rattan folk systematics underdifferentiation of species recognized in Western taxonomy tends to be more commonly observed. However, limited overdifferentiation is also encountered, particularly when juvenile stages are lexically distinguished from adult stages in specific folk taxa (see Table 6.5).

Understanding how local people both view and use their resources is an often greatly undervalued means of developing integrated, sustainable strategies for forest and resource management. When complemented with biosystematic data, the study of local classification systems of indigenous groups can provide considerable insights into which species are used and why, and how, in these functional respects, species relate to each other. In the case of African rattans, such an approach has been critical in providing clear information on the most widely utilized taxa and which species could be targeted for promotion in the context of sustainable development. This is particularly pertinent for the plethora of development and government agencies that are currently calling for the promotion of "rattan" to contribute to the improved livelihoods and wider conservation of the forests of West and Central Africa. It is essential that such initiatives first identify which species are of particular importance and then focus on the few species worthy of such investment. Now that a sound taxonomic base has been provided through the study of nomenclature and use and the utilization of each biological species is more comprehensively known, it is hoped that any future development of the rattan resource will take place in a coherent and structured manner.

Acknowledgments

The U.S. Forest Service, the Central African Regional Programme for the Environment, and the International Network for Bamboo and Rattan funded the work of the African Rattan Research Programme from 1997 to 2000; their support is gratefully acknowledged. I am also grateful to Phil Burnham, Barrie Sharpe, and Ruth Malleson for stimulating discussions on this subject. Finally, a great many thanks are extended to all the rattan harvesters, traders, artisans, and villagers with whom I have worked over the past few years and who have generously shared their knowledge and insights of the forest and the rattans on which many of them rely for their livelihoods.

Notes

1. Two other groups of palms confined to the New World have climbing representatives and are often mistaken as being closely related to true rattans. *Chamaedorea* (subfamily Ceroxyloxideae; tribe Hyophorbeae) and *Desmoncus* (subfamily Arecoideae; tribe Cocoeae) climb by means of reflexed terminal leaflets. Indeed, *Desmoncus* is often exploited for its canelike qualities and is used in the same way as the true rattans (Henderson & Chávez 1993). A climbing palm (*Dypsis scandens*) has also been discovered in Madagascar; it also is not related to the true rattans.

2. Spontaneous use indicates immediate short-term usage, regardless of species, such as tying a bundle of harvested leaves in the forest or repairing a cutlass handle on the farm. In this case, use is based on immediate, or "emergency," requirements and is not predetermined.

3. Material was studied from the Libreville, Yaoundé, Bata, Kumasi, Gold Coast, New York, Missouri, Hamburg, Geneva, Wageningen, Brussels, Florence, and Kew herbaria.

4. Disclaimer: This paper is not a taxonomic work and should not be considered the place of first publication for any new taxon or synonym it contains.

5. In the majority of Bantu languages the word for "tree" is polysemous for "wood," a lexical feature that is extremely widespread (Brown 1977). Interestingly, the Bantu proto-language had a word for "tree," *-ti-*, that was also the word for "medicine," indicating the historical importance of the plant-based approach to medicinal practice (Iliffe 1995) and suggesting that utility has played an important role in the development of plant-based vocabularies.

6. Where folk taxonomies have names for this category, they are usually polysemous with the oil palm (*Elaeis guineensis*), reflecting its symbolic and utilitarian importance.

7. Acaulescent members of this genus are also included under the palm life-form category.

8. These latter taxa, while possessing considerable utilitarian importance, are unrelated to other palms and to each other and can be classified as unaffiliated generics within most Niger-Congo folk classification systems in which they are recorded.

9. However, examples that point to functional utility as a major factor are also encountered. For example, within the genus *Laccosperma, L. opacum* and *L. laeve* are treated as generics within most non-Bantu classifications and, despite their close morphological similarities, are often distinguished based on their functional utility. In Twi, *L. laeve,* which is not used by that ethnic group, is called *tehan muhunu,* or "it lives in the world for nothing," whereas *L. opacum* is used for making specific baskets and is called *sayai.*

Literature Cited

Abbiw, D. 1990. The Useful Plants of Ghana. London: Intermediate Technology Press.

Ainslie, J. R. 1937. A List of Plants Used in Native Medicine in Nigeria. Oxford Forestry Memoirs, 7. Oxford: Clarendon Press.

Alexiades, M. N. (ed.). 1996. Selected Guidelines for Ethnobotanical Research: A Field Manual. Advances in Economic Botany, 10. Bronx: New York Botanical Garden.

Atran, S. 1983. The nature of folk botanical life forms. American Anthropologist 87(2): 298–315.

Balée, B. 1989. Nomenclatural patterns in Ka'apor ethnobotany. Journal of Ethnobiology 9(1): 1–24.

——— & D. Daly. 1990. Resin classification by the Ka'apor Indians. Pp. 24–34 *in* G. T. Prance & M. J. Balick (eds.), New Directions in the Study of Plants and People: Research Contributions from the Institute of Economic Botany. Advances in Economic Botany, 8. Bronx: New York Botanical Garden.

Berhaut, J. 1988. Flore illustrée du Sénégal: Monocotylédons et ptéridophytes. Dakar: Gouvernement du Sénégal.

Berlin, B. 1973. Folk systematics in relation to biological classification and nomenclature. Annual Review of Ecology and Systematics 4: 259–271.

———. 1977. The concept of rank in ethnobiological classification: Some evidence from Aguarana folk biology. American Ethnologist 3: 381–399.

———. 1992. Ethnobiological Classification: Principles of Categorization of Plants and Animals in Traditional Societies. Princeton, N.J.: Princeton University Press.

———, D. E. Breedlove & P. H. Raven. 1973. General principles of classification and nomenclature in folk biology. American Anthropologist 75(1): 214–242.

Bouquet, A. & M. Debray. 1974. Plantes médicinales de la Côte d'Ivoire. Travaux et documents de l'ORSTOM, 32. Paris: ORSTOM.

Brown, C. 1977. Folk botanical life forms: Their universality and growth. American Anthropologist 79: 317–342.

Burkill, I. H. 1935. A Dictionary of the Economic Products of the Malay Peninsula. 2 vols. London: Crown Agents.

———. 1997. The Useful Plants of West Tropical Africa. Vol. 4, Families M–R. Ed. 2. Kew: Royal Botanic Gardens.

Chevalier, A. 1934. Notes & actualities. Végetales Utiles 8: 342.

———. 1936. Sur quelques Palmiers de L'Angola donnant des produits utilisables. Revue de Botanique Appliqué 17: 895–898.

Conklin, H. C. 1962. The lexicographical treatment of folk taxonomies. International Journal of American Linguistics 28: 119–141.

Corner, E. J. H. 1966. The Natural History of Palms. London: Weidenfield & Nicolson.

Dalziel, J. M. 1937. The Useful Plants of West Tropical Africa. London: Crown Agents.

Davies, A. G. & P. Richards. 1991. Rainforest in Mende Life. London: ESCOR, ODA.

De Fretas, Y. 1992. Community versus company-based rattan industry in Indonesia. Pp. 142–149 *in* S. Counsell & T. Rice (eds.), The Rainforest Harvest: Sustainable Strategies for Saving the Tropical Forest? London: Friends of the Earth.

Defo, L. 1997. La filière des produits forestiers non-ligneux: L'Example du rotin au Sud-Cameroun. Yaoundé, Cameroon: University of Yaoundé, Department of Geography.

———. 1999. Rattan or porcupine? Benefits and limitations of a high-value non-wood forest product for conservation in the Yaounde region of Cameroon. Pp. 237–244 *in* T. C. H. Sunderland, L. E. Clark & P. Vantomme (eds.), The Non-Wood Forest Products of Central Africa: Current Research Issues and Prospects for Conservation and Development. Rome: FAO.

——— & T. C. H. Sunderland. 1999. L'Artisanat de rotin en milieu urbain au Cameroun: Rapport préliminaire. African Rattan Research Programme Technical Note No. 4. Kew: Royal Botanic Gardens.

Dransfield, J. 1992. The ecology and natural history of rattans. Pp. 27–34 *in* R. W. M. Wan, J. Dransfield & N. Manokaran (eds.), A Guide to the Cultivation of Rattan. Forest Record No. 35. Kuala Lumpur: Forest Research Institute.

——— & N. Manokaran (eds.). 1994. Rattans. Plant Resources of South-East Asia, 6. Wageningen, Netherlands: Pudoc Scientific Publishers.

Drude, O. 1895. Die Palmenflora des Tropischen Afrika. Engler's Botanische Jarhrbe 5: 108–136.

ECOSYN. 1999. Rotins. Unpublished manuscript for Lianes forestières d'Afrique de L'Ouest: Diversité, écologie et gestion. Coordination en Côte d'Ivoire.

Ekandem, M. J. 1955. The use of plants as symbols in Ibibio and Ibo Country. Nigerian Field 20(2): 53–64.

Ellen, R. 1998. Palms and the prototypicality of trees: Some questions concerning assumptions in the comparative study of categories and labels. Pp. 57–80 *in* L. Rival (ed.), The Social Life of Trees. Oxford and New York: Berg.

Falconer, J. 1994. Non-Timber Forest Products in Southern Ghana: Main Report. Chatham, U.K.: Natural Resources Institute.

Friedberg, C. 1974. Les Processus classificatoires appliqués aux objets naturels et leur mise en évidence. Journal d'Agriculture Tropicale et de Botanique Appliqué 21(12): 313–334.

Gianno, R. 1986. Resin classification among the Semelai of Tesak Bera, Pahang, Malaysia. Economic Botany 40(2): 186–200.

Given, D. R. & W. Harris. 1994. Techniques and Methods of Ethnobotany: As an Aid to the Study, Evaluation, Conservation and Sustainable Use of Biodiversity. London: Commonwealth Secretariat.

Godoy, R. 1990. The economics of traditional rattan cultivation. Agroforestry Systems 12: 163–172.

——— & T. C. Feaw. 1989. The profitability of small holder rattan cultivation in southern Borneo. Human Ecology 17: 347–363.

Grimes, B. F. (ed.). 1996. Ethnologue Language Name Index. Ed. 13. Dallas, Tex.: Summer Institute of Linguistics.

Guinea-López, E. 1946. Ensayo geobotánico de la Guinea continental española. Madrid: Dirección de Agricultura de los Territorios Españoles del Golfo de Guinea.

Guthrie, M. 1948. The Classification of the Bantu languages. London: SOAS.

———. 1969–1971. Comparative Bantu: An Introduction to the Comparative Linguistics and Prehistory of the Bantu Languages. 4 vols. Farnborough, U.K.: Gregg.

Hays, T. 1983. Ndumba folk biology and general principles of ethnobiological classification and nomenclature. American Anthropologist 85(3): 592–611.

Hédin, L. 1929. Les Rotins au Cameroun. Revue de Botanique Appliqué 9: 502–507.

Henderson, A. & F. Chávez. 1993. *Desmoncus* as a useful palm in the western Amazon basin. Principes 37: 184–186.

Holland, T. 1922. The useful plants of Nigeria. Kew Bulletin 9: 712–753.

Hunn, E. 1976. Towards a perceptual model of folk biological classification. American Ethnologist 3: 508–524.

———. 1982. The utilitarian factor in folk biological classification. American Anthropologist 84(4): 830–847.

Iliffe, J. 1995. Africans: The History of a Continent. Cambridge and New York: Cambridge University Press.

Irvine, F. R. 1952. Supplementary and emergency food plants of West Africa. Economic Botany 6(23): 23–40.

———. 1961. Woody Plants of Ghana. London: Oxford University Press.

ITTO. 1997. Bamboo & rattan: Resources for the 21st century? Tropical Forest Update 7(4): 12–13.

Jeffreys, M. D. W. 1960. Cordage among the Ibo. Nigerian Field 25(1): 42–43.

Kabuye, C. H. S. 1988. East African Plants Used in Basketry. Monographs in Systematic Botany of the Missouri Botanical Garden 25: 351–361.

Martin, G. J. 1995. Ethnobotany: A Methods Manual. London and New York: Chapman & Hall.

Mildbraed, J. 1913. Von den Bulus genutzte wildwachsende Pflanzen des Südkameruner Waldlandes. Notizblatte des Botanischen Gartend und Museums zu Berlin Dahlem Appendix 17: 1–16.

Minga, M. D. In press. L'Impact de l'exploitation du rotin sur la preservation de la forêt a Kinshasa, Republique Democratique du Congo. *In* Proceedings of the International Expert Meeting on African Rattans, Limbe, Cameroon, 1–3 February 2000.

Morakinyo, A. B. 1994. The ecology and silviculture of rattans in Africa: A management strategy for Cross River State and Edo State, Nigeria. M.S. thesis, University College of North Wales.

———. 1995. The commercial rattan trade in Nigeria. Forests, Trees and People Newsletter 25: 22–30.

Okigbo, B. N. 1980. Plants and food in Igbo culture. Annual address to the Ministry of Information, Culture, Youth and Sports. Lagos: Government Printers.

Oliver, R. 1999. The African Experience: From Olduvai Gorge to the 21st Century. Rev. ed. London: Weidenfield & Nicholson.

——— & J. D. Fage. 1975. A Short History of Africa. Harmondsworth, U.K.: Penguin.

Peluso, N. L. 1992. The rattan trade in East Kalimantan, Indonesia. Pp. 115–127 *in* D. C. Nepstad & S. Schwartzman (eds.), Non-Timber Products from Tropical Forests: Evaluation of a Conservation and Development Strategy. Advances in Economic Botany, 9. Bronx: New York Botanical Garden.

Profizi, J. P. 1986. Notes on West African rattans. Rattan Information Centre Bulletin 5(1): 1–3.

——— & J. P. Makita-Madzou. 1996. Utilisations des Arecaceae (Palmiers) au Congo par les Téké Tsaaya. Pp. 122–131 *in* A. Froment, I. de Garine, C. Binam Bikoi & J. F. Loung (eds.), Bien manger et bien vivre: Anthroplogie alimentaire et développement en Afrique intertropicale, du biologique au social. Paris: Coalition ORSTOM & l'Harmattan.

Randall, R. & E. Hunn. 1984. Do life forms evolve or do uses for life? American Ethnologist 11: 329–349.

Raponda-Walker, A. & R. Sillans. 1961. Les Plantes utiles du Gabon. Pp. 324–341 *in* Encyclopedie biologique. Paris: Paul Lechevalier.

Robyns, W. & R. Tournay. 1955. Flore du Spermatophytes du Parc National Albert. Vol. 3, Monocotyledons. Brussels: Musée National de Histoire Naturelle.

Siebert, S. F. 1997. Economically important rattans of Central Sulawesi, Indonesia. Principes 41(1): 42–46.

Staner, P. & R. Boutique. 1937. Matériaux plantes medicinales indigènes du Congo Belge. *Mem. l'Inst. Roy. Col. Belge.* pp 13.

Sunderland, T. C. H. 1998. The rattans of Rio Muni, Equatorial Guinea: Utilisation, biology and distribution. A report for the European Union Project No. 6 ACP–EG–020: Proyecto Conservación y Utilización Racional de los Ecosistemas Forestales de Guinea Ecuatorial (CUREF).

———. 1999a. The rattans of Africa. Pp. 227–236 *in* R. Bacilieri & S. Appanah (eds.), 1999. Rattan Cultivation: Achievements, Problems and Prospects. Kuala Lumpur: CIRAD–Forêt & FRIM, Malaysia.

———. 1999b. New research on African rattans: An important non-wood forest product from the forests of Central Africa. Pp. 87–98 *in* T. C. H. Sunderland, L. E. Clark & P. Vantomme (eds.), The Non-Wood Forest Products of Central Africa: Current Research Issues and Prospects for Conservation and Development. Rome: FAO.

———. 2001. The taxonomy, ecology and utilisation of African rattans. Ph.D. diss., University College, London and Royal Botanic Gardens, Kew.

——— & P. Tchouto. 1999. A participatory survey and inventory of timber and non-timber forest products of the Mokoko River Forest Reserve, SW Province, Cameroon. Unpublished report for USAID/CARPE.

Sylvic Sap. 1935. The West African mahogany trade. Nigerian Field 4 (3): 127–132.

Tenati, G. In press. The use of rattan for vocational rehabilitation in Zambia. *In* Proceedings of the International Expert Meeting on African Rattans, Limbe, Cameroon, 1–3 February 2000.

Thomas, D. W., J. M. Thomas, W. A. Bromley & F. T. Mbenkum. 1989. Korup Ethnobotany Survey: Final report to the World Wide Fund for Nature (U.K.).

Townson, I. M. 1995. Patterns of Non-Timber Forest Products Enterprise Activity in the Forest Zone of Southern Ghana. Report to the ODA Forestry Research Programme. Oxford: Oxford Forestry Institute.

Tuley, P. 1995. The Palms of Africa. Saint Ives, U.K.: Trendrine Press.

Uhl, N. W. & J. Dransfield. 1987. Genera Palmarum: A Classification of Palms Based on the Work of Harold E. Moore, Jr. Ithaca, N.Y.: L. H. Bailey Hortorium and International Palm Society.

Unwin, A. H. 1920. West African Forests and Forestry. London: T. Fisher Unwin.
Vansina, J. 1990. Paths in the Rainforest: Toward a History of Political Tradition in Equatorial Africa. London: James Currey.
Westermann, D. & M. A. Bryan. 1952. The Languages of West Africa. Handbook of African Languages. 2 vols. London and New York: Oxford University Press.
Wilkie, P. & A. Saridan. 1999. The limitations of vernacular names in an inventory study, Central Kalimantan, Indonesia. Biodiversity and Conservation 8: 1457–1467.
Williamson, K. & A. O. Timitimi (eds.). 1983. Short Izon-English Dictionary. Delta Series, 3. Port Harcourt, Nigeria: University of Port Harcourt Press.

Appendix 6.1: Indigenous Nomenclature and Utilization of African Rattans

Introduction

As discussed in the chapter, rattan canes are used extensively across West and Central Africa by local communities and play an important role in indigenous subsistence strategies for many rural populations. As such, the range of indigenous uses of rattan canes is vast. These uses are listed below by species; summary descriptions of the species can be found in Table 6.1. The vernacular names recorded are from the examination of herbarium records and notes, a literature review (cross-referenced to herbarium records), and my own field observations.

Calamus deërratus G. Mann & H. Wendl.

A clustering species of rattan with stems up to 30 m long, ecirrate leaves, and a long flagellum arising from, and adnate to, the sheath species. Plants are dioecious with male and female individuals. Widely distributed from Gambia and Senegal, south to Angola, and eastward to Uganda. Found predominantly in swamp forest.

Vernacular names

SENEGAL: *ki tid* (Balanta); *kintem* (Bainouk); *mantampa da sera* (Crioulo, Upper Guinea); *bu kètao bu ketav, fu fiaf, ka kèt, ka tay, ke hiya, kékiya* (Jola-Fogny); *tambem* (Fula-Pulaar); *tambi* (Tukulor); *tambo* (Mandinka); *tãbi* (Malinke); *e kapat* (Mandyak); *ratlan* (Wolof). GAMBIA: *tambo* (Mandinka). GUINEA–BISSAU: *quitite* (Balanta); *batanou* (Biafada); *mantampa de sera* (Crioulo, Upper Guinea); *tambem* (Fulfulde-Pulaar); *tambo* (Mandinka); *ecapate* (Mandyak); *quito* (Papel). GUINEA: *tambo* (Mandinka); *tâbi* (Malinke). SIERRA LEONE: *lumboinyo-lando* (Kisi); *kanga-mese* (Kono); *tambe* (Loko); *tambi* (Maninka); *tamba* (def. *tembui*) (Mende); *tambi* (Susu); *ra-gbet* (Themne); *tambu-na* (Yalunka). LIBERIA: *kpa kala* (Mano). CÔTE D'IVOIRE: *ailé-mlé* (Anyin); *gapapa* (Godié). GHANA: *demm̅eré* (Twi, also trade name); *néné,* (Akan); *ayeka* (Anufo); *ayeka* (Sehwi); *keteku* (Éwé); *ayeké* (Nzema). BENIN: *akete* (Defi); *dekun wéwé* (Gun-Gbe). NIGERIA: *erogbo, erugbo* (Edo); *ekwe-oji, iye* (Igbo); *apié* (the plant itself, or the cane rope made from it) (Ijo-Izon), *bwálàm* (a cane)

(Pero); *erogbo, erugbo* (Yoruba). CAMEROON: *nding* (Bulu). EQUATORIAL GUINEA: *nzing* (Fang). CENTRAL AFRICAN REPUBLIC: *bioh* (Banda-Yangere). DR CONGO: *kpude* (Zande); *ma-ndakele* (Ngbaka-Ma'bo); *ikonga* (Lombo); *babio* (Mongo-Nkundu); *lekwe* (BaMbuti). UGANDA: *bi-lekwe* (Amba).

Uses

The stems of *Calamus deërratus* are not as widely used as those of other species. Throughout the range of this species, the stems are considered inferior to those of other species of rattan, for they do not attain as great a diameter and hence are not as robust as the stems of many of the larger species of *Laccosperma*. Accordingly, they are often too thick and rather inflexible for weaving, for which a number of species of *Eremospatha* are preferred. However, the use of *C. deërratus* increases in the absence of other rattan canes (Profizi 1986; Morakinyo 1995; Tenati in press), and, aside from a wide range of construction and weaving applications, *C. deërratus* also contributes significantly to other household needs. For example, in Ghana the whole leaves are often used for thatching (Abbiw 1990). In Sierra Leone the apical bud (palm heart) is eaten by the Mende people as a dietary supplement, and they are in Ghana (Dalziel 1937; Irvine 1952, 1961). The young shoots are roasted whole and eaten in Ghana (Abbiw 1990). In lower Casamance of Senegal the leaves are grilled over a fire and then macerated; the liquid is drunk to promote weight loss in stout persons (Berhaut 1988), and the same preparation is also given for edema caused by vitamin deficiencies (Berhaut 1988). Ash from burning the roots is used as a kitchen salt in Guinea (Portères n.d., cited in Burkhill 1997) and Ghana (Abbiw 1990). In Sierra Leone the rind is twisted into small sponges and used to clean pans (Irvine 1961). Williamson and Timitimi (1983) state that the leaf sheath may be peeled off and twisted to make a rough rope.

Throughout West Africa in particular, *Calamus deërratus* has a wide range of uses for furniture construction and basketry to the extent that in the dry savannas of northern Nigeria, in the absence of other species, canes of *C. deërratus* were formerly supplied to prisons for workshop activities and the rehabilitation of prisoners (Burkhill 1997). Whole canes were widely used for fences and house building in other areas of Nigeria (Unwin 1920), as well as in Ghana (Abbiw 1990). Elsewhere in Africa, in Zambia and Uganda, in the absence of large-diameter canes, rattan furniture frames are made of two or three stems of *C. deërratus* joined together (Tenati in press). When the canes are split and woven, they have a multiplicity of purposes, such as the fabrication of "tie-ties" for staking yams (Morakinyo 1995). They are woven and used for fencing in Ghana (Abbiw 1990) and as binding materials for hut making and to tie thatch (Dalziel 1937; Berhaut 1988). In addition, both whole and split stems are used to make hammock suspension bridges (Dalziel 1937; Abbiw 1990) and foot loops for climbing palm trees (Dalziel 1937). More everyday uses include the fabrication of fish traps and weirs, kola baskets, bow strings, household matting and screens, and chair seats (Dalziel 1937; Irvine 1961; Abbiw 1990). In Nigeria the split stems were formerly used to tie timber trees together before they were floated downriver for sale (Sylvic Sap 1935), and in the continental region of Equatorial Guinea they were formerly used in the fabrication of temporary market baskets (Guinea-López 1946), although this is not the case today (Sunderland 1998).

References

SENEGAL: Dalziel 1937; Berhaut 1988; Burkhill 1997. GAMBIA: Anderson 131 (herb. MO!). GUINEA–BISSAU: Espirito Santo 1963 ex auctt., cited in Burkhill 1997. GUINEA: Dalziel 1937. SIERRA LEONE: Dalziel 1937; Deighton vocabulary 1957, cited in Burkhill 1997; Scott-Elliot 4738, 5121 (herb. K!); Thomas 2753 (herb. K!). LIBERIA: Dalziel 1937. CÔTE D'IVOIRE: ECOSYN 1999. GHANA: Dalziel 1937; Irvine 1961; Burkhill 1997; Vigne 1868 (herb. K!); Chipp 127 (herb. KUM!). BENIN: Profizi 1986; Aufsess 425 (herb. K!). NIGERIA: Dalziel 1937; Dunstan s.n. (herb. K!); Russell unpubl. notes; Williamson & Timitimi 1983; Bennet (herb. K!); Unwin 223 (herb. K!). CAMEROON: Mildbraed 1913. EQUATORIAL GUINEA: Guinea-López 1946; Tessmann 6 (herb. FI!); Sunderland 1998. CENTRAL AFRICAN REPUBLIC: Harris 820 (herb. K!). DR CONGO: Robyns & Tournay 1955; Evrard 1876 (herb. BR!); Louis 15541 (herb. BR!); Germain 210 (herb. BR!); Leonard 832 (herb. BR!); Troupin 296 (herb. BR!); Hart 633 (herb. MO!). UGANDA: Makombo s.n. (herb. K!).

Eremospatha barendii Sunderland

An uncommon species characterized by large, conspicuous bracts on the inflorescence. The dry, splitting ocrea is also an unusual character in this genus, shared only by the very distinct *Eremospatha wendlandiana.* Known only from one collection in a logging concession near Lolodorf, Cameroon.

Vernacular names

None recorded.

Uses

None recorded.

Eremospatha cabrae (De Wild. & Th. Dur.) De Wild.

A slender to moderate rattan species distinguished from the closely related *Eremospatha hookeri* by more rounded leaflets and dense pilose covering on the inflorescence. Restricted to Gabon, southward to Angola, and across to the lowland forests of the wider Congo Basin. Relatively common where it occurs, particularly in swamp forest and along rivers.

Vernacular names

GABON: *osono* (Tsogo); *osono* (Pinji); *ozono* (Myene); *li-bamba* (Vili); *nkolé* (Kélé); *nkolu* (Seki); *du-bamba* (Barama); *du-bamba* (Lumbu); *ivéta* (Duma); *iló-lóngo* (Kota); *u-lóngo* (Benga); *lé-mbumu* (Ndumu); *nlong* (Fang). DR CONGO: *li-findo* (Lombo);

lu-bambi (Kituba); *e-safa* (Mongo-Nkundu); *ki-sakata* (Kete). ANGOLA: *m'bamba* (Mbundu-Luanda).

Uses

In Gabon the split stems of this species are widely used the fabrication of temporary market baskets (Raponda-Walker & Sillans 1961). In the Congo Republic the complete stems of this species are used for furniture frames or are split and used for weaving, particularly by forest-dwellers (Profizi & Makita-Madzou 1996). The stems are also employed in the construction of cane bridges (Profizi & Makita-Madzou 1996). However, in the absence of *Eremospatha haullevilleana*, the split stems and epidermis of *E. cabrae* are used for weaving and basketry in DR Congo. In Angola, particularly in the absence of other species, the whole stems of *E. cabrae* were formerly used in the fabrication of armchairs that were "much appreciated by the colonists" and many of which were exported to Portugal (Chevalier 1936). Another main use of this species is utilization of the leaf sheath as a toothbrush by the Lombo of DR Congo.

References

GABON: Raponda-Walker & Sillans 1961. DR CONGO: Compère 2182 (herb. BR!); Germain 326 (herb. BR!); Hulstaert 1417 (herb. BR!); Jans 655 (herb. BR!); Kuasa (herb. BR!); Leonard 929 (herb. BR!); Louis 16797, 3804, 5656, 15169 (herb. BR!); Toussaint 2331 (herb. BR!). ANGOLA: Chevalier 1936.

Eremospatha cuspidata (G. Mann & H. Wendl.) H. Wendl.

A relatively rare species of rattan, characterized by the conspicuously apiculate leaflet apex. Restricted to the coastal forests of the Congo Basin; in open areas, especially on deep, white-sand savannas of Equatorial Guinea, Gabon, and DR Congo.

Vernacular names

EQUATORIAL GUINEA: *ndera* (Fang).

Uses

The split stems are used for light basketry and weaving, particularly in the absence of other species.

References

EQUATORIAL GUINEA: Guinea-López 1946; Sunderland 1998.

Eremospatha dransfieldii sp. nov.

Previously assigned to *Eremospatha hookeri*, this is a distinct taxon more closely related to both *E. laurentii* and *E. wendlandiana*. It is a robust species characterized by a conspicuous knee and a plasticity on leaflet form not previously recognized in the rattans of Africa. Restricted to the Upper Guinea forests from Sierra Leone to eastern Nigeria.

Vernacular names

SIERRA LEONE: *balu* (Kono); *mbalu* (def. *-ui*) (Mende); *ra-thamp* (Themne). NIGERIA: *epa-emele* (Yoruba); *inima ború* (Ijo-Izon); GHANA: Mfia (Twi).

Uses

The stem epidermis is easily peeled and is used in Sierra Leone by the Mende as a rope and for making baskets (Burkill 1997). The split cane itself is also a strong binding material (Small herb. K!). Used for basketry and furniture framework and traded, particularly in Ghana.

References

SIERRA LEONE: Deighton (vocabulary 1956, cited in Burkill 1997); Small 832 (herb. K!); Deighton 2591 (herb. K!). GHANA: Sunderland unpubl. notes. NIGERIA: Ainslie 1937; Okigbo 1980, cited in Burkill 1997); Williamson & Timitimi 1983.

Eremospatha haullevilleana De Wild.

A common and highly variable species across its range. This is a slender to moderate species of cane characterized by the long, somewhat striate, ocrea and the almost elongate-rhomboid leaflets. A species restricted to the lowland forests of the Congo Basin with outliers in southern Cameroon, Zambia, Tanzania, and Uganda.

Vernacular names

CENTRAL AFRICAN REPUBLIC: *pongbo* (Ngombe). CONGO: *mbaama* (Téké). DR CONGO: *li-findo* (Lombo); *mbowe* (Zande); *lu-popi* ((Nandi)); *n'kele* (Bangala); *m'bio* (Bangi); *lo-koli* (Kele); *ke-kele* (Lingala); *lu-kodi* (Luba-Shari); *lu-busi* (Tembo); *lu-bubi* (Lega-Mwenga); *yofoko* (Mungo-Nkundu); *lo-keko* (Lusengo); *kodi* (Luba-Kasai); *tukpuru* (Bhele). UGANDA: *bibbobbi* (Amba); *enga* (Luganda). TANZANIA: *urugage* (Ha).

Uses

The preferred species for basketry, weaving, and furniture manufacture throughout its range, even more so than *Eremospatha macrocarpa*. Profizi and Makita-Madzou (1996) state that this species is the best for weaving, for it possesses important qualities of strength, durability, and resistance to insect attack. Because of these qualities, the Téké Tsaaya of Congo-Brazzaville use the species widely, and it is considered second in importance only to the oil palm (*Elaeis guineensis*) (Profizi & Makita-Madzou 1996). The stems are used whole for a wide range of products, including cables for cane bridges, furniture framework, and building frames. The split stems are used in fabricating fish traps, noose-type snares to catch small terrestrial mammals, and handrails of river bridges made from the felled stems of *Musanga cecropioides*, as well as for a wide array of baskets. In the Bambama district, the apical bud of this species is widely consumed. During certain traditional rites, the stem of this species is respected as providing protection for the bearer. When the village expects malevolent spirits to visit it, the stems of this species are placed through the settlement and adorned with powerful fetishes (Profizi & Makita-Madzou 1996).

In the DR Congo the fruits are used by the Bangi for decoration, particularly in the manufacture of traditional collars (Dewevre herb. BR!). The Lombo use the acanthophylls as fish hooks (Louis herb. BR!) and the sap (the watery exudate collected when the stems are cut) as an abortifacient (Staner & Boutique 1937).

In Tanzania the split stems are used by the Ha for sewing bark (probably *Xylopia* sp.) beehives and food containers (Proctor herb. K!).

References

CENTRAL AFRICAN REPUBLIC: Carroll 115 (herb. MO!); Harris 2360 (herb. MO!). CONGO: Profizi & Makita-Madzou 1996. DR CONGO: Couteaux 471 (herb. BR!); Dewevre 581 (herb. BR!); Gathy 1638 (herb. BR!); Gerard 1432 (herb. BR!); Gillet 167 (herb. BR!); Gutzwiller 539 (herb. BR!); Herman 2138 (herb. BR!); Hulstaert 1418 (herb. BR!); Kitembo 60 (herb. BR!); Leclercq 736 (herb. BR!); Leonard 932, 933, 936, 1138 (herb. BR!); Liben 2603 (herb. BR!); Louis 3395, 1970, 772, 847, 3638, 11850, 9731, 12106, 9560, 8106, 7671, 9420, 16775 (herb. BR!); Nannan 117 (herb. BR!); Terashima 94 (herb. BR!); Troupin 9162, 2658 (herb. BR!); Gerard 1432 (herb. BR!); Minga in press. UGANDA: Makombo s.n. (herb. K!); Kabuya 1988. TANZANIA: Proctor 369 (herb. K!).

Eremospatha hookeri (G. Mann & H. Wendl.) H. Wendl.

A variable species with rhomboid to somewhat rounded leaflets. Distributed from Liberia to Gabon.

Vernacular names

NIGERIA: *itomi* (Ekit). CAMEROON: *ki-yince* (Balundu-Bima); *mbunden* (Bakundu-Balue). EQUATORIAL GUINEA: *alua-nlong* (Fang): GABON: *gigorula* (Sira).

Uses

In Nigeria the Ekit use the base of the leaf sheath as a chew stick (Morakinyo herb. K!).

References

SIERRA LEONE: NIGERIA: Morakinyo 1005 (herb. K!); Ayewoh 3852 (herb. K!). CAMEROON: Thomas et al. 1989; Sunderland 2001. EQUATORIAL GUINEA: Sunderland 1998. GABON: le Testu s.n. (herb. BR!).

Eremospatha laurentii De Wild.

This species is easily recognizable by the conspicuous knee, the straplike, linear, lowermost leaflets that are often highly reflexed and tightly clasping the stem and the entire leaflet apex. Found only in the Congo Basin, south of the Sanaga River in Cameroon but relatively common where it occurs.

Vernacular names

SIERRA LEONE: *bongei* (Mende). CAMEROON: *kpakpa* (Ewondo). CENTRAL AFRICAN REPUBLIC: *bo-kondi* (Banda-Yangere). EQUATORIAL GUINEA: *ebuat* (Fang). DR CONGO: *bo-ngale* (Mongo-Nkundu); *ikonga* (Lombo); *nkelele mo-none* (Lingala); *nkoli* (Bali).

Uses

This species is rarely used for furniture manufacture or basketry because the cane is known to be of somewhat poor quality.

References

SIERRA LEONE: Deighton 4117 (herb. K!). CAMEROON: Letouzey 11796 (herb. YA!). CENTRAL AFRICAN REPUBLIC: Harris & Fay 459 (herb. BR!). EQUATORIAL GUINEA: Sunderland 1998. DR CONGO: Leonard 55, 816, 980 (herb. BR!); Louis 15925, 15944 (herb. BR!); Dubois 912 (herb. BR!).

Eremospatha macrocarpa (G. Mann & H. Wendl.) H. Wendl.

This species is easily recognizable by the lack of a knee, the linear-lanceolate leaflets, with a narrowly praemorse apex. The leaves are bifid when juvenile and often accompanied by a distinct linear wrinkle on the ocrea. A very common and widespread species distributed from Senegal in West Africa to the western Congo Basin.

Vernacular names

SIERRA LEONE: *penden* (Kissi); *balu* (Kono); *mbalu* (Loko); *mbalu, koto mbalu* = juvenile (Mende); *ra-thamp* (Themne). LIBERIA: *bĕlĕ de bĕlĕ* (Mano). CÔTE D'IVOIRE: *ailè-mlé* (Anyin). GHANA: *mfia* (Akan-Asanti); *néné* (Nzima). BENIN: *dekon* (Defi); *dekun vovo* (Gun-Gbe). NIGERIA: *íkan* (Edo); *odu-aňã (Igbo); bórú* (Ijo-Izon); *ukan* (Yoruba); *ekakieri* = male (i.e. with no fruits), *irrumka* = female (with fruits) (Ekit); *iro* (Esan). CAMEROON: *filet* (Trade); *cane rope* (Pidgin); *echié* (Denya); *nlong* (indef.), *melong* (def.) (Bulu); *bana ndongo* = young cane (*bana* = child) (Balundu-Bima); *nloun* (Baasa). EQUATORIAL GUINEA: *nlong* (indef.), *mi-long* (def.) = juvenile stems, *ongam* = adult (Fang). GABON: *ke-gèma* (Lumbu); *nyèvila* (Sira); *ongam* (Fang); *ndètèse* (Kota); *iganga-tsungu* (Punu); *songu* (Vumbu); *tongo* (Tsogo); *mbubi* (Ndumu).

Uses

This species is reputed to be the best source of cane in Africa (Unwin 1920), and its quality is considered comparable to the small-diameter canes of Southeast Asia (Mildbraed 1913; Irvine 1961). *Eremospatha macrocarpa* is widely used for furniture construction, basketry, weaving, and tying wherever it occurs: Sierra Leone (Burkhill 1997), Côte d'Ivoire (ECOSYN 1999), Ghana (Irvine 1961; Abbiw 1990), Benin (Profizi 1986), Nigeria (Morakinyo 1994; Tuley 1995), Cameroon (Defo 1997, 1999; Sunderland 1999b, 2001), Equatorial Guinea (Guinea-López 1946; Sunderland 1998), and Gabon (Raponda-Walker & Sillans 1961). However, *E. haullevilleana* seems to be the preferred species for basketry and weaving in the DR Congo.

In Côte d'Ivoire the Anyi utilize the split stems for baskets, cages for transporting chickens to market, and bows for playing stringed instruments (ECOSYN 1999). In Nigeria the split stems of this species were formerly used to tie log rafts together for floating downstream to market (Sylvic Sap 1935), as well as for making a kind of string that the Yoruba used for tying lath pieces together in house building and for tying cloth in making coverings for canoes (Unwin 1920).

The powdered root is taken as a medicine for the treatment of syphilis by the Akan-Asanti in Ghana (Irvine 1961; Abbiw 1990) and the Yoruba of Nigeria (Ainslie 1937). The long, flexible stems of this species make it ideal for the construction of cane bridges, commonly encountered in Ghana (Irvine 1961) and among the many indigenous groups in Manyu Division, Sud-Ouest Province, Cameroon (Sunderland 2001).

References

SIERRA LEONE: Deighton vocabulary 1956, cited in Burkhill 1997. LIBERIA: Dalziel 1937. CÔTE D'IVOIRE: ECOSYN 1999. GHANA: Irvine 1961; Burkhill 1997; Sunderland 2001. BENIN: Profizi 1986. NIGERIA: Dalziel 1937; Williamson & Timitimi 1983; Russell unpubl. notes, cited in Burkhill 1997; Morakinyo 1003 (herb. K!); Nwambin 6942 (herb, K!). CAMEROON: Sunderland 2001. EQUATORIAL GUINEA: Guinea-López 1946; Sunderland 1998. GABON: Raponda-Walker & Sillans 1961.

Eremospatha quinquecostulata Becc.

A slender species of rattan with leaflets clustered in groups of 4 to 6 and conspicuously 5-costulate. Distributed from southwestern Cameroon to Gabon.

Vernacular names

CAMEROON: *calumé-echié* (Denya). GABON: *di-bula* (Sira).

Uses

Not used in Cameroon due to the presence of other, more desirable species of rattan. In Gabon, however, the stem is split and employed for basic weaving, particularly in the absence of other species.

References

CAMEROON: Sunderland 2001. GABON: le Testu s.n. (herb. BR!).

Eremospatha tessmanniana Becc.

A species of rattan characterized by rather glaucous gray-green leaflets and conspicuous forward- and rear-facing spines on leaflet margins. Distributed from Rio Muni, Equatorial Guinea, to the Takamanda region of the Cameroon-Nigeria border.

Vernacular names

CAMEROON: *calumé echié* (Denya). EQUATORIAL GUINEA: *ongam-akot* (Fang).

Uses

None recorded.

References

CAMEROON: Sunderland 2001. EQUATORIAL GUINEA: Tessmann 4 (herb. FI!); Guinea-López 1946.

Eremospatha wendlandiana Dammer ex Becc.

A highly distinctive rattan species with very triangular rhomboid leaflets and a characteristically tattering ocrea. The comparatively large size of the inflorescence and fruits is also unequaled in this genus. Distributed from southeastern Nigeria to Gabon, in the coastal Atlantic forest. This species is extremely common throughout its range, particularly in disturbed forest.

Vernacular names

NIGERIA: *eghounka* (Ekit). CAMEROON: *cane basket* (Pidgin); *mua-echié* (Denya). EQUATORIAL GUINEA: *akot* (Fang). GABON: *égoo* (Tsogo); *ngundju* (Punu); *ngundju* (Vumbu). CONGO: *ma-bulu* (Téké).

Uses

This species is not of much value commercially and is rarely traded. However, in the Cross River area of Nigeria the split stems are used to tie bamboo and stick-framed houses prior to plastering with clay, and the stem epidermis is used for tying yams (Morakinyo herb. K!). In Cameroon the split stems are also used for coarse basketry, particularly for temporary baskets used to transport produce to market and then discarded. The Balundu-Bima also used the base of the leaf sheath as a chew stick (Sunderland 2001). In the Congo the Téké use the inner stem for basketry and furniture construction and also consume the apex of the emerging stems (Profizi & Makita-Madzou 1996).

References

NIGERIA: Morakinyo 1001 (herb. K!). CAMEROON: Sunderland 2001; Sylvanus s.n. (herb. K!). EQUATORIAL GUINEA: Sunderland 1998. GABON: Raponda-Walker & Sillans 1961. CONGO: Profizi & Makita-Madzou 1996.

Laccosperma acutiflorum (Becc.) J. Dransf.

The largest of all the rattan species occurring in Africa, this species was previously assigned to *Laccosperma secundiflorum.* It has a distinct yellowish appearance, with the stems, petiole, rachis, and cirrus all being of that color. In mature individuals the ocrea is short and truncate, and the medium to dark green leaflets are linear-lanceolate, although they are shorter and held somewhat more horizontally on the rachis, rather than being conspicuously pendulous. Distributed from Sierra Leone to Cameroon, south to Gabon and DR Congo.

Vernacular names

NIGERIA: *ukpekpe* (Ekit). CAMEROON: *giant cane* (Pidgin). EQUATORIAL GUINEA: *ekwass* (Fang).

Uses

Despite its size, this species is reported to possess very poor quality cane, so it is rarely used.

References

CAMEROON: Sunderland 2001. EQUATORIAL GUINEA: Sunderland 1998.

Laccosperma laeve (G. Mann & H. Wendl.) H. Wendl.

This species possesses an erect inflorescence and ovoid fruits and seeds, the latter being rather smooth and flattened on one side. The absence of spines on the leaflet margin also distinguishes this species from *Laccosperma opacum.* Distributed from Liberia to Cameroon and south to Equatorial Guinea and Gabon in high forest.

Vernacular names

CÔTE D'IVOIRE: *ailé-mla* (Anyin). GHANA: *nguni* (Wasa); *tenan muhunu* = it lives in the world for nothing (Twi). NIGERIA: *itunibia* (Ekit). CAMEROON: *ge-nomé-echié* = slave to cane rope (Denya). CENTRAL AFRICAN REPUBLIC: *gao* (Banda-Yangeri). EQUATORIAL GUINEA: *ndele* (Fang). GABON: *munyengi* (Sira); *tèkè* (Tsogo).

Uses

The poor cane quality due to the extensive branching habit makes this species of little use. However, in Gabon *Laccosperma laeve* is sometimes used to make rope (Raponda-Walker & Sillans 1961). In the Central African Republic the BaBinga roast the roots of this species on axes and then eat the roots to improve virility (Carroll herb. K!).

References

CÔTE D'IVOIRE: ECOSYN 1999. GHANA: Kinlock 3237 (herb. KUM!). NIGERIA: Morakinyo 1001 (herb. K!). CAMEROON: Sunderland 2001. EQUATORIAL GUINEA: Sunderland 1998. CENTRAL AFRICAN REPUBLIC: Carroll 33 (herb. K!). GABON: Raponda-Walker & Sillans 1961.

Laccosperma opacum (G. Mann & H. Wendl.) Drude

The characteristic globose fruit and warty seed of this species, along with a somewhat wavy, as opposed to erect, inflorescence, distinguish it from *Laccosperma laeve* when fertile. The presence of spines on the leaflet margin of *L. opacum* also separates the two species in the absence of fertile material. Distributed from Liberia and Côte d'Ivoire to Cameroon, Gabon, and the Congo Basin, to eastern DR Congo, most commonly under a forest canopy.

Vernacular names

GHANA: *eholobaka* (Nzema); *sayai* (Akan-Asanti); *edem* (Kwawu). NIGERIA: *abu* (Edo); *ěkwě ǒya* = cane for tie-tie (Igbo). CAMEROON: *liko ko'ko* = close to cane (Mokpwe); *ge-nomé-echié* = slave to cane rope (Denyạ). EQUATORIAL GUINEA: *npue-nkan* (Fang). GABON: *ibulu* (Myene); *di-bulu* (Sira); *di-bulu* (Lumbu); *abulo* (Kele); *éboa* (Tsogo) *ulóngó-mwa-iki* (Benga). CONGO: *kimbana ki mukaana* (Téké).

Uses

The tendency for this species to produce aerial branches affects the cane quality significantly, and *Laccosperma opacum* is not an important source of cane. However, in Ghana the stem is sometimes split and used as binding material and for making the Akan-Asanti baskets, *kenten*. In Gabon forest workers drink the potable sap of this species, particularly during the long dry season, when many small streams dry up (Raponda-Walker & Sillans 1961). In the Congo the Téké eat the apical buds of this species; these are prepared by roasting the whole stems over, or next to, a fire (Profizi & Makita-Madzou 1996).

References

GHANA: Irvine 2300, 502 (herb. GC!); Irvine 1961. NIGERIA: Dalziel 1937; Burkhill 1997. CAMEROON: Cheek 5591 (herb. K!); Sunderland 2001. EQUATORIAL GUINEA:

Sunderland 1998. GABON: Raponda-Walker & Sillans 1961. CONGO: Profizi & Makita-Madzou 1996.

Laccosperma robustum (Burr.) J. Dransf.

A moderately robust rattan. The highly pendulous, regularly arranged, glaucous, blue-green leaflets of this species make it very distinctive. The stems are also often profusely armed; sheath with a very long, dry ocrea, often reflexed and longitudinally splitting. Distributed from southeastern Nigeria to the Congo Basin.

Vernacular names

CAMEROON: *eka* (Ewondo); *nkan, aka* = cleaned cane (Bulu); *dikah* (indef.), *mekah* (def.) (Bakundu-Balue); *gekwiya* (Denya); *makak* (Trade). CENTRAL AFRICAN REPUBLIC: *gao* (Banda-Yangere). EQUATORIAL GUINEA: *nkan, aka* = cleaned cane (Fang). GABON: *asperge* (nom forestier). DR CONGO: *ekpale-ekpale* (Bwa): *li-sele* (Lombo); *nkao* (Ngbaka-Ma'bo); *ikoonga* (Lombo).

Uses

This species is widely used throughout its range for its high-quality cane, which, in the main, is used whole. Previously assigned under the name, *Laccosperma secundiflorum*, the clear morphological and ecological differences between the species are marked, and as a result most local people distinguish between them in their folk classifications. The cane is used mainly for furniture and house frames, forms the basis for a thriving cottage industry, and is widely traded (Morakinyo 1994; Defo 1997; Sunderland 1998, 1999a, 1999b, 2001; Defo & Sunderland 1999).

The palm heart is eaten widely, from Cameroon through to Gabon (Rapona-Walker & Sillans 1961; Sunderland 2001) and contributes to the diets of many forest-dwellers. The young leaves are eaten in stews by the Fang in Rio Muni, Equatorial Guinea (Guinea-López 1946).

References

NIGERIA: Morakinyo 1000 (herb. K!). CAMEROON: Mildbraed 1913; Hédin 1929; Letouzey 8479 (herb. YA!); Thomas et al. 1989; Sunderland 2001. CENTRAL AFRICAN REPUBLIC: Fay 8236, 8254 (herb. MO!). EQUATORIAL GUINEA: Tessmann 2 (herb. FI!); Sunderland 1998. GABON: Wilks 1486 (herb. MO!). DR CONGO: Gerard 3218, 3933, 4951 (herb. BR!); Germain 8278 (herb. BR!); Leonard 1671 (herb. BR!).

Laccosperma secundiflorum (P. Beauv.) Küntze

Characterized by deep green, long, sigmoid leaflets and a long petiole. Shade-tolerant and often found under a forest canopy; distributed from Senegal to Cameroon, south to DR Congo.

Vernacular names

SENEGAL: *ka-likut* (Jola-Fogny). GUINEA–BISSAU: *tambem-hadje* (Fulfulde-Pulaar); *tambendjom* (indef.), *tambendjom-ô* (def.) (Mandinka). SIERRA LEONE: *lumboinyo-piando* (Kisi); *kangane* (Kono); *kafo* (Loko); *kavo* (def. *kavui*) (Mende); *ka-gbesu* = whole stems, *e-gbak* = leafless part of the stem (Themne). CÔTE D'IVOIRE: *kumh* (Attié); *agué* (Ebrié); *djoho, djolo* (Krumen); *ahika* (Anyin); *gblé* (Godié). GHANA: *willow* (Trade); *ayié* (Akan-Asanti); *ayikĕ* = large rattan (Nzema). BENIN: *kpanon* (Defi); *kpacha* (Gun-Gbe). NIGERIA: *ohwara* (Urhobo); *okankan* = whole cane, *ukwen* = when split (Edo); *òbóng* (Efik); *ukpé* = cane rope made of this species (Ijo-Izon); *iga* (Ekpeye); *añà* (Igbo); *epe-nla, ikan-ikó* = hook (Yoruba). CAMEROON: *ka-kawa* (Baka): *ekwos* (Balundu-Bima); *nde-gekwiya* (Denya). GABON: *nkan* (Fang); *nkanda* (Kélé); *ikandji* (Kota); *okana* (Ndumu); *mokangé* (Pinji); *mokangé* (Tsogo); *mukanda* (Sira); *mukanda* (Duma); *mukanda* (Lumbu); *nkogu* (Myene); *nkanyi* (Seki). CONGO: *mukaana a nguomo* (Téké). DR CONGO: *ma-kauw, bo-kauw* (def.) (Lingala); *bo-nganga* (Mongo-Nkundu); *nkau* (Kongo). ANGOLA: *mi-cau* (Mbundu-Luanda).

Uses

This species is the most desired large-diameter cane species in West Africa and, along with *Laccosperma robustum*, one of the two main commercial species in Central Africa (Hédin 1929; Irvine 1961; Berhaut 1988; Profizi 1986; Abbiw 1990; Tuley 1995; Burkhill 1997; Sunderland 1999a, 1999b, 2001).

Although the stems are mostly used whole throughout West Africa, the split stems are also used for binding house frames together and for constructing fish traps, baskets, and other woven products (Berhaut 1988; Profizi 1986; Abbiw 1990; Tuley 1995; Burkhill 1997). The leaves were formerly used by the Mende of Sierra Leone (Dalziel 1937), the Nzeme of Ghana (Abbiw 1990), and the Edo of Nigeria (Burkhill 1997) as a thatching material but have been replaced by metal roofing (Abbiw 1990; Davies & Richards 1991). In Côte d'Ivoire the leaves are also used for thatching, while the split stems are used for basketry and in the fabrication of traditional drums (securing the goatskin onto the wooden frame) (ECOSYN 1999). In the Congo the Téké use the split stems of this species as a mooring line for fastening their canoes to the riverbank. In much the same way, these split stems can also be suspended across a river to help with crossing on foot. The plant itself has significant symbolic value and is reputed to conjure up ghosts (Profizi & Makita-Madzou 1996).

The young emerging shoots are often cut and roasted whole before the inner soft pith is eaten (Raponda-Walker & Sillans 1961; Berhaut 1988; Abbiw 1990) or boiled and then fried in butter to remove the bitterness (Chevalier 1934). However, the Balundu-Bima of Cameroon state that consumption of the young shoots can lead to impotence, so males avoid it (Mallson pers. comm.).

The stem apex is widely eaten and is thought to contain an active heart stimulant that promotes a sense of well-being akin to that caused by kola nuts (Jordan herb. K!). Holland (1922) records how many local people in Nigeria transported a portion of the upper stem with them when they entered the forest for long periods to harvest wild rubber; the whole stems were roasted on a fire and the soft pith then removed and eaten. The cut stems also provide a potable sap drunk by forest travelers (Raponda-Walker & Sillans 1961; Berhaut 1988).

A tea made from the young shoots is used as an anthelmintic by the Akan-Asanti of Ghana (Abbiw 1990); a tea made from the young leaves, as a vermifuge in Gabon (Raponda-Walker & Sillans 1961). The Lombo from the Yangambi region of DR Congo use the sap from the cut stems of this species, mixed with the bark *Hua gabonii, Quassia africana,* and *Pterocarpus soyauxii,* to make a tea that is then used to treat dysentery (Staner & Boutique 1937).

References

SENEGAL: Berhaut 1988; Jordan 2064 (herb. K!). GUINEA–BISSAU: Burkhill 1997. SIERRA LEONE: Burkhill 1997. CÔTE D'IVOIRE: Adanohoun & Aké Assi 1972, cited in Burkhill 1997; Bouquet & Debray 1974; ECOSYN 1999. GHANA: Dalziel 1937; Sunderland 2001. BENIN: Profizi 1986; Aufsess (herb. K!). NIGERIA: Dalziel 1937; Otedoh 7251 (herb. K!); Burkhill 1997. CAMEROON: Letouzey 10605 (herb. YA!); Thomas et al. 1989. GABON: Raponda-Walker & Sillans 1961. CONGO: Profizi & Makita-Madzou 1996. DR CONGO: Leonard 815 (herb. BR!); Minga in press. ANGOLA: Chevalier 1936.

Oncocalamus macrospathus Burr.

A relatively robust cane with bright yellow rachillae, 3 central pistillate flowers, and a smooth seed. Distributed from southern Cameroon to northern Angola.

Vernacular name

CAMEROON: *eboti* (Ewondo).

Uses

None recorded.

References

CAMEROON: Letouzey 11889 (herb. YA!).

Oncocalamus mannii (H. Wendl.) H. Wendl.

A slender to moderately sized rattan cane characterized by the variable number of pistillate flowers in the cluster (1–7) and the conspicuously warty fruit. Distributed from southern Cameroon to Equatorial Guinea and Gabon.

Vernacular names

CAMEROON: *mfop n'lon* (Bulu). EQUATORIAL GUINEA: *asa-nlong* (juvenile), *ndoro* (adult) (Fang). CONGO: *mituo* (Téké).

Uses

The cane of this species is poor in quality, for it is rather inflexible and prone to breakage. However, particularly in the absence of other species, *Oncocalamus mannii* can be employed for coarse weaving. In the Congo the stem epidermis is used by the Téké for manufacturing coarse baskets and other local woven products. In Equatorial Guinea the juvenile stems are sometimes used for weaving in the same manner as the young stems of *Eremospatha macrocarpa*, which they resemble; hence the same name for the juvenile form (Sunderland 1998).

References

CAMEROON: Hédin 1929; van Gemerden s.n. (herb. K!). EQUATORIAL GUINEA: Sunderland 1998. CONGO: Profizi & Makita-Madzou 1996.

Oncocalamus tuleyi Sunderland

A robust cane with relatively base sheaths, bright crimson rachillae, three central pistillate flowers, and a smooth seed. A species with a limited distribution from southeastern Nigeria to southwestern Cameroon.

Vernacular names

NIGERIA: *iboh* (Ekit). CAMEROON: *madame* (Trade/Pidgin); *mo'ap* (Balundu-Bima); *edju* (Bakundu-Balue); *moa-echié* (Denya).

Uses

This cane is not used commercially at all due to its poor quality (Morakinyo 1994; Sunderland pers. obs.). It is rather weak, inflexible, and prone to breakage on bending. However, the base of the leaf sheath is often used by the indigenous communities of southwestern Cameroon as a chew stick (Thomas et al. 1989; Sunderland & Tchouto 1999; Sunderland 2001). In Nigeria the stem epidermis is often used for tying yams to their climbing poles ("yam-barns") because it rots fairly quickly and does not constrict the developing climbing stems (Morakinyo 1994).

References

NIGERIA: Morakinyo 1002 (herb. K!). CAMEROON: Thomas et al. 1989; Sunderland 2001.

Oncocalamus wrightianus Hutch.

A slender rattan with somewhat sigmoid leaflets. Distributed from Benin to the Niger Delta.

Vernacular names

BENIN: *hofle* (Defi); *gbe-dekun* (Gun-Gbe). NIGERIA: *akwal·* (Igbo); *pankéré* (Yoruba).

Uses

The Igbo in Nigeria split the stem to make tying material of different sorts: coarse cordage (*ekwelē* or *akwalà*), finer twine (*udo*), and string or thread (*elili*) (Jeffreys 1960). A specialized twine called *akwalà anya* is made for tying bulky packages (Jeffreys 1960) and for binding hut frames together. The leaf sheath base is also used by the Igbo as a chew stick (Dalziel 1937).

References

BENIN: Profizi 1986; Aufsess 430 (herb. K!). NIGERIA: Holland 1922; Dalziel 1937; Jeffreys 1960.

7

Crafts and Conservation: The Ecological Footprint of International Markets on an African Resource

A. B. Cunningham and S. K. Choge

Introduction

The concept of "ecological footprints" was developed by Rees (1992) to illustrate the large ecosystem area on which urban centers depend. The same concept can be applied at a national scale. The Netherlands, for example, requires 15 times its landmass to maintain the population it supports (Moffatt 1996). Despite its limitations (Moffatt 2000), the ecological footprint concept provides a useful way of illustrating the geographical scale and impacts of trade in plant products in response to international demand. Remote as many craft-production areas are, international market forces extend into them. The far-reaching effects of this trade are an important issue of which, unfortunately, many importers, retailers, and tourists from Europe, the United States, Canada, and Japan are completely unaware.

In this chapter we make three main points in examining the scale and complexity of a little-considered aspect of the ecological footprint of urban North America, Western Europe, and Japan on Africa. First, international trade in African crafts made by weavers and wood-carvers occurs on a geographical scale that is larger than most people realize and is growing year by year. Second, the ecological footprint resulting from international trade in African basketry and carving is having a serious impact on some species and habitats of local and, in some cases, global significance. As a result, there is a need in several African countries to shift from wild sourcing to on-farm cultivation or, in the case of dyes for Botswana baskets, to synthetic dyes of the same color as traditional dyes. Third, we discuss the opportunity that people in Western Europe and North America have, through selective buying and responsible sourcing, to play an important role in ensuring the sustainability of the crafts trade, rather than remaining contributors to rural resource depletion.

Weaving, Wood Carving, and Social Effects

Commercial craft production is an important means whereby poorer rural people, often from remote areas, are able to enter a cash economy. Although per capita income

Figure 7.1. Commercial trade in drums, including those made from *Sclerocarya birrea* wood and decorated with symbolic ochre, in the Binga district of western Zimbabwe.

from commercial crafts is normally very variable, depending on how much time individuals spend on weaving (usually women) or carving (usually men) versus other household, farming, or community activities, it often contributes significantly to the household income of rural people (Brigham 1994; Terry 1999). As a result, many people become involved in craft production in areas where few other economic opportunities exist. In Ngamiland, Botswana, for example, about half the female population of the villages of Etsha (ca. 1500 women) and Gomare/Tubu (400 women) were involved in basket making (Terry 1984, 1987). In western Zimbabwe, from a modest start in 1987, when just 25 women produced baskets for commercial sale, the Binga Crafts Association had grown to 3200 members by 1999. The commercial trade in Kenyan wood carvings started in 1919 through the remarkable efforts of Mutisya Munge, a Kamba carver (Elkan 1958). By 1954 the export value of the trade was worth U.K.£75,000 at 1950s rates (Elkan 1958). Today 60,000 Kenyan wood-carvers produce commercial carvings, primarily for export, providing household income for an estimated 300,000 dependents (Obunga 1995).

Crafts production also keeps traditional wood-carving and basket-weaving skills alive in a changing world, where enamel bowls and tin and plastic buckets replace baskets and wooden containers in household use. This is a positive attribute. Where craft development and marketing organizations need to take special care, however, is in avoiding the negative social and cultural impacts that can arise from a trade in artifacts that have sacred, ceremonial significance. In the African context, this most commonly applies to drums and masks. Nor is this unique to Africa. On the basis of many years of anthropological work in the Brazilian Amazon, for example, Ribeiro (1983) recommended that artifacts such as Tapirape masks and *ype* masks used in tribal ceremonies not be sold at all. Kenyan carvers commercially produce mock West African masks, poor-quality replicas of Buganda drums, and ersatz West African wooden fertility dolls that have no local cultural significance whatsoever, but in other cases culturally significant artifacts are in commercial trade. In northwestern Cameroon, for instance, Argenti (1998) documented the complex symbolism surrounding carved stools and masks and how "modern" carvings produced for "outside" sale compete with carvers' production of traditional objects for "inside" ceremonial masquerades. It is also tempting to suggest that the production of Gwembe Thonga drums—symbolically marked with ochre yet for commercial sale to tourists in the Zambezi valley, Zimbabwe and now being sold in large quantities (Figure 7.1)—may have two compounding negative effects, due not only to the commercialization of the drums that traditionally are objects of religious significance in this area but also to the use of *Sclerocarya birrea* (Anacardiaceae), among other species, to carve the drums (M. Mudenda pers. comm. 1998). This species is one of the most productive and nutritious sources of wild fruit, protein-rich kernels, and beer and commonly has traditional religious significance in southern Africa (Cunningham 1985; Peters 1988).

African Crafts: Commercial Trade and Resource Depletion

Weaving Fibers

Although a wide range of plant fibers, such as *Sansevieria* (Dracaenaceae) leaf fiber or the bark of many Leguminosae (*Acacia, Brachystegia*), Thymelaeaceae (*Gnidia, Lasiosiphon*), Malvaceae (*Hibiscus, Thespesia*), and Tiliaceae (*Triumfetta*), are used at a subsistence level, few species enter national trade; and even fewer, international export markets (see Appendix 7.1, Table 7.1). Nevertheless, several broad patterns of usage are evident. First, despite the small number of genera and species of African palms (19 genera, ca. 55 species) and woody bamboos (3 genera, 3 species), a similar pattern is seen in other parts of the Tropics, where the Arecaceae and Bambusaceae dominate commercially traded baskets. Second, rattans are widely used in the moist tropical forests of West Africa (particularly *Eremospatha macrocarpa* and *Laccosperma secundiflorum*) and Uganda (*Calamus deërratus*), with local overexploitation occurring in Nigeria (Morakinyo 1994) and Uganda in response to national urban demand. Third, stem fibers from the Marantaceae have an important role in basketry in West Africa (at least 4 genera used), Uganda (1 genus, 2 species in trade), and, probably, the Congo basin. Fourth, there are major regional differences in types of crafts produced: Africa's finest wood carvings are made in West Africa (northwestern Cameroon, Ghana, Mali), northern Mozambique, and southern Tanzania; some of the finest basketry in the world is made by rural women in southern and south central Africa.

The main basketry fibers used in southern Africa are *Hyphaene* (*H. coriacea, H. petersiana*) leaflets and, to a lesser extent, *Cocculus hirsutus* stems (Botswana, South Africa, Zimbabwe). The finest baskets of all, *mukenge* baskets of split *Combretum zeyheri* roots, are made by Mbundu and Lozi women from southern Angola and western Zambia. As a consequence of the high quality of this basketry and the support of rural development agencies, significant international export markets have been developed for southern African baskets in Europe (particularly Germany, Scandinavia, and the United Kingdom) and North America (both the United States and Canada). Baskets are also widely sold in southern Africa to overseas visitors or collectors. The largest volume of high-quality baskets made for export is produced in Botswana (Ngamiland), Zimbabwe (Binga area), and South Africa (KwaZulu/Natal). Although excellent, export-quality baskets are made in the northern communal areas of Namibia (from *H. petersiana*), these are mainly for household use, local barter, and sale in Namibia or South Africa. A similar situation existed in Botswana prior to 1970 and also exists today in the Turkana district of northern Kenya, where *H. compressa* woven items (baskets, mats, brooms) cater on a significant scale to a tourist trade in outlets near or in the capital, Nairobi (Hoebeke 1989).

Although supplies of *Hyphaene coriacea* leaves are abundant on the Mozambique coastal plain (Cunningham 1988), resource depletion of palm fiber (*H. petersiana*) has resulted in problems for rural women elsewhere in southern Africa (Botswana, Zimbabwe). Walking farther and paying more for palm leaves are consequences of a growing scarcity of palms in an area where palms were formerly common only along drainage lines and perennial rivers. In the Binga district of Zimbabwe, for example, the distance that women have to travel to collect palm leaves has increased with booming export sales and more people having become involved in basket production. In 1985 a reported 25 women were involved in basket production, traveling 3 km to collect *H. petersiana* leaves. By 1989 nearly 500 women were involved, but by that stage they were having to walk 18 km to collect palm leaves. In Kavango, Namibia, women are walking long distances to collect *Hyphaene* palm leaves. The majority (67%) of the 12 participants in a basketry course in Rundu, a small town in the Kavango region of northern Namibia, walked to collect palm leaves at Ngone, 30 km away (M. E. Terry pers. comm.). The digging up of edible palm hearts (the apical meristem of the palm) and, possibly, the felling of adult female palms for the edible pericarp around the fruits are considered to have had the greatest impact on wild palm populations. In 1983, 97% of 143 basket makers in Etsha and 55% of 69 basket makers in Gomare/Tubu, two villages in northwestern Botswana, complained about the scarcity of *H. petersiana* palm (Terry 1984, 1987). Less time is therefore spent on household activities, agriculture, and basket weaving itself.

Dye Resources

Although a wide range of dyes can be used to color basketry fiber, relatively few species are used for commercial basketry compared with species used on a local, subsistence scale. This subset of dye resources employed in commercial trade represents the most colorful and colorfast sources from plants. The use of silty mud to color weaving fiber a gray-black shade is also widespread in tropical Africa but is not further dealt with in this chapter. As seen in Appendix 7.1, Table 7.2, two patterns emerge: first, the importance of the Leguminosae as colorfast dye sources and the widespread use of the genus *Indigofera,* probably due to the anthropogenic dispersal of favored, fast-growing dye sources such as *I. arrecta;* and second, that the most important dye sources in commercial trade in southern Africa are from the bark of *Berchemia discolor* (Rhamnaceae), the roots of *Euclea divinorum* and *E. natalensis,* and the fruits of *Diospyros lycioides* (Ebenaceae).

Bark of the bird plum tree (*Berchemia discolor*) is a major source of dye for commercial basketry production in Botswana, Namibia, and Zimbabwe. This species is well known in southern Africa for its tasty edible fruits, the main reason why they are traditionally conserved in agricultural fields in this region. Formal monitoring of the impact of *Berchemia* bark (and root bark) use in the Etsha area of Botswana during the 1982–1993 period showed that *Berchemia* tree populations that were the main source of supply in 1982 had been killed off by 1992 (Cunningham unpubl. data). As a result, women basket makers walk farther and farther to obtain basketry dyes, spending less time in agriculture, making baskets, and looking after children. This has also resulted in changing patterns of dye use. Due to the local overexploitation of *Berchemia* and *Euclea* populations in some areas, increased use of *Indigofera tinctoria* and *I. arrecta* leaves and *Rubia cordifolia* roots as dye sources has taken place, as has use of a pink dye from fungus-infected *Sorghum* stalks.

Wood Carving: Species, Trade, and Impacts

Although many species can be carved, the wood qualities of a relatively small range of species have made them the favored sources for commercially traded carvings (see Appendix 7.1, Table 7.3). These can be divided into two broad groups. The first group comprises dense, slow-growing hardwoods, which are favored for their fine grain and for their ability to withstand cracking and take a fine polish. These are carved into bowls and figures. Many of them are from the Leguminosae, with localized overexploitation of several *Cord* species (Boraginaceae) also apparent. A growing scarcity of these preferred hardwood species has become a very relevant issue for wood-carvers in Malawi (*Dalbergia melanoxylon, Pericopsis angolensis*), Kenya (*Brachylaena huillensis,*

D. melanoxylon), and, increasingly, Namibia (*Pterocarpus angolensis*) and Zimbabwe (*Afzelia quanzensis, P. angolensis*). Significant from the perspective of conservation of the genetic resources of valued timber trees is the likely genetic erosion taking place due to overexploitation of some wood-carving species on a national or regional scale. Examples of this are likely in *A. quanzensis* (Zimbabwe), *B. huillensis* (Kenya), *D. melanoxylon* (savanna woodlands of Kenya and Malawi), *P. angolensis* (Malawi), and, probably, *Cordia platythrysa* (Ghana), which is listed as one of Africa's threatened tree species (WCMC 1998).

The second group of carving woods has less dense, lighter-colored wood. These are generally carved into stools or drums from trees in the Apocynaceae (*Holarrhena floribunda, Funtumia africana, F. elastica*), Araliaceae (*Polyscias fulva*), or Anacardiaceae (*Sclerocarya birrea*). These are faster-growing species and can better withstand demand for carving wood, yet there are cases in which depletion of wild populations has occurred. *Polyscias fulva*, for example, is a favored species for wood carving through most montane forest areas of Central and West Africa (Rwanda, Burundi, western Uganda, Cameroon). Due to its rapid growth rate and abundance in disturbed forests, supplies of this light-demanding softwood tree species easily meet the subsistence demand for production of carved stools, spoons, and blacksmith bellows in the densely populated (150–350 people per km^2) area of western Uganda. The same does not apply in the Bamenda highlands of northwestern Cameroon, however, where carvers in a similarly dense farming population are renowned for the large, innovative, and culturally significant carvings they make from *P. fulva*. Although Argenti (1998) only recorded 40–50 carvers in the Mount Kilum-Oku area of northwestern Cameroon (out of a total population of ca. 60,000), this species has been overexploited in the forest. As a result, many carvers in the area now plant *P. fulva* on their small-scale farms (Argenti 1998).

Depletion of tree species used in carving due to felling for timber, furniture, or carving itself can have a variety of negative social impacts. First, felling female *Sclerocarya birrea* trees results in loss of a fruit and nut source that is rich in nutrients commonly lacking in a starchy staple diet (vitamin C and proteins) and of major significance to poorer rural households (Cunningham 1985; Peters 1988). Second, reduction in some tree species favored for wood carving also represents reduced local self-sufficiency in preferred sources of medicinal bark, an issue that applies to some of the hardwood species as well. In terms of the faster-growing tree species, the possibility of localized depletion of *Holarrhena floribunda* (sese), the species most favored in Ghana for carving stools for international trade, is worth investigating. Extracts of *H. floribunda*, which contains steroidal alkaloids and trichthecenes, are recognized as having significant antitumor effects and as being moderately effective against leishmaniasis (Abreu et al. 1999; Loukaci et al. 2000). Similar concerns could be expressed about other species as well. *Pterocarpus angolensis* is effective against schistosome parasites (Ndamba et al. 1994), and other species in this genus (*P. marsipium*) have been shown to be effective in treating hyperglycemia (Manickam et al. 1997). The naphthoquinones in several *Cordia* species are known to be effective anti-inflammatory (Ficarra et al. 1995), antifungal, and larvicidal agents (Ioset et al. 1998, 2000). *Dalbergia melanoxylon* bark extracts contain strong antimicrobial compounds (Gundidza & Gaza 1993); *Spirostachys africana* is

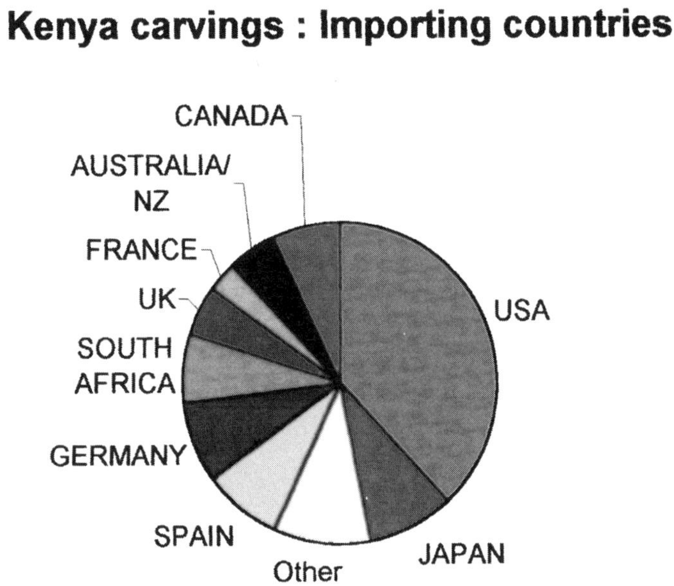

Figure 7.2. The proportion of Kenyan wood carvings imported by various countries, showing the importance of the United States as a market. *Source:* C. Alloo unpubl. data.

traditionally used as an antimalarial in Mozambique; and the burned bark of *Daniella olivieri* is an extremely effective mosquito repellent (Palsson & Jaenson 1999).

Even though we are unaware of any study carried out on the large wood-carving trade in Ghana, we suggest that the largest value and volume of African carvings in international trade comes from Kenya. The bulk of these carvings are exported to North America, primarily to the United States (Figure 7.2). With declining economic situations and rising unemployment in many African countries, income generation through carving and weaving is on the increase. For this reason, the Kenyan carving trade bears some useful lessons for organizations involved in craft development elsewhere in Africa.

Although the Kenyan wood-carving trade started 80 years ago with an emphasis on ebony (*Dalbergia melanoxylon*), over time it has become characterized by two features. The first is the successive overexploitation of favored species, beginning with *D. melanoxylon* and followed by *Brachylaena huillensis* and, more recently, *Spirostachys africana, Olea europaea* ssp. *africana,* and *Combretum schumannii.* The second feature is the movement of WaKamba wood-carvers from their home area of Wamunyu (Machakos district, Kenya) to the resource-rich frontier forests of coastal and central Kenya (Obunga & Sigu 1996). At present, seven species are found to be in use in significant quantities, but *B. huillensis* is by far the most important of these. Preliminary results from an ongoing study by S. Choge show that *B. huillensis* logs account for more than 89% of carving wood. In the first five months of 1998, for example, more than 36,726 log pieces (a standard 1.5 m long, on average 3 logs cut per tree) were delivered to Kenya's wood-carving centers. The Mombasa carving center alone received 21,106 log pieces of *B. huillensis;* Wamunyu received 8,325 pieces; and Lunga Lunga, on the Kenyan side of the border with Tanzania, received 5,897 log pieces from the resource-rich frontier of Tanzania. All other species combined stood at 4,220 log pieces (10.3%). Volumes and standing trees represented by these log quantities are currently being calculated, but the quantity of trees cut for production of carvings is certainly more than 40,000 trees per year. By contrast, carvers and furniture makers in the Bushbuckridge area of the eastern Transvaal lowveld used just 30 and 60 trees per year, respectively (Shackleton 1993).

Nearly all of the carving cooperative centers purchase wood from timber traders who act as intermediaries between the cooperatives and the wood sources. Much of this timber is illegally taken from state forests and national parks, as well as from private and communal lands. Trees are felled, cut to standard sizes, and transported to wood-carving centers, where they are graded. Experienced members of the centers mark each piece of wood with a unique reference code and determine the price at which it should be sold. The increasing number of wood traders may have helped to push the prices of timber supplied to the carving cooperatives down to levels far below market prices.

Ecological Footprints

The most widespread ecological footprint of the international trade in African crafts is generated by the wood-carving industry, rather than the basket-weaving industry. The impact of wood use for carving also needs to be compared with that of felling for timber. These relative impacts vary between sites and from country to country. In Zimbabwe, for example, in a study of wood carving along the Victoria Falls–Bulawayo road, an estimated 657 m^3 per year of carving wood, primarily *Afzelia quanzensis, Pterocarpus angolensis,* and *Kirkia acuminata,* was being cut (Matose et al. 1997). This represented only 13% of the 5000 m^3 per year of timber cut in the same area by concessionaires. However, the number of people in the wood-carving industry and the quantity of wood cut has been rising dramatically in response to drought and economic decline. In Kenya, there is no doubt that timber logging has a greater impact on montane forest such as Mount Kenya or Mount Elgon. In Kenyan coastal forests the reverse applies, and logging for carving woods has the greatest impact.

Basketry resource depletion nevertheless certainly poses a problem for rural women who walk farther or pay more for increasingly scarce resources. In addition, localized die-off of *Berchemia discolor* trees in Botswana represents loss of a popular and nutritionally valuable food source. Most eastern and southern African basketry species are widespread savanna species whose populations are affected by localized impacts on dye resources (*Berchemia, Euclea*) and, in a few cases, on *Hyphaene* palms; namely, near villages in western Ngamiland, Botswana (Cunningham & Milton 1987) and in the Binga area of Zimbabwe and part of Kavango, Namibia (Cunningham & Terry 1995). In other cases, harvesting of dyes is carefully managed by local women (Binga, Zimbabwe), or abundant palm resources and a low level of consumption of edible palm hearts mean that crafts production is sustainable (Cunningham & Terry 1995; Konstant et al. 1995; Sullivan et al. 1995).

Depletion of ebony (*Dalbergia melanoxylon*) and *Brachylaena huillensis* for wood carving has spread from Kenya to northern Tanzania in response to market demand. Most large to medium-sized *D. melanoxylon* trees have been felled in all but the most remote savanna woodlands of Malawi and Kenya, with normally vigorous resprouting heavily browsed by livestock. A similar widespread felling of *B. huillensis* has taken place in forests of Kenya and, to an increasing extent, Tanzania, initially for sawmilling (production of parquet flooring) but, more significantly in the past 30 years, for wood carving. This widespread loss of large, reproductively mature trees may erode the genetic base of these economically important tree species in the long term.

Also of concern is the ecological impact that the felling of more than 40,000 trees per year has on forest-dependent animals and on the structure of their East African coastal forest habitat, an endangered forest type and one of 200 priority ecoregions identified by Olson and Dinerstein (1998). Both *Brachylaena huillensis* and *Combretum schumannii* are forest tree species felled from (theoretically) protected areas of the last remaining forests in Kenya and Tanzania. These forests are also the habitat of endangered species such as the Sokoke Scops owl (*Otus irenae*) and the golden-rumped elephant shrew (*Rhychocyon chrysopygus*), which use hollow trees for nesting or shelter. A preliminary study in 1998 by Kyalo Solomon, who worked with woodcutters felling trees to supply wood-carving centers, recorded the proportion of felled trees that were hollow and the number occupied by birds, reptiles, and small mammals. On this basis, according to Solomon, more than 18,000 of the 40,000 trees cut per year would represent hollow-tree habitat. In just one year, the loss of hollow-tree habitat would effect more than 8000 golden-rumped elephant shrews and 2300 tropical girdled lizards, also one of the East African coastal forest endemics. Although no scientific record has ever been made of Sokoke Scops owl nests, this species certainly would nest in hollow trees, just as other Scops owl species do. In addition, Sokoke Scops owls are found mainly in *Cynometra-Brachylaena*–dominated forest in the Arabuko-Sokoke forest (the most important and—until the recent discovery of another site in Tanzania—the only known locality for this species in the world). For these reasons, it appears highly likely that the Kenyan carving industry poses the main contemporary threat to the endangered Sokoke Scops owl.

Agroforestry and Alternatives

Palms for Basketry

Palm cultivation has the potential to reverse the trend of declining availability of weaving material near basket-producing villages. Of the nine palm species in eastern and southern Africa, *Borassus aethiopum* and the three *Hyphaene* species (*H. petersiana, H. compressa,* and *H. coriacea*) have the best potential for agroforestry, due to their multiple-use potential and long, strong leaf fiber for basketry. *Phoenix reclinata,* though widespread and a source of popular edible fruits, is less suitable for basketry due to its short leaflets. Implementation of *Hyphaene* cultivation can bring several benefits to farmers. First, palms are well suited to the multiple-cropping systems of small farmers in southern Africa, for they provide a windbreak around fields that reduces wind erosion and evapotranspiration. Second, they provide a spiny barrier that discourages livestock from entering the fields and represent a multiple-use crop that yields leaf fiber, robust petioles that can be used as fuel, and edible fruits, in addition to lending itself to the more destructive uses for palm hearts and palm wine (which could be a by-product of thinning cultivated palm stands over time). In addition, unlike *Raphia,* which dies after terminal flowering at 25–30 years, *Hyphaene* palms do not die after flowering. They also sprout vigorously, providing a perennial crop that does not require replanting, and older specimens are resilient after the burning of brushwood that can accompany the clearing of fallow fields.

Pilot studies in Botswana have shown that *Hyphaene petersiana* cultivation is a viable proposition for agroforestry. This can provide a source of leaves close to basket-producing villages and an alternative to wild populations. What also needs to be taken into account, however, is that different planting systems may be required to fit local agricultural systems (Figure 7.3).

Wood-Carving Species

In Kenya, the on-farm cultivation of fast-growing introduced species such *Azadirachta indica* and *Jacaranda mimosifolia* also offers a viable alternative to harvesting in the wild, in terms of relieving shortages, promoting conservation in the overharvested wild populations, and providing an assurance of added income. In natural conditions, most indigenous tree species favored for carving are slow growing and may offer limited incentives to those who would wish to cultivate them. Carved animals from *Brachylaena huillensis* and *A. indica* are not easily distinguished by buyers, yet the two species differ greatly in growth rates. For example, *B. huillensis* trees reach 40 cm diameter at breast height (dbh) in 100 years, 45 cm dbh in 130 years, and 60 cm dbh in 175 years (Figure 7.4). By contrast, *A. indica* (neem), an introduced species that is being promoted as an alternative to *B. huillensis,* can reach 40 cm dbh in 16 years (Lemmens et al. 1995). On the Kenyan coast, *A. indica* trees in a Kenya Forestry Research Institute cultivation trial increased in height by 0.7–0.9 meters and 1.1–1.7 cm in diameter annually, with some individuals attaining a dbh of 43 cm in 21 years.

One of the ways around the problem of limited incentives to cultivate indigenous carving species due to their slow growth rates may be the cultivation of a range of wood-carving species (or multiple-use tree species) in mixed plantings. In Tanzania, for example, White (1988) recommended interplanting the fast-growing fuelwood species *Cassia siamea* with slow-growing *Dalbergia melanoxylon* as a means of offsetting the costs involved. What is lacking, however, is an economic comparison of the costs and benefits of fast-growing versus slow-growing wood-carving species, either individually or in mixed plantations. A study to provide information on values and volumes of wood-carving raw materials from different sources and on present levels of competition for the use of these resources is well advanced (Muturi & Choge 1998). This study is investigating the viability of cultivation as an alternative to harvesting

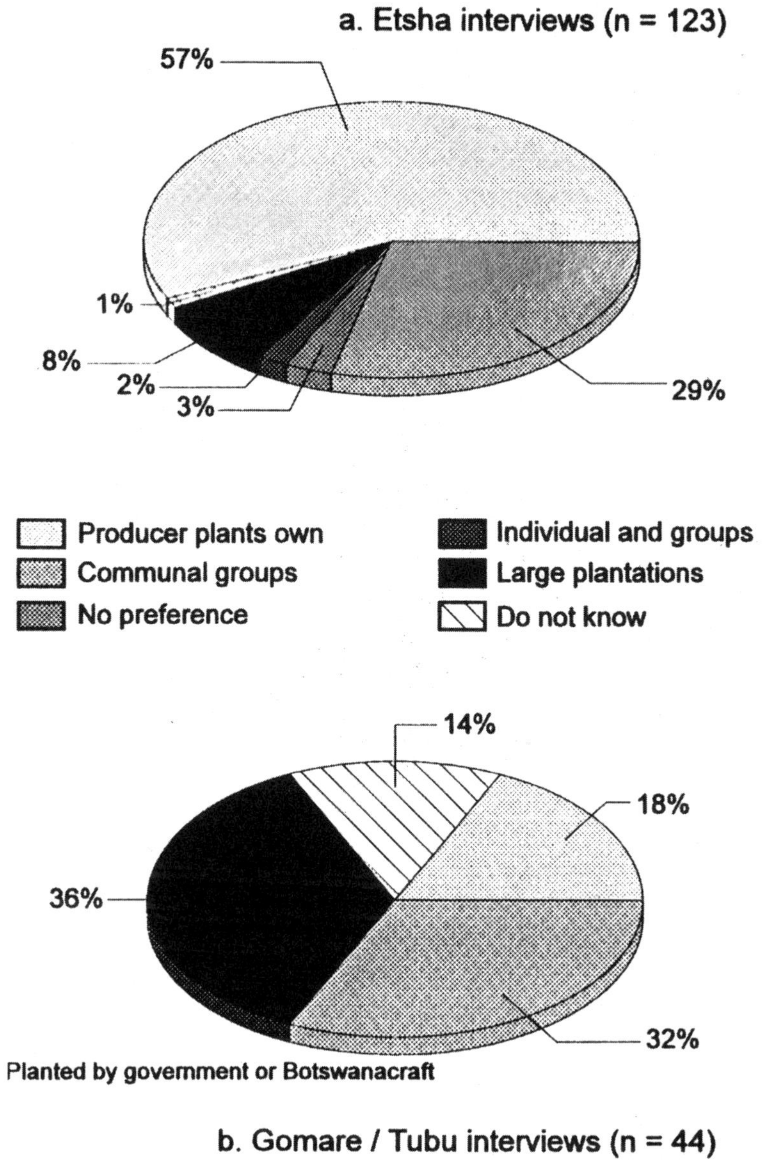

Figure 7.3. Perceptions of local subsistence farmers in two villages in northwestern Botswana as to the best methods of implementing *Hyphaene* palm cultivation. **a.** Etsha, primarily Hambukushu dryland farmers. **b.** Gomare/Tubu, mainly Bayei floodplain (*molapo*) agriculturalists. *Source:* After Terry 1986.

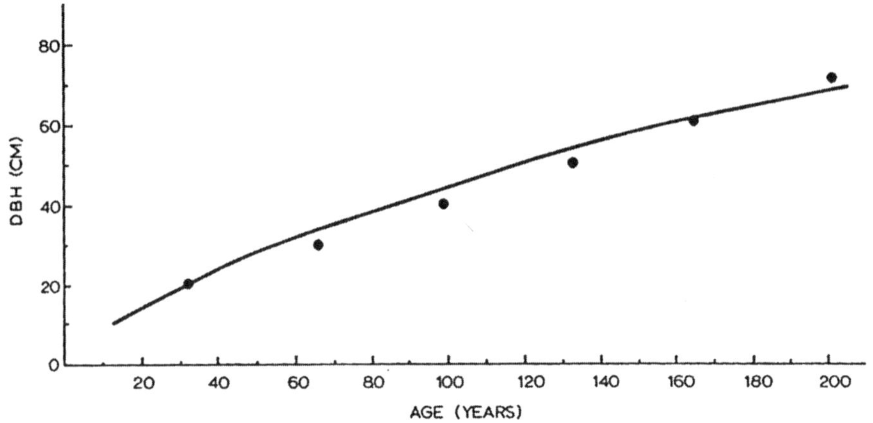

Figure 7.4. The relationship between diameter at breast height and age of *Brachylaena huillensis* (muhugu mahogany), the main species carved in Kenya for export. *Source:* Kigomo 1989.

from the wild, with an emphasis on the possibility of obtaining carving wood from on-farm plantations in order to alleviate current wood shortages.

An added incentive for on-farm forestry for carvings may come from carving under the certification system used by the Forestry Stewardship Council, and steps are being taken toward implementing this process (Schmitt & Cunningham 2002). Certification of wood carvings offers an opportunity for buyers and importers based in Europe and North America and tourists visiting Africa to play an important role in ensuring the sustainability of the crafts trade through selective buying and responsible sourcing, rather than remaining contributors to rural resource depletion. Certification of carvings faces complexities similar to those of small-scale timber certification in many parts of the world, particularly its high cost (Bass 1998; De Camino & Alfaro 1998; Tolfts 1998). The solutions to these problems are also similar, such as group certification schemes, developing local certification standards accredited by the council, and local capacity to reduce the cost of certification. If this can be done, then links with fair-trade organizations, coupled with greater awareness among importers and the buying public, can create two key components for an incentive for responsible sourcing of wood for carving: greater market share and a premium paid for certified carvings. If this works, and we certainly hope it will, then it offers an example of how the heavy ecological footprint of Europe, North America, and Japan on African craft resources can be turned into a positive opportunity for buyers to contribute to rural peoples' livelihoods yet tread lightly on the landscape.

Acknowledgments

We would like to thank Dr. P. K. Konuche and Dr. B. Kigomo, director and research director, respectively, of the Kenya Forestry Research Institute and the WWF/UNESCO/ Kew People and Plants Initiative (funded in Africa through the National Lotteries

Charities Board, based in the United Kingdom) for their support. The opportunity for one of us (A. Cunningham) to attend the International Botanical Congress in Saint Louis, Missouri, where the material in this chapter was presented, was made possible through the efforts of Nan Vance and a travel grant from the U.S. Forest Service. Comments on the chapter by the book's editors, Luisa Maffi and Tom Carlson, are appreciatively acknowledged.

Literature Cited

Abbiw, D. 1990. The Useful Plants of Ghana. London: Intermediate Technology Press.

Abreu, P. M., E. S. Martins, O. Kayser, K. U. Bindseil, K. Siems, A. Seemann & J. Frevert. 1999. Antimicrobial, antitumor and antileishmania screening of medicinal plants from Guinea-Bissau. Phytomedicine 6(3): 187–195.

Argenti, N. 1998. The role of carving, masquerades and metamorphosis in the hierarchy of Oku, North West Province, Cameroon. Ph.D. diss., University College London.

Bass, S. 1998. Forest Certification: The Debate about Standards. Rural Development Forestry Network Paper 23b. London: ODI.

Bowen, M. R. 1992. A Bibliography of Forestry in Somalia and Djibouti. London: ODA.

Brigham, T. E. 1994. Trees in the rural cash economy: A case study from Zimbabwe's communal areas. M.A. thesis, Carlton University.

Cunningham, A. B. 1985. The resource value of indigenous plants to rural people in a low agricultural potential area. Ph.D. diss., University of Cape Town.

———. 1988. Leaf production and utilization in *Hyphaene coriacea:* Management guidelines for commercial harvesting. South African Journal of Botany 54(3): 189-195.

———. 1996. People, park and plant use: Recommendations for multiple-use zones and development alternatives around Bwindi-Impenetrable National Park, Uganda. People and Plants Working Paper 4. Paris: UNESCO.

——— & G. W. Davis. 1997. Human use of plants. Pp. 474–506 *in* R. M. Cowling, D. M. Richardson & S. M. Pierce (eds.), Vegetation of Southern Africa. Cambridge and New York: Cambridge University Press.

——— & S. J. Milton. 1982. A survey of plant resource utilization by the Botswana basket industry in the CFDA, Ngamiland. Report for Veld Products Research, Gabane, Botswana.

——— & ———. 1987. Effects of basket-weaving industry on mokola palm and dye plants in north-western Botswana. Economic Botany 41(3): 386-402.

——— & M. E. Terry. 1995. Basketry, people and resource management in southern Africa. Pp. 363–375 *in* F. Ganry & B. M. Campbell (eds.), Sustainable Land Management in African Semi-Arid and Sub-Humid Zones. Montpellier, France: CIRAD.

De Camino, R. & M. Alfaro. 1998. Certification in Latin America: Experience to Date. Rural Development Forestry Network Paper 23c. London: ODI.

Elkan, W. 1958. The East African trade in wood carvings. Africa 57(2): 314–323.

Ficarra, R., P. Ficarra, S. Tommasini, M. L. Calabro, S. Ragusa, R. Barbera & A. Rapisarda. 1995. Leaf extracts of some *Cordia* species: Analgesic and anti-inflammatory activities as well as their chromatographic analysis. Farmaco 50: 245–256.

Gundidza, M. & N. Gaza. 1993. Antimicrobial activity of *Dalbergia melanoxylon* extracts. Journal of Ethnopharmacology 40: 127–130.

Hoebeke, P. 1989. Eongol: The doum palm (*Hyphaene compressa*) as a biological resource in Turkana district, Kenya. Cand. Scient. thesis, University of Trondheim.

Ioset, J. R., A. Marston, M. P. Gupta & K. Hostettmann. 1998. Antifungal and larvicidal meroterpenoid naphthoquinones and a naphthoxirene from the roots of *Cordia linnaei*. Phytochemistry 47(5): 729–734.

———, ———, ——— &———. 2000. Antifungal and larvicidal cordiaquinones from the roots of *Cordia curassavica*. Phytochemistry 53(5): 613–617.

Itoua, I., E. Underwood, D. M. Olson & E. Dinerstein. 1998. A Conservation Assessment of Terrestrial Ecoregions of Africa. Washington, D.C.: WWF–US Conservation Science Program.

Kigomo, B. N. 1989. Studies on the regeneration and growth characteristics of *Brachylaena huillensis* in semi-deciduous forests of Kenya. Ph.D. diss., University of Oxford.

Konstant, T. L., S. Sullivan & A. B. Cunningham. 1995. The effects of human population pressure on *Hyphaene petersiana* basketry resources in the palm savanna of Owambo, northern Namibia. Economic Botany 49(4): 345–356.

Lemmens, R. H. M. J., I. Soerianegara & W. C. Wong (eds.). 1995. Plant Resources of South-East Asia. Vol. 5(2), Timber Trees: Minor Commercial Timbers. Leiden, Netherlands: Backhuys.

Loukaci, A., O. Kayser, K. Bindseil, K. Siems, J. Frevert & P. M. Abreu. 2000. New trichothecenes isolated from *Holarrhena floribunda*. Journal of Natural Products 63(1): 52–56.

Manickam, M., M. Ramanathan, M. A. Jahromi, J. P. Chansouria & A. B. Ray. 1997. Anti-hyperglycemic activity of phenolics from *Pterocarpus marsupium*. Journal of Natural Products 60(6): 609–610.

Matose, F., M. Mudhara & P. Mushove. 1997. The woodcraft industry of the Bulawayo–Victoria Falls road. Institute of Environmental Studies Working Paper 2. Harare: University of Zimbabwe.

Moffatt, I. 1996. Sustainable Development: Principles, Analysis and Policies. Carnforth, U.K.: Parthenon Press.

———. 2000. Ecological footprints and sustainable development. Ecological Economics 32: 359–362.

Morakinyo, T. 1994. The ecology and silviculture of rattans in Africa: A management strategy for Cross River State and Edo State, Nigeria. M.S. thesis, University College of North Wales.

Muturi, G. M. & S. K. Choge. 1998. Economic Values of Woodcarving Timber in Kenya. Technical Programmes Report. Nairobi: Kenya Forestry Research Institute.

Ndamba, J., N. Nyazema, N. Makaza, C. Anderson & K. C. Kaondera. 1994. Traditional herbal remedies used for the treatment of urinary schistosomiasis in Zimbabwe. Journal of Ethnopharmacology 42(2): 125–132.

Obunga, R. 1995. Sustainable development of woodcarving industry in Kenya. Technical progress report. Unpublished report for the WWF/UNESCO/Kew People and Plants Initiative, National Museums of Kenya, Nairobi.

——— & G. Sigu. 1996. Sustainable development of woodcarving industry in Kenya. Draft Technical Progress Report (Phase 2). Unpublished report for the WWF/UNESCO/Kew People and Plants Initiative, National Museums of Kenya, Nairobi.

Olson, D. M. & E. Dinerstein. 1998. The Global 2000: A representation approach to conserving the Earth's most biologically valuable ecoregions. Conservation Biology 12: 502–515.

Palsson, K. & T. G. Jaenson. 1999. Comparison of plant products and pyrethroid-treated nets for protection against mosquitoes (Diptera: Culcidae) in Guinea-Bissau, West Africa. Journal of Medical Entomology 36: 144–148.

Peters, C. R. 1988. Notes on the distribution and relative abundance of *Sclerocarya birrea* (A. Rich.) Hochst. (Anacardiaceae). Pp. 403–410 *in* P. Goldblatt & P. P. Lowry II (eds.),

Modern Systematic Studies in African Botany. Monographs in Systematic Botany from the Missouri Botanical Garden, 25. Saint Louis: Missouri Botanical Garden.

Rees, W. E. 1992. Ecological footprints and appropriated carrying capacity: What urban economics leaves out. Environment and Urbanization 4: 121–130.

Ribeiro, B. G. 1983. Why and for whom are Indian handicrafts made? Pp. 29–45 *in* B. G. Ribeiro, M. R. B. Alvim, A. M. Heye, V. de Vives, J. S. D'Avila & D. L. M. Teixeira (eds.), The Traditional Artisan and His Role in Contemporary Society. Rio de Janeiro: FUNARTE.

Schmitt, S. F. & A. B. Cunningham. 2002. Reducing the ecological footprint of the "wooden rhino": The case for certification of Kenyan woodcarvings. Pp. 259–263 *in* A. Guillén, S. A. Laird, P. Shanley & A. R. Pierce (eds.), Tapping the Green Market: Certification and Management of Non-Timber Forest Products. London: Earthscan.

Shackleton, S. E. 1993. A situation analysis of the woodcraft industry in the Bushbuckridge district of the eastern Transvaal, with particular reference to resource use. Unpublished report, Wits Rural Facility, Klaserie, South Africa.

Sullivan, S., T. L. Konstant & A. B. Cunningham. 1995. The impact of utilization of palm products on the population structure of the vegetable ivory palm (*Hyphaene petersiana*) in Owambo, northern Namibia. Economic Botany 49(4): 357–370.

Sunderland, T. C. H. 2002. Status of rattan resources and use in West and Central Africa. Pp. 77–87 *in* J. Dransfield, F. O. Tesoro & N. Manokaran (eds.), Rattan: Current Research Issues and Prospects for Conservation and Development. Rome: FAO.

Terry, M. E. 1984. Handcraft Survey Ngamiland CFDA: A Survey of Basketmakers, Etsha, Ngamiland, Botswana. Gaborone: Botswanacraft Marketing Co.

———. 1986. Handcraft Survey Ngamiland CFDA: The Basket Industry of Gomare and Tubu. Gaborone: Botswanacraft Marketing Co.

———. 1987. Mokola Palm and Extension Activities in Ngamiland CFDA, Botswana. Gaborone: Botswanacraft Marketing Co.

———. 1999. The socio-economic significance of the handcraft industry in Botswana. Ph.D. diss., University of London.

Tolfts, A. 1998. How Appropriate Is Certification for Small-Scale Timber Processors in Melanesia? Rural Development Forestry Network Paper 23d. London: ODI.

WCMC. 1998. World List of Threatened Trees. Cambridge: World Conservation Monitoring Centre.

White, R. 1988. Report of the Consultant Mission to Tanzania: Mpingo (*Dalbergia melanoxylon*). Nairobi: UNEP.

Appendix 7.1: Species, Trade names, Densities, Uses and Ecoregions

The main commercially traded African basketry fibers (Table 7.1) dyes (Table 7.2), and carved woods (Table 7.3), showing the species of chief concern in terms of national (*), regional (i.e., cross-border trade within the region) (**), and international (***) trade and the ecoregions from which they are taken. Vegetation types and ecoregions follow those used by Itoua et al. (1998).

Key to parts used and main uses: ar = edible arils; ax = axe handles; b = bowls, bp = building poles; bpt = body paint; bt = boats; dr = drums; dy = dye; ch = charcoal; fu = furniture; fw = fuelwood; ins = insect repellent; mus = musical instruments; o = oil seeds; sp = spoons; st = stools; sti = sticks.

Key to tropical moist broadleaf forests: CC = Congolian coastal forests; CR = Cross-river coastal forests; NL = Nigerian lowland forests; NWCL = NW Congolian lowland forests; NECL = NE Congolian lowland forests; CCC = Central Congolian coastal forests; GL = Guinean lowland forests; VB = Victoria basin moist forests.

Key to tropical moist swamp forests: CSW = Congolian swamp forests; GSW = Guinean swamp forests.

Key to tropical montane and submontane forests: MCM = Mount Cameroon montane forests; CH = Cameroon [highland] forests; ARM = Albertine rift montane forests; EARC = Eastern arc montane forests; EAM = East African montane forests.

Key to tropical dry broadleaf forests: NZI = Northern Zanzibar-Inhambane coastal forests; SZI = Southern Zanzibar-Inhambane coastal forests; MPCF = Maputoland-Pondoland coastal forests.

Key to grasslands/savanna/shrublands: GFS = Guinean forest-savanna mosaic; WSS = Western Sudanian savannas; SACB = Somalian *Acacia-Commiphora* bushlands; MACB = Masai *Acacia-Commiphora* bushlands; ZMIO = Zambezian miombo woodlands; ZMOP = Zambezian and mopane woodlands; WMW = Western mopane woodlands; TVB = Transvaal bushveld; KAS = Kalahari *Acacia* savannas.

Key to flooded grasslands and savannas: ODFS = Okavango Delta flooded savannas; ZSM = Zambezian swamps and marshes.

Sources: 1 = Abbiw 1990; 2 = Cunningham & Milton 1982; 3 = Cunningham 1988; 4 = Cunningham & Milton 1987; 5 = Cunningham unpubl. field records; 6 = Sunderland 2002; 7 = Morakinyo 1994; 8 = Cunningham 1996; 9 = Cunningham & Davis 1997; 10 = Shackleton 1993; 11 = Obunga 1995; 12 = Bowen 1992.

Table 7.1
African basketry fibers

Family, genus, species	Part used	Ecoregion in which used	Source
Arecaceae			
Borassus aethiopum	lf	WSS, ZMIO	1, 5
*Calamus deërratus**	st	VB, not widely used in West Africa	5
Dendrocalamus strictus	st	GSW	1
*Eremospatha macrocarpa**	st	GSW, CR	6, 7
*Hyphaene compressa**	lf	NZI, MACB	8
*Hyphaene coriacea**	lf	SZI, MPCF	3, 4
*Hyphaene petersiana****	lf	ODFS	4
*Hyphaene thebaica**	lf	WMW, ZMIO	1
Laccosperma opacum	st	WSS	1
*Laccosperma secundiflorum**	st	GL, CR, GSW, CSW	6, 7
Raphia farinifera	lf	GL, GSW,	1, 5
Raphia hookeri	lf	VB, CC, NL, NWCL, CCC, CSW	1
Bombacaceae			
*Adansonia digitata***	bk	TVB, ZMIO	5
Celastraceae			
Loeseneriella apocynoides	st	ARM, EAM	8
Cyperaceae			
Cyperus articulatus	cl	Widespread, wetlands	1
Cyperus latifolius	cl	Widespread, wetlands	3
Cyperus natalensis	cl	MPCF, SZI wetlands	3
Cyperus papyrus	cl	Wetlands throughout tropical Africa	8
Cyperus sexangularis	cl	Widespread, wetlands	3
Gramineae			
Arundinaria alpina	st	ARM, EARC	8
Oxytenanthera abyssinica	st	EAM, EARC, CH	5
Phargmites australis	cl	Pan-tropical wetlands	1, 5
Juncaceae			
*Juncus kraussii**	cl	MPCF salt marshes	3, 9
Marantaceae			
Marantachloa leucantha	st	VB, CSW, GCW, ARM	1, 8
Marantachloa purpurea	st	VB, CSW, GSW	1, 5
Marantachloa ramosissimum	st	CSW, GSW	1
Sarcophrynium brachystachys	st	CSW	1
Thalia welwitschii	st	CSW	1
Thaumatococcus daniellii	st	CSW	1
Menispermaceae			
Cocculus hirsutus	st	TVB, ODFS, ZMIO	5, 9
Smilacaceae			
Smilax anceps	r	ARM, widespread but not widely used in other places	8

Table 7.2
African basketry dyes

Family, genus, species	Part used	Ecoregion in which used	Source
Anonnaceae			
Enantia polycarpa	b	GL	1
Cochlospermaceae	b	GL	1
Cochlospermum planchonii	r	GL	1
Cochlospermum tinctorium	r	GL	1
Ebenaceae			
Diospyros lyciodes	fr	MPCF, SZI, ZMOP, TBV	2, 3
*Euclea divinorum****	r	MPCF, SZI, ZMOP, TBV	3, 4
Euclea natalensis	r	MPCF, SZI, ZMOP, TBV	3
Guttiferae			
Harungana madagascariensis	br	GL,VB,ARM	1, 5
Leguminoseae			
Acacia nilotica	fr	GFS, WSS	1, 5
Adenanthera pavonina	Wd	GL	1
Baphia nitidia	wd	GFS	1
Haematoxylon campechianum	wd	GL	1
Indigofera arrecta	lv, r	MPCF, SZI, ZMOP, TBV, GL, GFS	1, 3
Indigofera spicata	lv, r	GL, GFZ	1
Indigofera suffriticosa	lv, r	GL, GFS	1
Indigofera tinctoria	lv, r	ODFS, KAS, ZMOP	1, 5
Lonchocarpus cyanescens	lv	GFS	1
Lonchocarpus laxiflorus	lv	GFS	1
Lonchocarpus sericeus	sd	GFS	1
Pterocarpus erinaceus	r	GL	1
Rhamnaceae			
*Berchemia discolor****	bk	ODFS, KAS, ZMOP	2, 4
Rubiaceae			
Rubia cordifolia	r	MPCF	3

Table 7.3
African wood-carving species

amily, genus, species	Items carved for commercial use	Other main uses	Ecoregion in which used	Source
nacardiaceae				
*Sclerocarya birrea***	dr	fr, b	MPCF, TBV	5
pocynaceae				
Alstonia boonei	dr, b	sp, med	GL, CC, CR, NL, NWCL	1, 5
Antiaris congensis	dr	bt, med	GL	1
Funtumia africana	dr, st	med	GL, CC, CR, NL, VB	1, 5
Funtumia elastica	dr, st	med, rubber	GL, CC, CR, NL, VB	1, 5
*Holarrhena floribunda****	st (Ghanaian)	med	GL, NWCL	1
Wrightia natalensis	fu	med	MPGF, SZ	3
raliaceae				
*Polyscias fulva****	st, figures, dr	ply, sp	CH, ARM, EAM	8
alanitaceae				
Balanites maughamii	trays	med bark	MPCP	3
oraginaceae				
Cordia africana	headrests	fu, fr, sp	MACB, EAM	5
Cordia millenii	b, sp	med, t, x	WSS	1
*Cordia platythyrsa****	dr, tools, sp	med, t, x	GL	1
Cordia sinensis	headrests	bp, fu, frs	MACB	5
ombretaceae				
Combretum imberbe	figures	bp, fw	ZMOP, TVB	5, 10
*Combretum schumannii****	figures	bp	NZI	11
Terminalia sericea	trays	bp, ax, fu	ZMIO, TVB, MPGF	10
Terminalia orbicularis	b	bp, ax	SACB, MACB	1
Terminalia spinosa	headrests	bp	MACB, NZI	5
Terminalia superba	st	t	GL, CC, CR, NL	1
ompositae				
*Brachylaena huillensis****	b, figures	t	NZI, SZI, EAM	5, 11
uphorbiaceae				
*Spirostachys africana****	figures, trays, b	med, t	NZI, SZI, TVB	5, 11
irkiaceae				
*Kirkia acuminata***	figures	med	ZMOP	5
eguminoseae				
*Afzelia quanzensis***	figures	t, fu	NZI, SZI, MPCF, ZMIO	3
*Dalbergia melanoxylon****	figures, b	t, mus	ZMIO, WSS	10, 11
Daniellia oliviera	st	t, med, ins (bark)	GL, CC	1
*Pericopsis angolensis****	st	t	ZMIO	5
*Pterocarpus angolensis***	st, b, figures	c, med, t, fu	ZMIO, TVB	5, 10
Pterocarpus erinaceus	st	t, med	GL	1

Table 7.3, *continued*

Family, genus, species	Items carved for commercial use	Other main uses	Ecoregion in which used	Source
Leguminoseae, *continued*				
Pterocarpus santalinoides	xylophone keyboards	med, fw	GL	1
*Pterocarpus soyauxii***	st	bp, c, t	CC, GL	5
Meliaceae				
Trichilia emetica	figures	med, shade, oil, ar	MPCF, SZI	3
Moraceae				
*Antiaris toxicaria****	dr, st	med, t		5
Oleaceae				
*Olea europaea ssp.africana****	b, sp	ch	TVB, ZMIO	10, 11
Rhamnaceae				
Berchemia discolor	b, st	sti, dy	ODFS, ZMOP	10
*Berchemia zeyheri**	b	fr	TVB, MPCF	3
Salvadoraceae				
Dobera glabra	mortar & pestles	st, med	SACB, MACB	5, 12

8

Effects of Economic Liberalization on Dao Women's Traditional Knowledge, Ecology, and Trade of Medicinal Plants in Northern Vietnam

Jennifer C. Sowerwine

Introduction

Vietnam is undergoing rapid economic change, embracing a market-oriented economy after more than 40 years of socialist central planning. It has tremendous natural resource capital, including valuable medicinal plants that contribute significantly to affordable primary health care. Much of the anthropological literature on biocultural diversity reports a rapid disappearance of indigenous botanical knowledge and healing practices associated with the demise of tropical forest ecosystems in the Third World (Greaves 1994; Brush & Stabinsky 1996; Carlson 2001; Florey 2001; Lizarralde 2001; Zent 2001). Yet in the highlands of northern Vietnam, despite the rapid decline of forest cover due to war, extensive logging, fires, and agricultural expansion, traditional herbal medicine persists.[1] In fact, traditional botanical medicine is becoming increasingly important in meeting urban and rural health care needs and in supporting the livelihoods of rural minority people. It is estimated that more than 80% of Vietnamese depend on traditional botanical medicine as their primary means of health care. This challenges claims that medicinal plant ecological habitats and knowledge are becoming historical relics in the twenty-first century.

In this chapter I analyze the effects of specific political-economic transformations on the knowledge, ecology, and trade of medicinal plants by Dao women in the mountains of northern Vietnam.[2] Specifically, I explore how these changes affect the long-term sustainability of medicinal plant biodiversity, the knowledge systems about those plants, and the landscapes within which this biocultural diversity is produced and reproduced. Rather than merely document the effects, I critically analyze the multiple and contradictory macrolevel policies. My focus is on Dao women because they are widely recognized as reputable healers and active suppliers of herbal medicine throughout northern Vietnam.[3]

Specific liberalization policies, intended to improve the health, socioeconomic, and ecological conditions in the mountainous regions, are having variable effects on the sustainability of Dao traditional botanical medicine in different locales. I demonstrate how this system of medicine endures, despite the emergent threats. Central to this argument is the premise that Dao traditional botanical medicine is not some static or coherent entity that must be preserved but a set of processes that are continually in flux across space and time.

Dao Traditional Botanical Medicine

I begin my discussion of the ways in which economic liberalization, or *doi moi,* policies are affecting the sustainability of Dao traditional botanical medicine by defining terms and phrases. Among the myriad "economic liberalization policies," I refer only to those that either directly or indirectly affect the sustainability of Dao traditional botanical medicine. These include changes in health care policy and programs, changes in land and forest policy, including intensification of national resettlement and reforestation programs, and opening of the market economy. "Dao traditional botanical medicine" refers to the dynamic and inseparable link between medicinal plants (biological diversity) and the culturally unique and gendered knowledge about those plants (cultural diversity). In order to generate and maintain knowledge about herbal medicine, there must be a healthy plant base; and in order for a plant to be or become a medicine, people must generate knowledge about its medicinal use through experimentation, practice, and transference to future generations. This inextricable link between medicinal plant diversity and cultural knowledge of the Dao constitutes a central component of the "biocultural diversity" of northern Vietnam (see Maffi 2001). "Sustainability" of Dao traditional botanical medicine refers to maintaining the conditions under which both the ecology of medicinal plants and the gendered knowledge about those plants can be produced (created, innovated) and reproduced or sustained (enabled to regenerate, transferred to younger generations) over space and time in the context of a rapidly changing political economy (Table 8.1).

The biocultural diversity of Dao traditional botanical medicine (both medicinal plants and cultural knowledge) is produced (created or innovated) and reproduced (maintained or transferred) over time through specific social and ecological processes. "Medicinal plants" are produced, or grow, in diverse microecological habitats. They are reproduced, or regenerate, through varying degrees of natural or anthropogenic disturbance such as fire, cultivation, grazing, harvesting, and protection. "Cultural knowledge" about medicinal plants is produced, or generated, through the treatment of everyday physiological processes such as becoming ill, fatigue, and childbirth. Many herbalists are also experimenting with new uses for botanical medicines. This knowledge is reproduced, or transferred from one generation to the next, through the cultivation of and movement through many different land-use types. Through the acts of walking to and from, and working within, swidden agricultural fields, knowledge is transferred from the older female generation to the younger. Knowledge about everyday treatments for the common cough, backache, and swollen or aching limbs, as well

Table 8.1
Sustainability matrix of Dao traditional medicine

	Biocultural diversity	
Sustainability	Medicinal plants (biological diversity)	Cultural knowledge (cultural diversity)
Production (creation / innovation)	Medicinal plant diversity is created or produced in diverse microecologies, including old forest, river, rock forest, swidden and fallow fields, and "bare hills."	Medicinal plant knowledge is innovated or produced to treat illness, facilitate the birth process, enhance general health, and through everyday experimentation.
Reproduction (maintenance / transference)	Medicinal plant diversity is sustained or reproduced through various disturbance and management regimes of cultivation, fire, grazing, harvesting, and regeneration, as well as protection.	Cultural knowledge of medicinal plants is maintained or reproduced through active movement across diverse landscapes. This knowledge is sustained through transference from mother to daughter or daughter-in-law.

as for pregnancy and postpartum cleansing, is typically transferred from mother-in-law to daughter-in-law. In-house residence of the mother-in-law is important for this link to be maintained.

The Socioecological Landscape

Dao traditional medicine is historically rooted in a dynamic socioecological landscape, a landscape in which ecological patterns and processes inherently shape and are shaped by agricultural, agroforestry, and other livelihood processes over time and space. Medicinal plants and knowledge about those plants are produced and reproduced by means of the creation of, and movement across, diverse landscapes over time, through a particular gendered labor process based on swidden agriculture and agroforestry practices. Because, according to my data, women and children are the primary agriculturists, medicinal plant knowledge is innovated and transferred to younger generations through the practice of walking to and from the fields, clearing and planting diverse swidden and wet rice fields, collecting firewood, harvesting non-timber forest products, and grazing buffalo. Through active management of these landscapes, resulting in varying degrees of disturbance and regeneration, the Dao produce a rich mosaic of microecologies uniquely suited to a broad spectrum of medicinal plant species.

Dao traditional medicine comprises both shamanistic and botanical healing practices, depending on the perceived cause of disease. If an ailment is believed to be caused by a spirit, shamans are summoned to perform ceremonies to exorcise the malignant spirit by appeasing it, often through satiating its "hunger" by butchering a pig or a chicken. Among the Dao Do (Red Dao)[4] there is only one spirit, La Hu Quay, whose illness manifestations can be successfully treated with herbs. This spirit is thought to cause the majority of chronic ailments, including those with cancerlike symptoms. Acute, common ailments such as a cough, stomachache, bruise, sprain, snake bite, or leech bite, may be treated first by ceremonial prayer and then by administration of a botanical medicine if symptoms persist. General postpartum healing for the mother and child involves herbal treatments, including bathing, drinking tea infusions, and applying poultices from a wide variety of plants. Depending on the disease, herbal remedies may comprise a single plant or may involve the gathering of more than 200 different species from many different habitats.

Traditional Dao settlement patterns and agricultural practices have historically fostered the development and regeneration of diverse ecosystems rich in biodiversity[5] and have enabled the sustained production and transference of traditional knowledge of botanical medicine. Dao settlement patterns can be characterized as permanent, whereas agricultural practices are a combination of rotational swidden agriculture within a fixed territory and permanent wet rice agriculture.[6] For example, a typical Dao family may actively cultivate 3–5 upland rice and corn fields at a time for 3–5 years, along with 4–5 wet rice fields, after which they leave the swidden land fallow for 4–20 years to allow regeneration before they return to cultivate that land again. Rather than "abandoning" former swidden fields, as is commonly assumed, the customary land-tenure rules allow families to maintain tenure over former swidden fields. Some families may have up to 15 swidden plots "in reserve" at various stages of regeneration, in which they graze animals and continue to harvest medicinal plants, fuelwood, edible fruits, tubers and bamboo shoots, bamboo, and wood for housing and pigsty construction. Customary rules that clearly define land-tenure boundaries and rights, based on ancestral land-clearance claims, serve as a means of ensuring this diversity of vegetation types and ecosystems across the montane landscape. Without these social sanctions of resource use and control, outsiders or other community members would have an incentive to clear the "abandoned" fallow fields as soon as they become fertile, after about 3–4 years.[7] Ironically, most of these swidden fields are officially classified as "barren land" under government classification schemes.

Contrary to common belief, the production and reproduction of medicinal plant diversity is not necessarily confined to old forests but is intricately linked to this diversity of land-use practices. Most medicinal plants used for home treatment are harvested from disturbed habitats: near the home, along the trail to the swidden or wet rice fields, or in the agricultural fields themselves. Ethnobotanical transects and participant-observation exercises revealed that different medicinal plants are found in conjunction with different types of land use. For example, some important medicinal plants grow only in freshly burned swidden fields. Others are found only in regeneration forests. Others grow only in heavily disturbed habitats, such as along trails or in intensively grazed areas. Some semi–shade-tolerant medicinal species require an overstory canopy that has been thinned of underbrush, whereas certain climber species require fairly significant forest structure for development. Others are found only along the edges of wet rice fields. It is crucial to understand the specific habitats in which medicinal plants grow, and how they regenerate, when designing and implementing conservation and development programs that may alter the landscape, as is currently happening in Vietnam.

The innovation and transference of medicinal plant knowledge is intricately linked not only to the creation of diverse landscapes but also to movement through them. Walking from the home to the swidden field one crosses many different habitats in which medicinal plants are typically harvested, such as in the forest around the home, along the edges of wet rice fields, in riparian ecosystems, in fallow corn or rice swiddens, on grassy hills, and in forests, growing on limestone karsts. As groups of family members and neighbors go to work the fields or collect firewood or fodder, children and grandchildren observe and acquire medicinal knowledge from their mothers, grandmothers, and/or mothers-in-law (Figure 8.1). For example, a cut on the finger produced while trimming the radicles off a *Canna* tuber is treated with the juice of Mia chuay (*Ageratum conyzoides* L.) to stop the bleeding. Bellyaches are treated by chewing on the young shoots of Gim bua (*Rubus parvifolius* L. or *R. triphyllus* Thumb.) or the young leaves of Ken tay tra (*Maesa montana* A. DC.).

The acquisition and transference of medicinal knowledge also occurs through everyday physiological experiences. When one falls ill, either a family member or a well-known herbalist is summoned to gather plants for treatment. While being treated, sometimes for several months, the patient becomes acquainted with the herbs and often masters the remedy. Upon successful treatment and exchange of knowledge, gifts of rice, distilled corn, rice or cassava "wine," silver coins, or chickens are offered to the healer as a form of reciprocity. Fatigue, backaches, and weakness caused by everyday labor can be treated with a blood-toning plant such as Tam kha mhay (big health vine; *Ventilago leiocarpa* Benth.) or Dia tan (Veronica sp.). Aching joints and rheumatism are treated with Hau dang (*Homalomena aromatica* Schott); sore legs, with Chu tay mhay (*Illigera* sp.). After a woman gives birth, a family member, typically the child's grandmother, is responsible for gathering the herbs needed to treat the new mother.[8] Plants are administered in the form of a bath or tea infusion, including Dia ui (*Millettia* sp.) and Gim tiu (*Uncaria* sp.), for the mother to regain her strength, appetite, and color as well as to induce lactation and reduce numbness in the extremities. Thus a woman usually learns medicine from her mother-in-law (or her mother, if she lives nearby) when she is of birthing age. Residence patterns also affect the knowledge acquisition process. Young families who move out of their parents' home typically know less about medicinal plants than do those families with a mother-in-law in the home. Families in which the mother-in-law died at an early age also possess less knowledge of even basic herbal remedies.

Contrary to common assumptions, not only is traditional botanical medicine not spatially confined to the old forest, it is not temporally confined to ancestral knowledge. Discussions with the Dao revealed that although herbal remedies are passed down from one generation to the next, many new medicines are continually being discovered through daily movement between homes and fields. For example, people who have a stomachache while walking to the swidden field gather herbs with known efficacy. However, if those herbs are unavailable or if the symptoms persist, they will experiment with other

Figure 8.1. Walking home from the neighboring village, Chao Lo May of Sapa transfers ethnobotanical knowledge on uses of medicinal plants to her children.

plants, carefully avoiding known toxic plants such as Yang meu (*Oldenlandia capitellata*), to find one with the desired therapeutic effect.

Sapa and Ba Vi

Two Dao villages were selected for case studies. The first was among the Dao Do in a highland village belonging to Sapa District, near the Chinese border. The topography is marked by peaked mountains, long, cultivated slopes, and wide, rambling valleys intricately terraced in wet rice agriculture (Figure 8.2). Mist surrounds the forested mountain tops, and hamlets of 20–30 households dot the patchwork landscape. The average elevation is approximately 1200 m. The village center is a two-hour drive over an undeveloped road from the district capital or a full day's trip from Hanoi by train. Most of the 10 hamlets in the village are accessible only by foot, approximately a 1–4-hour walk from the village center. Access to new technologies and services is limited. Only within the last few years has electricity been introduced through minihydroelectric generators. The Dao in Sapa are engaged in a mixture of terraced wet rice production and rotational rice, corn, canna, and cassava swidden production to meet their subsistence needs. Women are the primary keepers and practitioners of herbal medicine. They are not, however, involved in the trade of medicinal plants to the same extent as are the Dao women of Ba Vi.

Since Vietnam opened its borders with China, the cash economy in the Dao village in Sapa has boomed. The sale of cardamom (*Amomum aromaticum* Roxb.), an understory agroforestry crop in the Zingiberaceae plant family, soared in the last decade. At an average price of 25,000–30,000 VN dong per kilo, some families can earn up to U.S.$2000 per annum selling cardamom,[9] more than ten times the average annual per capita income of lowland ethnic Kinh. Requiring 50–75% canopy cover for optimum production, cardamom provides incentives for Dao villagers to protect the remaining old forest. It also provides incentives to villagers to allow their swidden lands to regenerate sufficiently, in order to create the ecological conditions necessary for cultivating large fields of cardamom. Widespread cultivation of cardamom is encouraged by the state as a source of income for highland minorities. However, land preparation involves significant clearing of understory vegetation, which may affect not only plant diversity but also the long-term integrity of the forest. Compared with the Dao in Ba Vi, the Dao in Sapa are living on the frontier; a borderland region where the effects of national policy implementation are limited. They have not been resettled, as have the Dao in Ba Vi, and reforestation programs have not yet begun. Thus, the farmers continue to engage in traditional forms of agricultural production, maintaining a landscape of diverse vegetation types.

The second case study was among the Dao Quan Chet in Ba Vi District, Ha Tay Province, adjacent to the buffer zone of Ba Vi National Park, 70 km outside Hanoi. Encompassing an area of 7377 ha, the park rises sharply to 1296 m above the agricultural plains. It was formally established as a national park in 1992; however, it has been managed as a forest reserve, a state forest enterprise, and a protected area since the early 1900s, first by the French and, since the 1940s, by the Vietnamese government. Park management zones are delineated according to elevation (AREA 1993). The strictly protected zone, which extends upward from the 400-m contour, contains all of the primary forest remaining in the park, totaling 1544 ha. Within this area, fewer than 950 ha are undisturbed or only slightly affected by human use. The particular use zone extends from the 400-m to the 100-m contour level and is designated primarily for rehabilitation and development. Extending for 1 km below the 100-m elevation is the buffer zone, which is designated as the habitation and agricultural area for the populations surrounding the park. The main functions in this zone include reforestation, agricultural production,

Figure 8.2. From the forested mountain ridges to the terraced valley bottoms, the Sapa landscape is a rich mosaic of land-use types, including swidden and fallow fields, that harbor diverse numbers of medicinal plants. Except for forests and wet-rice terraces, all land is classified as barren by the government and is scheduled to be reforested by 2010.

agroforestry, integrated home garden systems, and firewood plantations. However, land is scarce and poor in this area, and the results of these programs have been limited.

Formerly, the Dao practiced shifting agriculture in the park, as they continue to do in Sapa. Resettlement of the Dao began in 1965 and was completed in 1991. The area to which they were resettled is heavily eroded and lacks sufficient land and water to yield adequate crops for subsistence, not to mention surplus for sale. Due to lack or poor quality of wet rice land, some families continue to engage in swidden agriculture in the park, while others have intensified the collection of medicinal plants from the old forest within the park boundaries for sale. Park officials have criminalized both sets of activities. Village officials estimate that, since the opening of the market in the early 1990s, more than 80% of the local households rely on income from medicinal plant trade (Sowerwine et al. 1998). It is primarily Dao women who collect, process, and sell the herbal medicines. Contrary to many studies on the effect of economic reforms on women in Vietnam and other postsocialist countries (Aslanbeigui et al. 1994; Beresford 1994; Desai 1995; Goodkind 1995), the Dao women have reported an improvement in their standard of living in recent years from their increased trade of medicinal plants.

Since the early 1980s national park officials have contracted with local government officials to reforest former swidden land (100–800 m) within the park with eucalyptus (*Eucalyptus alba* Reinw. ex Blume), acacia (*Acacia mangium* Willd. and

A. auriculiformis A. Cunn. Ex Benth.), pine (*Pinus massoniana* D. Don.), and fruit trees in an attempt to regreen "the face" (oriented toward Hanoi) of the park. To compensate for the lost agricultural land, farmers have negotiated short-term subcontracts with their local leaders to grow cassava (*Manihot esculenta)* as an understory crop for three years, within the lower elevations of the park, until the canopy of those plantations closes and prohibits further cultivation. Uncertainty as to future food supply, due to insecure tenure regimes, is resulting in heavy "mining" of the soil through intensive cassava production and in increasing signs of resistance through "illegal" tree harvesting, destroying seedlings, and uprooting saplings (Figure 8.3). In sum, the establishment of the national park and the resettlement of the Ba Vi Dao, followed by widespread forest plantations, intensive cassava cultivation, and increased trade, are placing mounting pressure on the reproduction of Dao traditional botanical medicine.

The Effects of Economic Liberalization

Officially beginning in 1986, the Vietnamese government launched doi moi, a series of "economic renovation" strategies. This shifted the economy away from central planning toward a more market-oriented approach in an effort to achieve rapid economic development and modernization. This transition involved a range of economic liberalization policies and programs, including: decollectivization and allocation of forest and agricultural land to individual households; economic investment for development of montane areas, which included support for fixed cultivation and sedentarization of ethnic minorities, as well as reforestation of so-called barren lands; liberalization of domestic and foreign markets; and restructuring of public finance, including the reduction of health care subsidies. In this section I compare the effects of four doi moi agendas—health care reform, resettlement programs, forest land reform, and market reform—in two Dao villages on the long-term sustainability of Dao traditional botanical medicine. Although the intent of these programs was to improve the social, economic, health, and ecological conditions of the country, the results were often variable and contradictory. The Vietnamese government supports traditional medicine wholeheartedly. However, certain elements of the reform policies threaten the reproduction of the knowledge, ecology, and trade of medicinal plants, while others enhance it. Ethnobotanical data from this research suggests that even though the processes that constitute Dao traditional medicine are sustained, well-intentioned conservation and development policies must be critically evaluated to ensure that the objectives are indeed met.

Health Care Reform

Public finance restructuring has involved a significant reduction in government subsidies for health care. This has resulted in a rise in consumer demand for traditional botanical medicine due to its efficacy and affordability. A study commissioned by the World Bank in 1999 and conducted by the Vietnam Institute of Traditional Medicine (VITM) in Hanoi, found that 60–81% of the respondents considered traditional

Figure 8.3. This 7-year-old *Acacia* sp. plantation may indeed function to "regreen" Ba Vi National Park. However, cassava production below the canopy denudes all vegetation, including valuable medicinal plants, and depletes the soil of nutrients essential for natural regeneration.

medicine to be cheaper than "modern" medicine. One to 19% of the respondents said that traditional medicine cost nothing, because they collect or grow their own herbal medicines (Tran V. H. pers. comm. 2000). Another study, conducted in 1997–1998, by the VITM found that the expenditure for one bed in the traditional medicine hospital is 45% of the expenditure for one bed providing modern medicinal treatment (Tran V. H. pers. comm. 2000).

Despite the recent influx of Western pharmaceutical companies, many doctors and scientists in Hanoi estimate that more than 80% of the Vietnamese population, both rural and urban, still use botanical medicines regularly. In 1997 it was estimated that the proportion of herbal medicines consumed relative to Western pharmaceuticals increased from 20–30% in 1987 to 40–60% in 1997 (Thoi Bao Kinh Te 1997). The perception that botanical medicine has fewer side effects than pharmaceuticals has also enhanced its popularity. With the open policy on private enterprise, traditional healers began to open (or reopen) private practices, regaining the ability to provide greater service to the general population. In 1995 there were 350 registered traditional healers (S. Johnsen & K. E. Anderson pers. comm.), yet demand continues to exceed supply as people can be seen lining up outside the healers' offices.

Dao women herbalists in Ba Vi have reaped tremendous economic benefits from this intensified demand for traditional medicine by increasing their sale of herbal medicines to ethnic Kinh in the lowland areas of northern Vietnam. With improved transportation and communication networks, Dao women have been taking their herbal medicines to the road, selling as far north as Quang Ninh and as far south as Thanh Hoa Provinces. They often sell their medicines in marketplaces as well as directly in people's homes. Interestingly, during the American war in Vietnam, ethnic Kinh soldiers stationed in Ba Vi were treated successfully by Dao women herbalists. After the war the soldiers returned to their home provinces and told their friends and family of their successful treatment. Over the years Dao women have been invited to the former soldiers' villages to sell their medicine, eventually developing complex trade networks that continue to expand by word of mouth. With the advent of doi moi, travel and sales have increased significantly.

As the patient load increases, Dao women have intensified their harvest of medicinal plants in the national park. However, after the establishment of Ba Vi National Park in 1992, forest-product collection rights were terminated, essentially criminalizing former traditional land-use activities of the Dao. In the past, when the resource base was near their homes, Dao women harvested only the stems and leaves of plants (which allowed for regrowth). Now that the women herbalists must travel greater distances, they try to maximize the returns on their labor by harvesting the greatest substance-to-volume ratio, even if this compromises the regenerability of the plants. Moreover, the women report, someone else may collect the plant before they have a chance to return. A combination of factors, including an increase in demand for Dao herbal medicine, lack of rights to collect from the park, and spatial displacement from the resource base, results in unsustainable and overharvesting practices.

A second element of health care reform, which has variable effects on Dao traditional medicine, is the government program to provide health care to everyone. Each village has a local health station, typically staffed by an ethnic Kinh, that administers 96 medications approved by the Ministry of Health. Only one of these medications is composed of medicinal plants. Although extensive government policies are designed to integrate traditional botanical medicines with modern medicines, including Article 39 in the Constitution,[10] local community health stations often fail to do so. As a result, many in the younger generation are finding biomedical health care more convenient and more "modern." They are becoming less interested in collecting, preparing, and using local traditional medicines. Unfortunately, many diseases are not treatable with pharmaceutical medicines provided by the health stations. In both Ba Vi and Sapa, traditional botanical medicine practitioners persist. However, during this era of "modernization" and development, international support for strengthening local health stations without supporting the integration of local traditional botanical medicine may result in declining local interest in learning about and being treated with traditional medicine. How these dynamics may affect perceptions, attitudes, and behavior toward Dao traditional botanical medicine in the future remains an important issue, particularly if the resource base, in the case of Ba Vi, becomes increasingly threatened.

Lastly, with public-finance restructuring, subsidies for state pharmaceutical enterprises over the last ten years have declined significantly. Many of these companies, in an effort to survive, have shifted their mission from production of medications (largely based on medicinal herbs) to trade in Western pharmaceuticals and pharmaceutical equipment with overseas partners. This has reduced the demand for and pressure on several wild medicinal plant populations around Sapa, such as Hoang tinh (*Polygonatum* sp.), Hoang lien gai (*Berberis wallichiana* DC.), Dang sam (*Codonopsis javanica* (Blume) Hook. f.), and Cau dang (*Uncaria* sp.). It has also reduced the demand for many cultivated species, such as Tam that (*Panax bipinnatifidus* Seem.), Duong quy (*Angelica sinensis* (Oliv.) Diels), Xuyen khung (*Lingusticum wallichii* Franch), and Thao qua (*Amomum aromaticum*).

Resettlement and Reforestation

State ideological portrayal of highland landscapes as degraded, "barren," or empty and as populated by "backward" ethnic minority populations has served for centuries to legitimize state interventions into highland provinces along international borders to control and manage people and resources. This kind of spatial ordering by the state is what Vandergeest and Peluso call "internal territorialization" (1995: 388). The Vietnamese state's expression of internal territorialization has involved the use of maps, censuses, cadastral surveys, and land-use classifications to justify resettlement of ethnic groups engaged in various forms of shifting agriculture. It has also included heavy public investment in timber reforestation programs to create a source of revenue for the state. Government reforestation programs in Vietnam, borrowing from the nineteenth-century German model of scientific forestry, emphasize sustainable-yield monocrop plantations. This model reflects an ideology that disregards culture, biodiversity, and the intricate relationship between the two.

I argue that state strategies for classifying nearly 40% of the national territory of Vietnam as barren and attribution of that "barrenness" to ethnic minority land-use practices not only render invisible but ultimately undermine local land-use and land-management practices that have produced and reproduced medicinal plant diversity for centuries (Figure 8.4). Ethnobotanical transects and in-depth interviews provide evidence that these lands are not barren but, in fact, are rich repositories of botanical medicines and important sources of income for both the rural minorities and the ethnic majority Kinh population.

Since the 1960s the Vietnamese authorities have made every effort to sedentarize ethnic minority populations in order to stabilize volatile border regions, create a stable taxable population, arrest deforestation, which they incorrectly attribute primarily to swidden agriculture, and create a "modern" society, with wet rice symbolizing the most advanced stage of agricultural production. The Center for Ethnic Minorities and Mountain Areas (CEMMA) was established in 1992 to facilitate the "stabilization" of ethnic minority populations and the intensification of highland development. From 1993 to 1996 the program's budget increased more than fourfold (CEMMA/UNDP 1995). In 1995 the United Nations Development Programme pledged tremendous financial support for strengthening CEMMA's capacity and program implementation. In 1993 the government established Program 327, known as the "regreening of open land and bare hills," which earmarked a portion of the funds to encouragement of fixed cultivation and settlement through forest allocation to ethnic minority households,

Figure 8.4. Despite the government's classification of this hillside as "barren," it is an intensively managed patchwork of swidden or fallow fields in various stages of regeneration or protected for understory *Amomum aromaticum* production. Medicinal plants used locally for primary health care are found in all land-use types depicted.

infrastructure development, and increased provision of services. From 1993 to 1998 alone, this portion of the funds totaled nearly U.S.$28 million (MARD 2000). In 1997 the government established the Five Million Hectare Program (government decision no. 661/CT), with a primary focus on reforestation and resettlement programs and special emphasis on the montane and border areas. Policies and programs for upland resettlement are almost invariably linked to reforestation and development programs and are thus obtaining increasing support from multilateral and bilateral aid agencies interested in the latter. This international financing increases the capacity for implementing policy and, therefore, accelerates the processes that threaten the sustainability of Dao traditional medicine.

The effects of these resettlement policies on the knowledge, ecology, and trade of medicine among the Dao are multifaceted. In Ba Vi, Dao populations living within Ba Vi National Park were moved out of the strictly protected zone (elevation 400 m and above) in the initial phase of the sedentarization programs in 1968. By 1991 most Dao families were permanently resettled outside the national park, in the buffer zone. The alienation of Dao people from their ancestral land has created a spatial barrier to the reproduction of knowledge. While they were living on the mountain, medicinal plant collection was part of women's daily routines, but now it takes them nearly half a day to reach the old swidden fields and forest. Thus only those actively involved in the trade of medicinal plants, typically older women whose labor is no longer primary, can afford the time to go. Younger women are too busy cultivating their fields at lower elevations to follow their mothers and mothers-in-law to the medicinal plant collection grounds. Because people no longer cultivate diverse swiddens, they no longer move through diverse environments.[11] Thus traditional botanical knowledge acquisition, innovation, and transference are threatened.

Contrary to the conservation objectives of Ba Vi National Park, the resettlement of the Dao and delineation of strict boundaries around the park have actually led to unsustainable harvesting practices and increased degradation within the park. Some women have shifted their harvesting to neighboring hills in search of particular medicinal species, such as Binh voi (*Stephania rotunda* Lour.), that are increasingly difficult to find in Ba Vi. Many of these women are returning to their birth villages to gather plants, essentially undermining conservation objectives by transferring unsustainable harvesting practices to less visible areas. Ironically, the establishment of rigid national park boundaries and exclusion of local populations from its management appear to increase degradation of the resource base, not only within the park but beyond it as well.

In Sapa, on the other hand, although permanent settlements are encouraged, the Dao have been able to maintain their traditional agricultural and habitation patterns. Thus resettlement policies appear to have had little or no effect on the acquisition and transference of traditional botanical medicinal knowledge. However, current government land-planning maps reveal clear state intentions to regulate traditional agricultural lands and forests. The stated goal is to ensure stability and income opportunities for the Dao, as well to enhance watershed protection. This would be done primarily through the reforestation of highly productive swidden fields and fallow lands.

Of the 33 million ha of land in Vietnam, 19 million are designated as "forest land"; 12–13 million, as "barren land." Currently, about 60–65% of the land in the northern montane area is officially classified as "barren" (Poffenberger 1998). Despite official figures, which depict low forest cover in the highlands, a steady stream of medicinal plants from these regions continues to supply an increasing domestic and international demand for herbal medicines. It is estimated that 70% of the medicinal plants traded in the market originate in this region (Rake et al. 1993). It is also estimated that hundreds of thousands of people depend on the medicinal plant industry for employment (Van Tien 1991: 13). Income generated from trade of non-timber forest products, which include medicinal plants, often exceeds income derived from the sale of agricultural products such as rice (de Beer 1993). One village outside Hanoi produces 1700–2000 tons of dried medicinal plants per year, generating an income that is 2–3 times higher than that from the rice produced (de Beer 1993). Today the highlands of Vietnam supply more than 557 phytomedicinal products and 458 essential oils to the Vietnamese economy (Nguyen 1995). Results from my study also indicate that so-called barren land (actually under various stages of swidden regeneration, grazing land, or used in other productive ways) is high in medicinal plant diversity. Studies elsewhere in Southeast Asia (Agarwal 1994; Caron 1995) and other regions of the world (Hoskins 1982; Fortmann & Rocheleau 1985; Hecht et al. 1988) also document the high productivity and biodiversity found on so-called barren lands.

As international donors enter Vietnam and provide increasing financial support for conservation and development initiatives, state reforestation programs to "regreen" "barren land" are gaining momentum. The first of those efforts began in the mid-1980s with international support from the World Food Program (WFP). The goals of this program were to establish 68,000 ha of forest plantations and 16,500 ha of fruit orchards (Frankenfort 1994). It entailed a "food for work" plan in which villagers received grain in exchange for planting tree seedlings. In 1992 government decision no. 327/CT, regarding the use and reforestation of barren hills, combined resettlement, sedentarization, and reforestation programs under one umbrella. U.S.$230 million have since been invested in this project (Christ & Kloss 1998). Beginning in 1997 the Five Million Hectare Program was launched in an effort to establish 2 million ha of protected forest and 3 million ha of production forest between 1998 and 2010, with the potential for even greater foreign and domestic investment. The total estimated budget for the program is 37.6 trillion VN dong, or U.S.$2.68 billion (MARD 2000). The effects of these reforestation programs on the ecology of, and subsequent access to, medicinal plants may ultimately threaten the reproduction of medicinal plants, and the knowledge thereof, among local populations. In both field sites ethnobotanical transects, conducted in different ethnoecological land-use types, demonstrate the importance of "barren lands" and suggest potential degradation of medicinal habitats through reforestation.

In Sapa District, 66% of the Dao village land was classified as "barren" by the Land Administration Department of Sapa in 1998. Through in-depth interviews, landscape mapping, and ethnobotanical transects, however, I found that all of the land—other than rocky outcrops—is actively in use. I also found that the majority of medicinal plant species consumed on a frequent basis are collected within this so-called barren land. On landscape maps sketched by the villagers, the Dao identified 12 distinct ethnoecological land-use categories. In each of the three hamlets we conducted one 50-by-10-m transect through each land-use type—a total of 32 transects—to determine species richness of medicinal and useful plants as an indicator of diversity (Table 8.2; in some cases, one of the land-use categories did not exist in the social ecology of the hamlet).

Working with local healers, we collected both social and ecological data for each transect. Social data included information about the place-name and its meaning, distance from village, historical and current uses and disturbance of the land (cultivation, grazing, firewood gathering, medicinal collection, fire, logging, fallow, etc.), and land tenure (de facto versus de jure access to and control over land). Ecological data included location information (elevation, slope, aspect), canopy cover and height estimate, percentage cover of life-form (tree, shrub, vine, herb, grass), and visual estimates of the most abundant species in each transect. Each transect was sketched and photographed. Soil samples were collected at the 0, 25, and 50 m marks to be tested for pH, moisture, and texture, along with local classification of soil. A list of medicinal and other useful plants was generated and frequency of occurrence was noted by walking through the transect systematically with the healer. Detailed information about the use, distribution, relative importance, frequency of harvest, and availability of each plant was collected. One-meter plots at 10-m intervals on alternating sides of the transect

Table 8.2
Land-use types arranged by the highest average numbers of useful and medicinal plant species per 500 square meters in three Sapa hamlets

Land-use type	Hamlet 1	Hamlet 2	Hamlet 3	Average
Regeneration forest (3–5 years)	38	40	42	40
Old forest (without cardamom)	33	49	27	36
Around the home	37	22	35	31
Old forest (with cardamom)	24	30	39	31
River edge	28	33	26	29
Regeneration forest (10–20 years)	20	N/A	32	26
Rock forest	N/A	22	29	26
Regeneration forest (6–9 years)	23	25	24	24
Wet rice land	24	24	22	23
Regeneration forest (1–2 years)	22	23	24	23
Swidden	N/A	22	24	23
Barren	17	N/A	21	19

line were fully sampled as a further source of data on percentage of useful, relative to total, plants in the plot. Voucher specimens of each medicinal plant were collected and deposited in the Hanoi College of Pharmacy herbarium. Botanical determinations, soil analyses, and transect data analysis are currently under way.

Analysis of the transects, following local land-use classification, demonstrates the rich diversity and medicinal importance of various transitional habitats (Table 8.2). This is contrary to government classification of these lands as barren. Regeneration fields of 3–5 years have the highest average number of species per 500 m^2 (40), followed by the old forest without cardamom (36), around the home (31), and old forest with cardamom (31). Wet rice land, swidden land currently in use, and 1-year swidden regeneration lands have equal average numbers of useful and medicinal species per transect (23). Locally classified "barren land" falls last in terms of species richness (19), but the difference between the most diverse land-use type (40) and least diverse land-use type (19) is not immense. This suggests that useful and medicinal plants are not confined to the old forest alone but are widely distributed across the landscape.

Although no reforestation programs have yet been implemented in Sapa, plans are under way to reforest all land-use categories collectively classified as "barren" under the Five Million Hectare Program. This includes everything but old forests with and without cardamom, rock forests, and wet rice lands. These programs could lead to rapid and significant alteration of the existing mosaic of diverse ecosystems that harbor rich biodiversity of useful and medicinal plants for the local communities.

Reforestation programs in Ba Vi have a different history. Beginning in the 1980s the WFP program reforested large tracts of "degraded land" in the park, largely using introduced *Eucalyptus alba, Acacia mangium,* and *A. auriculiformis.* Other commercial species, such as *Pinus massoniana, Cinnamomum camphora* L., *C. cassia* Presl, *Cinchona ledgeriana* Moens, *Euphoria longan* Lour. Steud, and *Camelia sinensis* L., were also planted (AREA 1993). According to local Dao farmers, many of these plantations are on former swidden fields and settlements, areas where women used to collect medicinal plants on a regular basis. Since resettlement and the conversion of their former swiddens into monocrop plantations, however, women must collect most medicinal plants from the remaining "old forest" and regeneration forests in the strict protection zone of the park—above the 400-m contour level—or from neighboring mountains.

As in Sapa, 12 distinct land-use types were identified on Ba Vi mountain (Table 8.3): two natural forest types (old forest and regeneration forest), three plantation forests (pine, eucalyptus, and acacia), two swidden types (cassava and canna), riparian zone, "bare hill," wet rice field, and holy forest. Three transects were conducted in each land-use type in order to determine species richness.[12] Again as in Sapa, the regeneration forest has the highest average number of medicinal and useful species (33), followed by the old forest (31) and active cassava swidden field (30).[13] The least rich land-use types are holy forest (21), followed by wet rice field (22) and eucalyptus plantation (23). Interestingly, the pine-plantation transects have a relatively high average number of species (27). This may be attributed to the age of the three pine plantation—more than 80, 35, and 15 years old, respectively—in which natural regeneration has occurred.

Table 8.4 demonstrates the relative importance of the medicinal and useful species in each land-use type, ranked by healers as most important, of average importance, or less important. The highest average number of "most important" medicinal and useful species are in the regeneration forests (19). Old forests provide the second highest number of most important species (16), followed by "grass forests" or "bare hill" (15.3). The land-use types with the lowest number of most important species are wet rice fields (5.3), holy forest (6),[14] and eucalyptus plantation (6). Although pine plantations were third highest in terms of medicinal and useful species richness (in Table 8.3), they rank eighth in terms of harboring "most important" species (7.3) in Table 8.4. Contrary to national land-use classification schemes, "bare hills" are in fact not barren but, as demonstrated here, important sources of medicine. Some of the species considered "most important" by the Dao for medicine, found in the so-called barren land, are Dia senh mieng (*Psychotria reevesii*), Lay go mieng (*Clerodendron crytophyllum* Turcz.), Nham nho nh'ha (*Ficus palmatiloba*), Sat tong mhay (*Tetracera scandens* L.), Ta chem dieng (*Litsea citrata*), and Khuc Khac (*Smilax glabra* Roxb.).

The highest percentage of "most important" medicinal plants per land-use type follows a pattern which is similar to that of the total number of most important plants. In the regeneration forest transects 57% of the species were considered "most important" by the Dao, followed by old forest and bare hill, both at 51%. The combined percentage of species with most and average importance indicates that the regeneration forest has the highest (95%), followed by wet rice fields (91%). Of all plant species found in the bare hills, along the rivers, and in *Canna* swidden fields, 90% are considered "most important" or of "average importance." The least important land-use type in terms of valuable medicinal plants appears to be pine plantations, in which only 55% of the plant species are considered of highest or average importance, followed by eucalyptus plantations (60%). Tables 8.3 and 8.4 clearly indicate that medicinal plants again are not confined to old forests alone but occur in diverse habitats. As in Sapa,

Table 8.3
Land-use types arranged by the highest average numbers of useful and medicinal plant species per 500 square meters in four Ba Vi sites

Land-use type	Site 1	Site 2	Site 3	Site 4	Average
Regeneration forest	30	29	39	–	33
Old forest	32	30	33	30	31
Cassava swidden	28	37	24	–	30
Pine plantation	32	23	27	–	27
Riverside	31	29	22	–	27
Acacia plantation	18	25	35	–	26
Old settlement	30	24	21	–	25
Canna swidden	25	22	26	–	24
"Grass forest" or bare hill	33	14	25	–	24
Eucalyptus plantation	20	25	–	–	23
Wet rice field	25	23	18	–	22
Holy forest	21	–	–	–	21

Table 8.4
Land-use types arranged by relative importance of species

Land-use type	Average number of most important species	Percentage of most important species /total	Percentage of species of most importance and average importance /total
Regeneration forest	19	57	95
Old forest	16	51	86
"Grass forest" or bare hill	15.3	51	90
Cassava swidden	10.6	36	69
Riverside	10.3	38	90
Acacia plantation	9	35	76
Canna swidden	8.6	36	90
Pine plantation	7.3	27	55
Old settlement	7	28	67
Eucalyptus plantation	6	27	60
Holy forest	6	29	76
Wet rice field	5.3	24	91

so-called barren lands in fact prove to have relatively high species richness and relatively high importance of medicinal plants for the health care needs of the Dao. Regeneration forests ranked highest in terms of both number and percentage of most important species. Both pine and eucalyptus plantations rank relatively low in terms of relative importance of medicinal and useful species. Thus old forests and natural regeneration forests—former swidden fields—harbor a greater diversity of medicinal and other useful species than do reforestation plantations of eucalyptus and pine.

Lastly, I examined each transect to determine the number of most important species that local healers considered rare or difficult to find. The land-use types with the highest average number[15] of rare/most important species were the regeneration forests (3.6) and old forests (2.75). The land-use types in which only one rare/important species was found were eucalyptus plantation, old settlement, river transect, and wet rice field. *Canna* swidden fields had no rare/most important species. The number of important rare species that were found in only one land-use type were pine plantation (3),[16] old forest (2), regeneration forest (2), acacia plantation (2), and cassava swidden (1). Some of those habitat-specific species were Bay la mot ha (*Daiswa* sp.), found only in the old forest, Mu chau (*Dracontomelum duperreanum* Pierre), and Pieu lam (*Canaria pimela*), found only in an eight-year-old acacia plantation that had not been intercropped for five years, Sung ken (*Polygonatum tonkinense*), found only in the regeneration forest, and To dau dieng (*Sapium discolor* (Champ.) Maell. Arg.), found only in the old pine plantation. This indicates that rare and important medicinal plants for the Dao are not confined to one particular habitat but exist within a diverse set of habitats in transition. The data suggest that natural regeneration forests, rather than plantations, would be better policy prescriptions to "regreen" barren land and conserve medicinal biodiversity simultaneously. The low number of important/rare species in the cassava and canna swiddens may be due to the fact that these lands are now under permanent cultivation, with little or no period of regeneration, due to restrictions on swidden agriculture on Ba Vi mountain.

In summary, the establishment of the national park, resettlement, and reforestation programs, supported by international conservation and development aid, may actually have undermined (in Ba Vi) or threatened to undermine (in Sapa) the conservation of both medicinal plant biodiversity and knowledge among the Dao. Resettlement of Dao populations to marginal lands outside the park in Ba Vi has created a spatial barrier to the reproduction of knowledge. No longer having formal rights to the resource, Dao women lack incentive to harvest medicinal plants sustainably, placing further pressure on the resource. Reforestation with single timber and fruit tree species, which belong to the national park, has converted diverse ecosystems, represented by various stages of swidden regeneration, into habitats that hinder important medicinal plant reproduction and community economic stability.

Market Reform

Liberalization of both international and domestic markets has also had variable effects on the knowledge, ecology, and trade of medicinal plants. Open borders have enabled increased foreign investment, including conservation and development aid. This aid may be strengthening government policies that often undermine the processes that sustain Dao traditional botanical medicine. New trade relations have resulted in increased exports of herbal medicines from Vietnam to mainland China, Taiwan, Korea, and some European and U.S. markets (Brevoort 1998), and U.S. tea and drug development companies have already begun to bioprospect in Vietnam. The liberalization of domestic markets has intensified the trade of medicinal plant materials from the highlands in lowland consumer markets. Within the last few years several domestic pharmaceutical companies have begun to produce traditional botanical remedies in "modern" form, appealing to the new generation of consumers. These new developments raise three issues: long-term prospects for sustainability of medicinal plant diversity in the context of rapidly opening markets, intellectual property rights and proper benefit sharing for ethnic minorities who make their knowledge of botanical medicine available to individuals, companies, or state entities, and the effect of long-term availability of medicinal plants for local health care needs.

With the more fluid border with China, the fruits of the cardamom (*Amomum aromaticum*), which the Dao have produced in small quantities for generations as a stomachache treatment, are fetching increasingly higher prices. Large volumes are being sold across the border to be used as a central spice in many Chinese dishes. As of September 2000 the price for cardamom had skyrocketed to 120,000 VN dong per kilo, or approximately U.S.$8.50.[17] As a result, the Dao are now investing in cardamom cultivation through natural swidden regeneration. Ironically, the rise in cardamom trade may result in increased forest cover, independent of state reforestation efforts. However, cardamom cultivation requires the clearing of underbrush, potentially undermining medicinal plant diversity.

As household incomes rise from the sale of *Amomum*, the Dao in Sapa sell fewer medicinal plants to the local market, since the average price of 2000 VN dong (less than U.S.$0.15) per kilo is not worth their labor. This reduces pressure on the medicinal plant resource base. However, some villagers are still actively involved in the trade of more than 20 species of plants, including Chim yang doi (*Smilax glabra*), Gim tieu (*Uncaria* sp.), Gia bi gim (*Acanthopanax* sp.), Phan chiet (*Panax pseudo ginseng* Wall.), and Chong phim lau (*Schefflera* sp.). The Dao in Sapa are particularly concerned that some special plants used for postpartum cleansing are being overharvested by migrant Kinh for other purposes, due to the new market demand. The market changes are thus having limited effects on the reproduction of medicinal plants and do not appear to have any immediate significant impact on medicinal plant knowledge of the Dao.

In Ba Vi, however, with the rise of domestic markets for traditional botanical medicine, tension is beginning to emerge between biodiversity/knowledge conservation and sustained income. Open markets and reduction in government subsidies for health care are resulting in an increase in domestic demand for herbal medicine. On one hand, this has resulted in enormous economic benefits for Dao women—and some men—in Ba Vi, who have expanded the collection and trade of medicinal plants. In an effort to capture some of those economic benefits, many women, previously not involved in the trade, quickly learned from their neighbors. Thus, in the early 1990s the medicinal knowledge base among women in the village expanded. On the other hand, this intensified trade has increased pressure on the resource base, potentially threatening its long-term sustainability.

Newly introduced cash crops are also affecting the ecology and knowledge of medicinal plants among the Dao of Ba Vi. With the contemporary devolution of agricultural decision making to the household, the Dao began to cultivate cassava in 1998, in order to supply raw material to biscuit and candy factories in and around Hanoi. Cassava production is changing household labor relations and apparently affecting the reproduction of knowledge and medicinal plant diversity. Because it is extremely labor intensive, it consumes more of men's and women's time than do other agricultural activities. As a result, women in families that are short on labor—such as younger families—have less time to follow their mothers and mothers-in-law to the forest and market and thus miss opportunities to learn how to identify, harvest, process, and treat with medicinal plants. As mentioned above, because of their inability to recognize medicinal plants, these women tend to weed them out during cassava production. Older women, on the other hand, who are knowledgeable about herbal medicine, retain the medicinal plants while weeding. No longer participating in the trade of herbal remedies, the young Dao women forgo an important source of personal income, because money earned from cassava production is controlled and managed by men.

For the last few years the Dao in Ba Vi have been intensively cultivating cassava, not only in the allowed buffer zone but also within the boundaries of the park, through informal arrangements with local government leaders. Cassava cultivation sometimes extends above the 800-m contour into the strictly protected zone, thus undermining the conservation objectives of the national park. However, the cash earned from sale of cassava helps compensate for Dao farmers' lack of wet rice land (a typical family grows only enough rice for 2–3 months of consumption).[18] In order to ensure high yields, cassava fields must be carefully maintained, including weeding 2–3 times per season. This intensive weeding involves the removal of all other growth, including medicinal plants, ultimately reducing the seed bank, root stock, and nutritional quality of the soil. Rather than enhanced conservation of its biological resources, current conservation and reforestation policies in Ba Vi National Park are resulting in increased ecological degradation.

In response to the changes in supply and demand of medicinal plants, some older women have begun to plant medicinal plants in their home gardens. This innovative response may help sustain the gene base and raw material for Dao health care. Based on a study conducted by the Hanoi College of Pharmacy, however, very few young families grow medicinal plants in their home gardens (Tran V. O. pers. comm. 1999). This may reflect the erosion of medicinal knowledge caused by resettlement policies and changing land-use practices.

Discussion and Conclusions

Vietnam has an incredibly rich diversity of both peoples and plants and of innovative knowledge systems connecting the two. Although Dao women's knowledge of traditional medicine continues to thrive, current macroeconomic reform policies and programs, strengthened by emerging development and conservation aid, may begin to undermine possibilities for the sustainable reproduction of those knowledge systems and the biocultural landscapes within which they have been sustained for centuries. However, rather than portraying Dao traditional medicine as on a unilinear trajectory toward demise, I argue that the processes of knowledge acquisition and transference and of medicinal plant production and reproduction will be sustained, albeit in new and variable forms in different locales.

Because Sapa continues to be on the margins with regard to implementation of national policies and market liberalization (no resettlement, no reforestation, less integration into the market economy), the effects of these changes have been less prominent there than in Ba Vi. Processes of innovation and transference of medicinal knowledge continue to thrive. Rotational swidden agriculture maintains a landscape of varying and diverse habitats, fostering medicinal plant diversity. The immediate threat, challenging the sustainability of medicinal knowledge in Sapa, is the intensification of Western health care ideology channeled through local health stations. In an effort to become more "civilized,"[19] many younger Dao visit the local health station regularly to buy "modern" medicines. As foreign aid projects continue to invest heavily in the health sector, local health stations will become more powerful and may undermine the value, and thus the practice, of safe, effective, and affordable traditional botanical medicine.

In Ba Vi, although knowledge of traditional botanical medicine is strong, economic liberalization policies are having a more profound effect. Several interconnected processes are threatening the long-term sustainability of Dao traditional medicine. Exotic monocrop plantations of pine, acacia, and eucalyptus have undermined rich repositories of medicinal plants and other useful biological resources. Resettlement programs have created spatial and temporal barriers to access to the resource base, thus hampering the reproduction of Dao women's knowledge of medicinal plants. The lack of rights to the resources in the newly established park, provides disincentives for the sustainable harvesting of medicinal plants. Market reforms, which enable rapid economic development, are placing disproportionate demands on young Dao women's labor, thus reducing the time they can devote to learning the medicinal plant trade from their elders.

Despite these threats, Dao traditional botanical medicine is persisting in innovative and flexible ways. In Ba Vi, older women, familiar with the mountain landscape, continue to collect medicinal plants for sale and treatment, despite sanctions against doing so. Several young women have apprenticed themselves to these older women, learning the art and science of collecting, processing, selling, and administering herbal medicines. One can also observe innovation in response to the settlement process, with the cultivation of key medicinal plants in home gardens. This may not produce enough for sale, but it can provide the basis for local health care needs. We may even witness future innovation or discovery of new medicinal remedies by Dao women through experimentation with, cultivation in, and movement through the new "resettlement ecologies" in which they live today.

Innovation in processing and storage technologies is reducing postharvest loss and helping to reduce pressure on the resource base. In response to expanding long-distance markets, the Dao have begun making *cao,* a sticky extract or substrate of more 200 medicinal plants, used as the basis for many remedies (see Figure 8.5). A small piece of cao can substitute for a large volume of fresh or dried plant matter. Smaller packages of materials enable women to trade over greater distances and for longer periods of time. Medicinal plants in the form of cao can also be stored much longer than before, without concern for mold or insect loss. The Dao have also recently begun to store dried plant matter in plastic, rather than canvas, bags, in order to prevent damage from moisture and insects.

Well-intentioned conservation and development policies in Vietnam are having multiple and often contradictory effects on the long-term sustainability of medicinal plant biodiversity, the knowledge systems about the plants, and the landscapes in which they are produced and reproduced. For example, nationwide efforts to improve health care by strengthening local health stations are commendable. Yet these health care initiatives could be more effective and affordable if they incorporated local traditional botanical health care systems into the overall health care strategy. Working with local women healers to identify specific health care needs that are not currently met by local traditional medicine could better inform programs geared to improve local health stations.

State programs to reforest "barren" hills may appear to be an excellent conservation objective, but ethnobotanical research reveals that highland landscapes, which may appear barren to the naked eye, are in fact rich repositories of medicinal plants in various stages of swidden cultivation and fallow. Reforestation of these swidden lands may actually undermine existing medicinal plant habitats and local health systems based on women's knowledge of those plants.

Establishment of national parks for conservation of biodiversity, without taking into account the interdependence between local peoples and those resources, undermines both conservation objectives and the socioeconomic well-being of the adjacent communities. Introduction of high-yield cash crops improves the local economy, but at the expense of the reproduction of traditional knowledge and ecology of botanical medicines, as well as of sustained sources of income for women.

The future of Dao traditional botanical medicine will be shaped by constraints and opportunities that arise from interactions among biocultural knowledge, institutions at the local level, and political-economic processes at the state and international levels. Efforts to conserve traditional botanical medicine must move beyond locally specific solutions, such as education programs and home gardens, to consider greater political-economic forces effecting change. Development and conservation policies must be coordinated to facilitate, rather than undermine, the sustainability of the processes that constitute the biocultural diversity of Dao traditional botanical medicine. Rigorous policy analysis, based on rich, contextualized ethnobotanical data, can provide the basis for better informed macrolevel policy decisions. It is this link between microlevel processes and macrolevel change that must be critically analyzed in order to ensure the provision of safe, effective, and affordable health care for rural and minority populations now and in the future.

Acknowledgments

This research would not have been possible without logistical support from the Hanoi College of Pharmacy, especially Tran Van On, Pham Gia Hue, and Dinh Hien Van, and my team of field research collaborators: Hoang Van Lam, Nguyen Thi Thanh Huong, Do Quyen, Nguyen Manh Tuyen, Pham Thu Nga, Vu Thu Ha, and Nguyen Thi

Figure 8.5. Trieu Thi Bay of Ba Vi prepares a batch of cao, a sticky black substrate used as the basis for many herbal remedies, from a decoction of more than 200 medicinal plant species. This innovative way of processing, which concentrates the plant medicine, enables it to be stored longer and also lightens the packs of Dao women traders.

Hien. I thank the Ministry of Health of the Socialist Republic of Vietnam for granting me permission to conduct research and the Botany Department at the Hanoi College of Pharmacy for processing, storing, and determining my voucher specimens. The research was funded by a Social Science Research Council International Dissertation Research Fellowship and the Fulbright-Hays Doctoral Dissertation Research Abroad

Program. Special thanks to Suraya Afiff, Lea Borkenhagen, Tom Carlson, Luisa Maffi, Jennifer Sokolov, and Emily Yeh for providing comments on earlier drafts of the chapter. Most important, I am indebted to the people of Ba Vi and Sapa for sharing their time, experience, and wisdom with me.

Notes

1. According to the Vietnamese Forest Inventory Planning Institute, forests covered about 45% of Vietnam's total national territory in 1945. Since then, the area under forest cover has been reduced to approximately 26% of the national territory. In the northwest, forest covered only about 13% of the land in 1995 (Vu et al. 2000).

2. The Dao (pronounced *Zao*)—otherwise known as Yao (in China) or Mien (in Laos and Thailand)—are the ninth largest of the 54 ethnic groups (including the ethnic Kinh majority) in Vietnam. Numbering about 500,000, they reside primarily in the northern highlands at middle to high elevations. The Kinh are the largest ethnolinguistc group in Vietnam, making up roughly 87% of the population. Their primary language is Vietnamese, and they reside principally in the wet rice delta and coastal regions of the country.

3. The data presented in this chapter are based on 16 months (February 1999–June 2000) of fieldwork in two Dao villages in northern Vietnam. I take sole responsibility for the ideas offered.

4. Based on cultural and linguistic differences, the Dao in Vietnam are classified into at least six subgroups: Dao Trang (white Dao), Dao Quan Chet (tight-trouser Dao), Dao Tien (hanging-coin Dao), Dao Do (red Dao), Dao Thanh Y (green-shirt Dao), and Dao Lo Gan (no known translation) (Dang et al. 1993).

5. Although biodiversity can be defined in a number of ways, I use species richness—the number of species per unit area—as an indicator of medicinal diversity.

6. Swidden agriculture is a highly sophisticated form of rotational cultivation on sloping hill lands that farmers have developed to suit particular ecological conditions. A typical swidden agricultural cycle can be described as follows: clearing a plot of sloping land, allowing the brush to dry, setting fire to the brush to release nutrients into the soil, cultivating corn, rice, cassava, or some other crop—often intercropping—for 3–5 years, then leaving the land fallow for 1 to 20+ years, depending on population size, land availability, and cultural practices, before returning to the plot.

7. There is some evidence of attempted "encroachment" on Dao fallow lands in the village in Sapa by the ethnic Hmong, which was promptly stopped by leaders of the Dao clan responsible for that land.

8. Residence after marriage is patrilocal. The young couple typically lives with the husband's parents until their children are old enough for the family to establish an economically independent household. It is usually the eldest son who stays with the parents and ultimately takes over household responsibility. The wife's village is often several hours away from the husband's, so it is typically the husband's mother, rather than hers, who collects medicines for childbirth.

9. This figure is based on October 1999 prices, when U.S.$1 equaled 14,000 VN dong.

10. Article 39 states: "The State mobilizes and organizes all social forces in the building and development of Vietnamese medicine following a farsighted orientation; prevention shall be combined with treatment, traditional medicine and pharmacology with modern medicine and pharmacology, State health services with peoples' health services" (Socialist Republic of Vietnam 1993).

11. The existing canna and cassava swidden fields are now under permanent cultivation, due to the changes in access and tenure. They will be cultivated to the point at which natural regeneration will be compromised, thus disrupting the cycles and patterns of medicinally diverse regeneration fields.

12. Only one transect through "holy forest" was conducted because only one holy forest exists in the village; only two eucalyptus plantation transects were conducted; and an extra old forest transect was conducted.

13. It is important to note that, although cassava fields are typically intensively weeded, women healers who happen to manage these particular cassava fields informed us that they are careful not to remove medicinal plant species. This may help to explain the relatively high species richness.

14. The low diversity in the holy forest can be explained by its history of use. It was cultivated continually from 1963 to 1990. The cooperative then reforested the area with acacia trees intercropped with cassava. In 2000 the trees were harvested, and the forest was recently designated as cemetery land where all other land uses are forbidden.

15. I calculate averages rather than total numbers of medicinal and useful species per land-use type because the total number of transects per land-use type differ.

16. All three rare and most important species were found in the transect of the 80-year-old pine plantation, where grazing and collecting fuelwood and medicinal plants are the primary disturbances.

17. The winter of 1999 was extremely cold—Sapa seeing its first snow in 16 years—and many of the cardamom plants died. Farther north, in China, the freeze destroyed most of the crop as well, thus driving up the price of cardamom. Although the yield of fruits in Sapa dropped significantly, because of the inflated prices many Dao families earned more income than they had in the previous year.

18. Actually, men spend much of the cassava money on showy objects such as motorbikes or concrete homes instead of investing it in long-term reproduction of the household such as education, food processors, or livestock.

19. An issue worth mentioning, but beyond the scope of this chapter, is the widespread government effort to "civilize" the "backward" ethnic minority peoples in the mountainous areas. Government radio stations broadcast programs daily in the "deep and far regions" of Vietnam, encouraging ethnic minorities to adopt "modern" lifestyles and habits.

Literature Cited

Agarwal, B. 1994. A Field of One's Own: Gender and Land Rights in South Asia. Cambridge and New York: Cambridge University Press.

AREA. 1993. Ba Vi National Park Management Plan. Hanoi: Association for Research and Environmental Aid.

Aslanbeigui, N., S. Pressman & G. Summerfield (eds.). 1994. Women in the Age of Economic Transformation: Gender Impact of Reforms in Post-Socialist and Developing Countries. London and New York: Routledge.

Beresford, M. 1994. Impact of Macroeconomic Reform on Women in Vietnam. VIE/93/w02. Hanoi: UNIFEM.

Brevoort, P. 1998. The Booming U.S. Botanical Market: A New Overview. HerbalGram 44: 33–46.

Brush, S. B. & D. Stabinsky (eds.). 1996. Valuing Local Knowledge: Indigenous People and Intellectual Property Rights. Washington, D.C.: Island Press.

Carlson, T. J. S. 2001. Language, ethnobotanical knowledge, and tropical public health. Pp. 489–502 *in* L. Maffi (ed.), On Biocultural Diversity: Linking Language, Knowledge, and the Environment. Washington, D.C.: Smithsonian Institution Press.

Caron, C. 1995. Magic, *wind-borne fear, and nature's nerf ball: Ethnobiology in Siruvattukkadu Kombai.* Institute of Current World Affairs, May 5; Pp 1–20.

CEMMA/UNDP. 1995. Framework for External Assistance to Ethnic Minority Development. Hanoi: Center for Ethnic Minorities and Mountainous Areas and the United Nations Development Programme.

Christ, H. & D. Kloss. 1998. Consultancy Report No. 16: Land Use Planning and Land Allocation in VN with Particular Reference to Improvement of Its Process in the Social Forestry Development Project (Song Da). Hanoi: Ministry of Agriculture and Rural Development, Social Forestry Development Project.

Dang, N. V., T. S. Chu & L. Hung. 1993. Ethnic Minorities in Vietnam. Hanoi: Gioi.

De Beer, J. 1993. Nonwood forest products in Indochina, focus: Vietnam. Working paper submitted to the Food and Agriculture Organization of the Untied Nations, Rome.

Desai, J. 1995. Vietnam Through the Lens of Gender: An Empirical Analysis Using Household Survey Data. Hanoi: United Nations.

Florey, M. 2001. Threats to indigenous knowledge: A case study from eastern Indonesia. Pp. 325–342 *in* L. Maffi (ed.), On Biocultural Diversity: Linking Language, Knowledge, and the Environment. Washington, D.C.: Smithsonian Institution Press.

Fortmann, L. & D. Rocheleau. 1985. Women and agroforestry: Four myths and three case studies. Agroforestry Systems 2: 253–272.

Frankenfort, D. 1994. World Food Programme Wood Market Survey for Proposed WFP: Assisted Project 5322. Hanoi: World Food Program.

Goodkind, D. 1995. Rising gender inequality in Vietnam since reunification. Pacific Affairs 68(3): 342–359.

Greaves, T. (ed.). 1994. Intellectual Property Rights for Indigenous Peoples: A Sourcebook. Oklahoma City: Society for Applied Anthropology.

Hecht, S., A. B. Anderson & P. May. 1988. The subsidy from nature: Shifting cultivation, successional palm forests, and rural development. Human Organization 47(1): 25–35.

Hoskins, M. 1982. Social forestry in West Africa: Myths and realities. Paper presented at the annual meeting of the American Association for the Advancement of Science.

Lizarralde, M. 2001. Biodiversity and loss of indigenous languages and knowledge in South America. Pp. 265–281 *in* L. Maffi (ed.), On Biocultural Diversity: Linking Language, Knowledge, and the Environment. Washington, D.C.: Smithsonian Institution Press.

Maffi, L. (ed.). 2001. On Biocultural Diversity: Linking Language, Knowledge, and the Environment. Washington, D.C.: Smithsonian Institution Press.

MARD. 2000. National Five Million Hectare Reforestation Programme, 1998–2010. Hanoi: Ministry of Agriculture and Rural Development.

Nguyen, T. D. 1995. Geography of Vietnam. Hanoi: Gioi.

Poffenberger, M. (ed.). 1998. Stewards of Vietnam's Upland Forests: A Collaborative Study. Research Network Report No. 10. Berkeley: University of California, Berkeley, Center for Southeast Asia Studies, Asia Forest Network.

Rake, C., G. Meyer, V. T. Luong, Q. L. Nguyen, D. Q. Vu, V. T. Nguyen & C. A. Do. 1993. Markets of Important Forest Products, Non Timber Forest Products and Agricultural Products in the Provinces Hoa Binh, Son La and Lai Chau in the North West of Vietnam. Baseline Study No. 4. Hanoi: Social Forestry Development Project Song Da.

Socialist Republic of Vietnam. 1993. The Constitution of 1992. Hanoi: Gioi.

Sowerwine, J., H. D. Nguyen & M. Poffenberger. 1998. Ba Vi National Park and the Dzao. Pp. 71–89 *in* M. Poffenberger (ed.), Stewards of Vietnam's Upland Forests: A Collaborative Study. Research Network Report No. 10. Berkeley: University of California, Berkeley, Center for Southeast Asia Studies, Asia Forest Network.

Thoi Bao Kinh Te (Economic Times). 1997. Quoted in Vietnam News, December 3.

Van Tien, L. 1991. Country Paper Vietnam: Contribution to the Regional Expert Consultation on Non-Wood Forest Products in the Asia-Pacific Region. Bangkok: RAPA.

Vandergeest, P. & N. L. Peluso. 1995. Territorialization and state power in Thailand. Theory and Society 35: 385–426.

Vu H. M., V. T. Nguyen & H. Warfvinge. 2000. Task Force I: Clarification of the National Five Million Hectare Reforestation Programme. Draft Report. Hanoi: Partnership for the National Five Million Hectare Reforestation Programme.

Zent, S. 2001. Acculturation and ethnobotanical knowledge loss among the Piaroa of Venezuela: Demonstration of a quantitative method for the empirical study of TEK change. Pp. 190–211 *in* L. Maffi (ed.), On Biocultural Diversity: Linking Language, Knowledge, and the Environment. Washington, D.C.: Smithsonian Institution Press.

9

Medical Ethnobotany and Therapeutic Uses of *Croton lechleri* Müell.-Arg. for Human Health

Thomas J. S. Carlson

Introduction

In ecosystems throughout the world, ethnolinguistic groups have selected and developed bioactive plants into therapeutic botanical medicines. Botanical medicines from thousands of different plant species contribute to the primary health care needs of 80% of people on earth (Farnsworth et al. 1985). Therapeutic pharmacological activity has been confirmed in medicinal plant species from around the world in experimental biology studies on a spectrum of different disease groups, including respiratory disease, cardiovascular disease, Type 2 diabetes mellitus, viral infections, fungal infections, and malaria (Balick 1990, 1994; Carvalho & Krettli 1991; Bierer et al. 1995; Carlson et al. 1997, 2001; Luo et al. 1998; Fort et al. 2000). Numerous studies have also demonstrated that plants used as medicines have a higher rate of bioactivity than do plants collected randomly (Carlson 2001).

Numerous ethnolinguistic groups in the western Amazon basin use the bark latex of the Amazonian tree *Croton lechleri* Müell.-Arg. to treat a variety of skin, gastrointestinal, and respiratory diseases. In this chapter I first discuss how ethnobotanical field research generated hypotheses on the therapeutic benefits of the bark latex of *Croton lechleri*. Next come descriptions of numerous in vitro cellular and in vivo animal studies and of human clinical studies that were designed to test these hypotheses and to confirm bioactivity of this medicinal plant for the treatment of diarrhea, respiratory infections, and herpes simplex skin infections and as a skin wound healer.

Bioactivity of Ethnobotanical versus Random Plant Collections

In recent decades, plants have been evaluated for medicinal qualities by large pharmaceutical companies by means of random or taxonomic collection of large numbers of plants for testing in standardized high-throughput in vitro screening systems. In this high-throughput approach, plant species are evaluated for activity against a spectrum of different diseases without consideration of whether the plants are used to treat any of them. It has been demonstrated that this approach can lead to identification of effective and safe bioactive molecules (Cragg et al. 1997). However, a more efficient method of screening for medicinal bioactivity is to select plants used by humans to treat specific diseases. Plant species are selected by conducting careful ethnobotanical field research with traditional botanical healers. A high percentage of biologically active plants can be generated through the utilization of sound medical ethnobotanical field research methods (Carlson & King 1997). For example, 40 of the 70 plants (57%) used ethnomedically to treat Type 2 diabetes demonstrated statistically significant antihyperglycemic activity in a diabetic mouse model compared with the vehicle control group (Carlson et al. 1997).

A significantly higher rate of pharmacological activity in plant extracts used ethnomedically compared with extracts from randomly collected plants has been demonstrated in several studies. Multiple published studies on plant extracts assessed in anti-cancer in vitro assays at the National Cancer Institute (NCI) were evaluated by Spjut and Perdue (1976). This survey reported anti-cancer activity in a cellular in vitro model in 10.4% of randomly collected plant species, 29.3% of plant species used to treat nematodes, 38.6% of plant species used as fish poisons, and 52.2% of plant species used as dart or arrow poisons. Anti-infective activity of ethnomedically collected plants versus randomly collected plants has also been compared in numerous studies. A study conducted in Brazil found anti-malarial activity in 18% (4 of 22) of the plants used ethnomedically, compared with 0.07% (2 of 273) of the plants collected randomly (Carvalho & Krettli 1991). Anti–HIV activity was assessed in NCI assays that compared extracts of randomly collected plants from Belize with extracts of "powerful plants" used by a traditional Maya healer in Belize. Tannins are present in many traditional botanical medicines. When the tannins in the plant extracts from Belize were not selectively removed, 25% (5 of 20) of the ethnobotanically indicated plants were active, compared with 6% (1 of 18) of the randomly collected plants (Balick 1990). A larger group of plants from Belize were assessed in a similar way in the same NCI anti–HIV assay. This study showed that 1.6% (1 of 61) of the randomly collected plants were active, compared with 15% (11 of 73) of the plants presented by the traditional Maya healer (Balick 1994). Vlietnick and Vanden Berghe (1991) assessed antiviral activity in plants and found that 5% of randomly collected plants demonstrated antiviral activity, compared with 25% of the ethnobotanical plant collections.

In another study, the frequency of isolating antiviral molecules was assessed in plants identified through ethnobotanical research to treat viral infections, compared with natural products collected randomly (Carlson et al. 1997). The plants collected ethnobotanically were assessed against three different antiviral assays (cytomegalovirus, influenza, and respiratory syncytial virus); the randomly collected natural products were screened against an antiviral assay for herpes simplex virus. The same virologist supervised the screening programs for each of the four viruses. The results are as follows: 97 plants used ethnomedically to treat viral infections screened against respiratory syncytial virus yielded 8 active molecules with 8.2% isolation frequency; 231 plants used ethnomedically to treat viral infections screened against cytomegalovirus yielded 5 active molecules with 2.2% isolation frequency; 123 plants used ethnomedically to treat viral infections screened against influenza virus yielded 2 active molecules with 1.6% isolation frequency; and 15,000 randomly collected natural products screened against herpes simplex virus yielded 2 active molecules with 0.013% isolation frequency. The frequency of isolating pure molecules with antiviral activity from plants used ethnomedically to treat viral infections is, respectively, 630 times, 169 times, or 123 times higher, depending, respectively, on the virus (respiratory syncytial virus, cytomegalovirus, or influenza) being tested when compared with the frequency of isolating anti–herpes simplex virus compounds from natural products that were randomly collected.

The above studies show that ethnobotanical knowledge of local people enables scientists to more successfully identify pharmacologically active plants. This chapter will describe how the ethnobotanical and ethnopharmacological uses of *Croton lechleri* by a variety of ethnolinguistic groups in the western Amazon basin helped guide ethnobotanists, laboratory scientists, and human clinical scientists evaluate the plant's therapeutic pharmacological qualities.

Methods

The study of the pharmacological properties of the stembark latex of *Croton lechleri* was based on medical ethnobotanical field research and experimental biology, pharmacology, and human clinical studies. Ethnobotanical field research was conducted in both rural and urban areas of Peru. A physician/ethnobotanist field research approach was used to obtain data on the diseases being treated, plant parts used, method of preparation, mode of administration, dose size, dose interval, and length of treatment. These methods generate detailed ethnomedical and ethnopharmacological information on the use of medicinal plants.

A multidisciplinary team of scientists collaborated from the laboratories of Shaman Pharmaceuticals, the Stanford University Medical Center, the University of California at San Francisco Medical Center, the University of California at Berkeley, and the Children's Hospital at Oakland Research Institute. This team included scientists from the disciplines of ethnobotany, chemistry, experimental biology, pharmacology, human clinical medicine, drug development, and regulatory affairs. Based on hypotheses generated by our ethnobotanical field research on the uses of the latex of *Croton lechleri,* in vitro cellular and in vivo animal experimental biology and pharmacology studies were designed and conducted. These investigations were followed by comprehensive human clinical studies.

Ethnomedical Uses of *Croton lechleri* Stembark Latex

Numerous ethnolinguistic groups in the western Amazon basin treat a variety of ailments with red latex from the stembark of the medium-sized tree *Croton lechleri,* in the Euphorbiaceae plant family (Table 9.1). Common names for this tree include *sangre de drago, sangre de grado, masujuain* (Cofan, in Ecuador/Colombia), *lan huiqui*

Table 9.1
Ethnomedical uses of *Croton lechleri* bark latex

- Topical uses
  - Preparation
    - Raw bark latex applied topically
  - Dosage
    - Enough latex to cover surface of skin lesion 3 times a day
  - Diseases treated
    - Skin/mouth/gums/nose, in adults and children
      - On lacerations to stop bleeding
      - Wound healing
      - Herpes simplex
      - Bacterial skin infections
      - Gingivitis
      - After tooth extraction to stop bleeding
      - Nosebleeds
      - Acne
- Oral uses
  - Preparation
    - Bark latex mixed in cold water, warm water, or alcohol
  - Dosage (10 drops = 0.5 ml)
    - 0–2 years old: 1 drop orally 1–3 times a day
    - 2–4 years old: 2 drops 1–3 times a day
    - 5–12 years old: 2–3 drops 1–3 times a day
    - Teenagers and adults: 3–10 drops 1–3 times a day
  - Diseases treated
    - Upper respiratory infections in both adults and children
      - Tonsillitis
      - Pharyngitis, throat infections
    - Lower respiratory infections in both adults and children
      - Lung infections, pneumonia
      - Tuberculosis
      - Influenza
      - Asthma
    - Gastrointestinal/hepatological disease in both adults and children (Carlson & King 2000)
      - Diarrhea without blood
      - Cholera
      - Traveler's diarrhea
      - Infant diarrhea
      - Diarrhea with blood
      - Gastric and duodenal ulcers
      - Gastritis
      - Hepatitis
      - Hemorrhoids

(Quichua, in Ecuador), *uruchnum* (Untsuri Shuar, in Ecuador), and *conéwé* (Waorani, in Ecuador). In South America the bark latex of this tree is used to treat a spectrum of diseases, including topical administration to treat herpes simplex virus lesions, and cuts and to heal wounds and oral administration for lung problems, flu, cough, diarrhea, stomach ulcers, hepatitis, and diabetes (Ubillis et al. 1994). The bark latex of this plant is also reported to be used to treat the following ailments: topically in Ecuador to treat inflammations (Cai et al. 1993); topically as a wound healer in South America (Pieters et al. 1993); to treat cancer and applied topically for wound healing in Brazil and Ecuador (Pieters et al. 1992; Cai et al. 1993); topically as a hemostatic and wound healer, as a vaginal bath before childbirth, and orally for intestinal and stomach ulcers in Peru (Duke & Vásquez 1994: 58); orally to treat rheumatism in Peru (Persinos et el. 1979); and orally to treat leucorrhea, fractures, and hemorrhoids in Peru (Rutter 1990: 349).

Additional species (*Croton draco* Schlect. and *C. urucurana* Baill.) with red latex are also used as medicines. *Croton draco* is used in the following ways: aerial parts are used to treat cancer in Panama (Gupta et al. 1996); infusion of dried aerial parts is taken orally in Mexico to treat fevers (Hernández & Delgado 1992); and latex is applied topically in Mexico to treat bleeding (Zamora-Martínez & Pola 1992). The red stembark latex of *C. urucurana* is taken for the following ailments: orally to treat diarrhea, respiratory tract infections, and urinary tract infections in Argentina; and topically as a wound healer and to treat cancer and orally to treat rheumatism in Brazil (Peres et al. 1997).

In the western Amazon basin of South America, our primary ethnobotanical field research demonstrated the use of the stembark latex of *Croton lechleri* to treat a variety of adult and child diseases of the skin, mouth, respiratory tract, and gastrointestinal tract (Carlson & King 2000).

Ethnomedical Uses of *Croton lechleri* Latex to Treat Diarrhea

Throughout the world, diarrhea is a major cause of both morbidity and mortality; each year an 6 million people die of it (Guerrant 1995). In many developing countries, diarrhea is the major cause of death in children less than 5 years of age (Guerrant et al. 1990; Guerrant 1995). It is also a major problem in people living with HIV in both developed and developing countries. Throughout the tropical world, botanical medicines are used to treat diarrhea. In 1993 and 1994 our primary medical ethnobotany field research conducted in Peru found that the bark latex from *Croton lechleri* is taken orally as a medicine by indigenous and mestizo children and adults to treat a variety of types of dysentery and diarrhea, including cholera (Carlson & King 2000). A survey of the ethnomedical use of the bark latex of *C. lechleri* was conducted in the outdoor markets of Iquitos, Peru, in 1996 by a Peruvian registered nurse, Dina Ayala, and an ethnobotanist, Franklin Ayala. Fifty-two randomly selected shoppers were interviewed, and 57% (30) of them reported using *C. lechleri* latex for the treatment of diarrhea (Carlson & King 2000). This information indicates that even in urban areas of the western Amazon basin this botanical medicine is commonly used for the treatment of diarrhea.

Chemistry of *Croton lechleri*

The stembark latex of *Croton lechleri* is the source of the proanthocyanidin molecule SP–303, which has a molecular weight of 2100 (Ubillis et al. 1994). A standardized botanical extract, SB–300, was developed from the crude bark latex of *C. lechleri.* Sixty-seven percent by weight of this standardized botanical extract consists of the proanthocyanidin molecule SP–303. The remaining 33% of the weight of the SB–300 standardized extract contains additional proanthocyanidin molecules that have not been specifically characterized.

In Vitro and In Vivo Studies on Formulations of *Croton lechleri* Stembark Latex

In Vivo Animal Safety Studies

Extensive in vitro cellular and in vivo animal studies have been conducted on SP–303. A common safety evaluation conducted in animals is the LD50 study, which measures the dose that is lethal in 50% of the study animals. Safety was demonstrated in very high doses of SP–303 in LD50 studies in rodents and dogs and in a six-month chronic consumption study in dogs. Safety was also evaluated in SB–300, a standardized botanical extract from *Croton lechleri* that contains 67% of SP–303 by weight. In these studies SB–300 demonstrated safety at very high doses in LD50 studies in rats.

In Vitro and In Vivo Efficacy Studies on Wound Healing

Species of *Croton* with red stembark latex, including *Croton lechleri,* were studied for wound-healing properties (Table 9.2). In a rat model, the red stembark latex was shown to stimulate all of the following: contraction of wound; crust formation; new collagen formation; and epithelial-layer regeneration (Pieters et al. 1995). In a mouse model, both the *C. lechleri* latex and the alkaloid taspine that is present in the latex demonstrated wound-healing activity (Vaisberg et al. 1989). A dihydrobenzofuran lignin compound isolated from the stembark latex was shown to stimulate human endothelial cells in a cellular model (Pieters et al. 1993).

In Vitro and In Vivo Efficacy Studies on Herpes Simplex Virus

In vitro studies on SP–303 have shown antiviral activity in viral CPE assays and plaque-reduction assays (Table 9.2) against herpes simplex virus–1 (HSV–1) and herpes simplex virus–2 (HSV–2), including human isolates of acyclovir-resistant HSV and foscarnet resistant HSV (Safrin et al. 1993; Ubillas et al. 1994). Studies demonstrated that SP–303 interferes with herpes simplex virus infection of cells (Ubillas et al. 1994). SP–303 also demonstrated antiviral activity when applied topically in a mouse HSV–2 model and a guinea pig HSV–2 model (Ubillas et al. 1994).

Table 9.2
In vitro and in vivo activity of formulations of *Croton lechleri* bark latex

- Skin wound healing activity
  - In vitro cicatrizant activity
    - Dihydrobenzofuran lignin compound: human umbilical vein endothelial cells (Pieters et al. 1993)
    - Taspine compound: human foreskin fibroblast migration (Vaisberg et al. 1989)
  - In vivo cicatrizant activity in rodent models
    - Stembark latex: rat model (Pieters et al. 1995)
    - Stembark latex (10%): mouse model (Vaisberg et al. 1989)
    - Taspine-HCl compound: mouse model (Vaisberg et al. 1989)
- Herpes simplex virus skin infection
  - In vitro anti-herpes activity of SP–303 molecule
    - Herpes simplex virus–1 and herpes simplex virus–2 (Safrin et al. 1993; Ubillas et al. 1994)
  - In vivo anti-herpes activity / mouse and guinea pig models of SP–303 molecule
    - Herpes simplex virus–2 (Ubillas et al. 1994)
- Respiratory infections
  - In vitro antiviral activity of SP–303 molecule (Wyde et al. 1993b; Ubillas et al. 1994)
    - Respiratory syncytial virus–1, respiratory syncytial virus–2, parainfluenza–1, parainfluenza–3
    - Influenza–A, influenza–B
  - In vivo antiviral activity of SP–303 molecule
    - Respiratory syncytial virus: cotton rat (Gilbert et al. 1993; Wyde et al. 1993a)
    - Parainfluenza: cotton rat (Wyde et al. 1993a)
    - Influenza: mouse (Gilbert et al. 1993)
- Gastric ulcers
  - In vivo anti-ulcer activity liquid stembark latex
    - Enhanced gastric ulcer healing in a rat model (Miller et al. 2000)
- Diarrhea
  - In vitro anti-diarrhea activity of liquid stembark latex
    - Anti-secretory mechanism by inhibiting chloride ion secretion across human intestinal epithelial cells in Ussing chamber (Miller et al. 2000)
  - In vitro anti-diarrhea activity SP–303 molecule and SP–300 extract
    - Anti-secretory mechanism by inhibiting chloride ion secretion across human intestinal epithelial cells in Ussing chamber (Gabriel et al. 1999; Illek et al. 2000, in prep.)
  - In vivo anti-diarrhea activity SP–303 molecule
    - Anti-secretory efffect in cholera mouse model (Gabriel et al. 1999)

In Vitro and In Vivo Efficacy Studies on Respiratory Infection

In vitro studies were conducted on the antiviral activities of SP–303 in CPE assays and plaque-reduction assays on a spectrum of different viruses that cause respiratory infection in humans (Table 9.2). SP–303 demonstrated in vitro antiviral activity against respiratory syncytial virus–A (RSV–A), respiratory syncytial virus–B (RSV–B), influenza–B (FLU–B), parainfluenza–1 (PIV–1), parainfluenza–3 (PIV–3), and influenza–A (FLU–A)

(Wyde et al. 1993b; Ubillas et al. 1994). Antiviral activity was shown with SP–303 in vivo studies in an RSV cotton rat model (Gilbert et al. 1993; Wyde et al. 1993b), a PIV–3 cotton rat model (Wyde et al. 1993b), and a FLU–A mouse model (Gilbert et al. 1993).

In Vitro and In Vivo Efficacy Studies on Diarrhea

Opiate-derived anti-diarrhea medicines, including lomotil, imodium, and paregoric of opium, act by reducing intestinal motility. A common side effect of these medicines is constipation. There is a medical need for a safe and effective anti-diarrhea agent that does not have constipation as a side effect. Our medical ethnobotanical research demonstrated that bark latex from *Croton lechleri* is used to treat cholera and other types of secretory diarrhea. Based on this observation, we generated the hypothesis that this botanical medicine works by an anti-secretory mechanism.

To test this hypothesis, cellular in vitro, animal in vivo, and human clinical studies were designed and conducted. Studies in rodents demonstrated that SP–303 does not affect intestinal motility. Studies were conducted to assess activity of these substances in models of secretory diarrhea. An Ussing chamber cellular assay was used to evaluate mechanism-of-action, and a cholera mouse model was used to assess efficacy. In the cellular Ussing chamber model, cAMP–mediated chloride ion secretion is measured as current across human intestinal epithelial cells. Chloride ion secretion mediates some forms of secretory diarrhea in humans. Studies in the Ussing chamber demonstrated that chloride-ion secretion is inhibited by SP–303 (Gabriel et al. 1999). Another study compared the SP–303 molecule with the SB–300 standardized extract and found that both had comparable significant inhibition of cAMP–mediated chloride ion secretion in the Ussing chamber (Illek et al. 2000, in prep.). Research in the cholera mouse model (Figure 9.1) demonstrated that SP–303 has a significant inhibitory effect on fluid secretion into the intestinal lumen when given simultaneously by mouth with cholera toxin as well as when given 3 hours after the cholera toxin was given (Gabriel et al. 1999). The whole liquid stembark latex of sangre de grado (*Croton* spp.) used in traditional medicine in the western Amazon basin demonstrated gastric ulcer healing in rats and showed an anti-secretory anti-diarrheal effect in a cellular model through action on sensory afferent nerves effecting intestinal epithelial-cell electrolyte movement (Miller et al. 2000). The above studies demonstrate that the SP–303 molecule, the SB–300 standardized botanical extract, and the whole liquid stembark latex all have anti-secretory anti-diarrheal mechanisms of action (Table 9.2).

Human Clinical Studies with SP–303 and SB–300

Human clinical studies were conducted to test the safety and efficacy of SP–303 and SB–300. Phase I human clinical studies evaluate safety; Phase II and Phase III human clinical studies evaluate efficacy as well as safety. In the double-blind, randomized, placebo-controlled studies, patients were randomly assigned to either a treatment group or a placebo group. The patients did not know whether they received the medicinal treatment or a placebo sugar pill. Also, the physicians and nurses evaluating

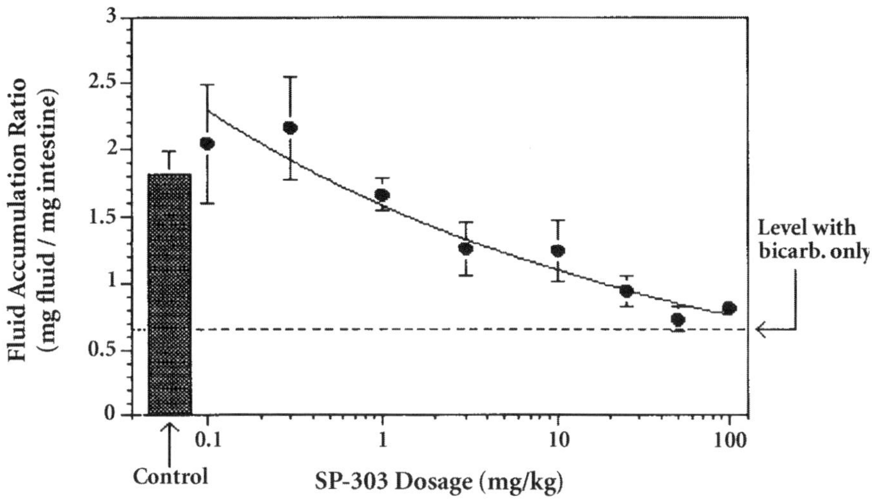

Figure 9.1. Dose response effect of SP–303 in intestinal fluid accumulation ratios (mg fluid / mg small intestine) when orally administered to mice 3 hours after cholera toxin. This ability demonstrates that the fluid-reducing effect of SP–303 is not attributable to the direct inactivation of CT by SP–303. *Source:* Gabriel et al. 1999.

the patients during the study did not know which patients received the medicinal treatment and which received the placebo sugar pill.

Human Studies on Pharmacokinetics and Safety of SP–303 and SB–300

Pharmacokinetic studies in infants and adult humans demonstrated that there is no significant absorption of the SP–303 molecule into the bloodstream. This is not surprising, because SP–303 is a large molecule with a molecular weight of 2100. Phase I human safety clinical studies demonstrated safety with orally administered SP–303 in both human infants and adults. More than 1000 humans who had taken SP–303 orally were evaluated in Phase I safety, Phase II efficacy, and Phase III efficacy placebo-controlled human clinical studies. No significant adverse effects were demonstrated in the SP–303 treatment groups, compared with the placebo groups (Carlson & King 2000). In a small study of 10 patients living with AIDS and chronic diarrhea, no adverse effects were demonstrated from taking the standardized extract SB–300 daily over a 12-month period (Koch et al. 2000).

Human Clinical Studies in Herpes Simplex Virus Infection with SP–303

A topical formulation of SP–303 was studied in a multicenter, randomized, double-blind, placebo-controlled Phase II clinical study for safety and efficacy against recur-

Table 9.3
Human efficacy studies of formulations of *Croton lechleri* bark latex

Skin disease
- Anti-herpes simplex efficacy: SB–303 molecule (double-blind, randomized, placebo-controlled study)
  - AIDS–associated genital herpes simplex virus in Mexico (Phase II study) (Orozco-Topete et al. 1997)

Diarrhea
- Anti-diarrhea efficacy: SB–303 molecule (double-blind, randomized, placebo-controlled studies)
  - Travelers' diarrhea in United States tourists in Jamaica and Mexico (Phase II study) (Dicesare et al. 1998)
  - AIDS–associated diarrhea in the United States (Phase II study) (Holodniy et al. 1999)
  - AIDS–associated diarrhea in the United States (Phase III study) (Koch 2000)
- Anti-diarrhea efficacy: SB–300 extract
  - AIDS–associated diarrhea in the United States (Koch et al. 1999a, 1999b, in prep.)

rent genital and anal herpes simplex lesions in people living with AIDS (Table 9.3). The SP–303 treatment group experienced complete healing of their herpes simplex virus lesions at a faster rate than did the placebo group ($p = 0.053$) (Orozco-Topete et al. 1997).

Human Clinical Studies on Traveler's Diarrhea with SP–303

A double-blind, randomized, placebo-controlled study was conducted in American travelers to Mexico or Jamaica who experienced traveler's diarrhea (Table 9.3). The study demonstrated a statistically significant reduction in diarrhea in the 48 people in the 250 mg SP–303 treatment group, compared with the 44 people in the placebo group (Dicesare et al. 1998). Compared with the placebo control group, the treatment group that received 250 mg of SP–303 for 2 days had a significantly faster resolution of diarrhea when assessed at 48 hours ($p = 0.0004$) and 72 hours ($p = 0.0002$) after initiation of treatment. More than 90% of the people in the treatment group showed partial or complete improvement of diarrhea in the first 24 hours, which was significantly higher than the placebo group ($p = 0.003$). In the subjective symptom scoring of relief from cramping and urgency, there was a significant improvement in the SP–303 treatment group, compared with the placebo group.

SP–303 Compound Study in People Living with HIV and Chronic Diarrhea

Frequent bouts of watery diarrhea are common in a large percentage of people with HIV. This is caused by the HIV and is also a side effect of anti–HIV drugs. A popular anti–HIV protease inhibitor medication, Nelfinavir, has been demonstrated to cause secretory diarrhea in both in vitro and human clinical studies (Andrade et al. 2000).

A Phase II double-blind, randomized, placebo-controlled study was conducted to evaluate the safety and efficacy of orally administered SP–303 for the symptomatic treatment of diarrhea in people living with HIV (Table 9.3). This study was conducted at San Francisco General Hospital, at the University of California at San Francisco Medical Center, and at the Palo Alto Veterans' Administration Hospital at Stanford University Medical Center. A 500 mg dose of SP–303 taken orally 4 times a day for 4 days by 51 people (25 placebo, 26 SP–303) demonstrated a significant reduction in stool weight ($p = 0.008$) and stool frequency ($p = 0.04$) (Holodniy et al. 1999).

A double-blind, randomized, placebo-controlled Phase III study was conducted to test for the efficacy and safety of orally administered 500 mg SP–303 4 times a day for 7 days in people living with HIV and chronic diarrhea. Compared with the placebo group (100 people), the SP–303 treatment group (98 people) revealed a statistically significant reduction in stool weight in all people ($p = 0.033$) and in people with severe diarrhea ($p = 0.008$) (Koch 2000).

Human Clinical Study with SB–300 Standardized Botanical Extract in People Living with HIV and Chronic Diarrhea

The standardized botanical extract SB–300 was evaluated in a 20-patient, open-label efficacy and safety study in people living with HIV and chronic diarrhea (Table 9.3). The study was conducted at San Francisco General Hospital and the University of California at San Francisco Medical Center. One or two 350 mg tablets of SB–300 were taken 4 times a day by mouth for 2 weeks by people in the study. The stool weights on day 13 and day 14 were compared with the pretreatment two-day baseline stool weights, and a statistically significant ($p = 0.0026$) reduction in stool weight was found (Koch et al. 1999a).

Another study was conducted on quality-of-life issues in a combined group of 42 people in the SB–300 study and the Phase III SP–303 study (Koch et al. 1999b). This study demonstrated significant improvement in the following measures of quality of life: time spent resting ($p = 0.03$); ability to leave home ($p = 0.03$); sexual activity ($p = 0.01$); and sum score of daily living activities (0.024).

Both SP–303 and SB–300 offer numerous advantages over the anti-diarrheal agents that are currently available as pharmaceuticals:

- Clinical studies have proved them to be safe and effective;
- No significant absorption of principal ingredient (SP–303) into blood was found;
- The anti-secretory mechanism normalizes excess chloride ion and fluid flow in the bowel and promotes normal stool formation;
- Unlike opiate-derived medicines, they do not affect intestinal motility and do not cause constipation;
- Unlike antibiotics, they do not cause drug-resistant strains of bacteria; and
- Unlike bulking agents, they do not cause bloating or uncomfortable intestinal fullness.

Table 9.4
Skin disease treatment by *Croton lechleri* formulations: ethnomedical uses, experimental biology studies, and human clinical studies

Ethnomedical uses
- Topical dose of bark latex (10 drops = 0.5 ml)
  - Topical administration: apply enough latex to cover surface of lesion 3 times a day.
- Skin/mouth/gums/nose diseases in adults and children
  - On lacerations to stop bleeding and enhance wound healing
  - Wound healing
  - Herpes simplex
  - Bacterial skin infections
  - Gingivitis
  - After tooth extraction to stop bleeding
  - Nosebleeds
  - Acne

Experimental biology studies
- Skin wound healing activity
  - In vitro cicatrizant activity
    - Dihydrobenzofuran lignin compound: human umbilical vein endothelial cells (Pieters et al. 1993)
    - Taspine compound: human foreskin fibroblast migration (Vaisberg et al. 1989)
  - In vivo cicatrizant activity in rodent models
    - Stembark latex: rat model (Pieters et al. 1995)
    - Stembark latex (10%): mouse model (Vaisberg et al. 1989)
    - Taspine–HCl compound: mouse model (Vaisberg et al. 1989)
- Herpes simplex virus skin infection
  - SP–303 molecule in vitro anti-herpes activity
    - Herpes simplex virus–1 and herpes simplex virus–2 (Safrin et al. 1993; Ubillas et al. 1994)
  - SP–303 molecule in vivo anti-herpes activity in mouse and guinea pig
    - Herpes simplex virus–2 (Ubillas et al. 1994)

Human clinical studies
- Anti-herpes simplex efficacy: SP–303 molecule
  - Double-blind, randomized, placebo-controlled study
  - AIDS–associated genital herpes simplex virus in Mexico (Phase II study) (Orozco-Topete et al. 1997)

Discussion and Conclusions

The ethnobotanical knowledge of ethnolinguistic groups in tropical communities around the world contributes significantly to affordable, safe, and effective health care in rural communities. The numerous experimental biology studies on a variety of disease categories described in this chapter demonstrate a higher rate of bioactivity in plants used medicinally by local people, compared with plants collected randomly. These studies confirm the wisdom of the local ethnomedical systems that identify and use safe and

Table 9.5
Respiratory disease treatment by *Croton lechleri* formulations: ethnomedical uses and experimental biology studies

- Ethnomedical uses
  - Respiratory disease
    - Upper respiratory in both adults and children
      - • Tonsillitis
      - • Pharyngitis, throat infections
    - Lower respiratory in both adults and children
      - • Lung infections / pneumonia
      - • Tuberculosis
      - • Influenza
      - • Asthma
- Experimental biology studies
  - Respiratory viral infections
    - SP–303 molecule in vitro antiviral activity against
      - Respiratory syncytial virus–1
      - Respiratory syncytial virus–2
      - Parainfluenza–1
      - Parainfluenza–3
      - Influenza–A
      - Influenza–B
        - (Wyde et al. 1993b; Ubillas et al. 1994)
    - SP–303 molecule in vivo antiviral activity against
      - Respiratory syncytial virus: cotton rat (Gilbert et al. 1993; Wyde et al. 1993a)
      - Parainfluenza: cotton rat (Wyde et al. 1993a)
      - Influenza: mouse (Gilbert et al. 1993)

therapeutic medicinal plants growing in their ecosystems. Comprehensive ethnobotanical research on the uses of the bark latex of the Amazonian tree species *Croton lechleri* has demonstrated a spectrum of uses to treat diseases of the gastrointestinal tract, respiratory tract, and skin (Table 9.1). This ethnobotanical research has provided valuable information that enabled hypotheses to be generated on potential therapeutic activities. Experimental biology studies (in vitro cellular and in vivo animal studies) and human clinical studies were designed and conducted to test these hypotheses. The experimental biology studies (Table 9.2) and human clinical studies (Table 9.3) conducted on formulations derived from the stembark latex of *C. lechleri* confirmed therapeutic bioactivity consistent with ethnobotanical uses for treating a spectrum of diseases of the skin (Table 9.4), respiratory tract (Table 9.5), and gastrointestinal tract (Table 9.6). Preparations of *C. lechleri* latex have been shown to safely and effectively enhance wound healing and treat herpes simplex virus infection, respiratory virus infections, and different types of watery diarrhea, including traveler's diarrhea and HIV-associated diarrhea. These studies have contributed to the development of formulations of *C. lechleri* latex that are now available to people in northern countries for the treatment of watery diarrhea, wound healing, and skin infections. The *C. lechleri* tree grows quickly

Table 9.6
Gastrointestinal disease treatment by *Croton lechleri* formulations: ethnomedical uses, experimental biology studies, and human clinical studies

Ethnomedical uses
- Gastrointestinal/hepatological disease in adults and children (Carlson & King 2000)
    - Diarrhea without blood
    - Cholera
    - Traveler's diarrhea
    - Infant diarrhea
    - Diarrhea with blood
    - Gastric and duodenal ulcers
    - Gastritis
    - Hepatitis
    - Hemorrhoids

Experimental biology studies
- Anti-diarrhea
    - In vitro anti-diarrhea activity of liquid stembark latex
        - Anti-secretory mechanism by inhibiting chloride ion secretion across human intestinal epithelial cells in Ussing chamber (Miller et al. 2000)
    - In vitro anti-diarrhea activity of SB–303 molecule and SB–300 extract
        - Anti-secretory mechanism by inhibiting chloride ion secretion across human intestinal epithelial cells in Ussing chamber (Gabriel et al. 1999; Illek et al. 2000, in prep.)
    - In vivo anti-diarrhea activity of SB–303 molecule
        - Anti-secretory effect in cholera mouse model (Gabriel et al. 1999)
- Gastric ulcers
    - In vivo anti-ulcer activity of bark latex
        - Enhanced gastric ulcer healing in a rat model (Miller et al. 2000)

Human clinical studies
- Anti-diarrhea efficacy: SB–303 molecule (double-blind, randomized, placebo-controlled studies)
    - Travelers' diarrhea in United States tourists in Jamaica and Mexico (Phase II study) (Dicesare et al. 1998)
    - AIDS–associated diarrhea in the United States (Phase II study) (Holodniy et al. 1999)
    - AIDS–associated diarrhea in the United States (Phase III study) (Koch 2000)
- Anti-diarrhea efficacy: SB–300 extract
    - AIDS–associated diarrhea in the United States (Koch et al. 1999a, 1999b, in prep.)

and abundantly in the western Amazon basin and provides local people with an inexpensive, safe, and effective treatment for a spectrum of health care needs. This is yet another example of the timeless value of a medicinal plant species to human health.

As ecosystems are degraded and languages are diminished around the world, valuable ethnobotanical systems that provide locally available, safe, and effective health care are threatened. It is crucial not only to conserve ecosytems but also to support the ethnolinguistic communities who live within these ecosystems and manage and use their biological resources. The conservation of biological, ethnolinguistic, and ethnobotanical knowledge has important public health benefits for local rural communities.

Acknowledgments

I thank Steven King, Lisa Conte, and Shaman Pharmaceuticals for their support of comprehensive research on the medicinal qualities of *Croton lechleri.* I also want to thank Elsa Mesa, Franklin Ayala, Dina Ayala, Beto Borges, and Carina Romero for their contributions to understanding the ecology and management of this species. Luisa Maffi, Chuck Peters, and Jennifer Sowerwine offered valuable comments on this manuscript. Finally, I want to express my deepest gratitude to the Cofan, Quichua, Untsuri Shuar, and Waorani ethnolinguistic groups and all the people and communities of the western Amazon basin who manage, use, and share this medicinal plant to enhance human health.

Literature Cited

Andrade, A., C. Sears, P. Rufo, W. Lencer & C. Flexner. 2000. Characterization of nelfinavir (NFV)-associated diarrhea: Secretory versus osmotic. Poster presented at the 7th Conference on Retroviruses and Opportunistic Infections, San Francisco, Calif., January 30–February 2, 2000.

Balick, M. J. 1990. Ethnobotany and the identification of therapeutic agents from the rainforest. Pp. 22–39 *in* D. J. Chadwick & J. Marsh (eds.), Bioactive Compounds from Plants. CIBA Foundation Symposium 154. Chichester, U.K., and New York: Wiley.

———. 1994. Ethnobotany, drug development and biodiversity conservation: Exploring the linkages. Pp. 4–24 *in* G. T. Prance, D. J. Chadwick & J. Marsh (eds.), Ethnobotany and the Search for New Drugs. CIBA Foundation Symposium 185. Chichester, U.K., and New York: Wiley.

Bierer, D. E., R. E. Gerber, S. Jolad, R. P. Ubillas, J. Randle, E. Nauka, J. Latour, J. M. Dener, D. M. Fort, J. E. Kuo, W. D. Inman, L. G. Dubenko, F. Ayala, A. Ozioko, C. Obialor, E. Elisabetsky, T. Carlson, T. V. Truong & R. C. Bruening. 1995. Isolation, structure elucidation, and synthesis of irlbacholine, 1,22–Bis[[[2–(trimethylammonium)ethoxy]–phosphinyl]oxy]docosane: A novel antifungal plant metabolite from *Irlbachia alata* and *Anthocleista djalonensis.* Journal of Organic Chemistry 60: 7022–7026.

Cai, Y., Z. P. Chen & J. D. Phillipson. 1993. Diterpenes from *Croton lechleri.* Phytochemistry 32(3): 755–760.

Carlson, T. J. S. 2001. Language, ethnobotanical knowledge, and tropical public health. Pp. 489–502 *in* L. Maffi (ed.), On Biocultural Diversity: Linking Language, Knowledge, and the Environment. Washington, D.C.: Smithsonian Institution Press.

——— & S. R. King. 1997. Ethnomedical field research methods to assess medicinal plants. Pp. 152–165 *in* M. M. Iwu, E. N. Sokomba, C. O. Okunji, C. Obijiofor & I. P. Akubue (eds.), Commercial Production of Indigenous Plants as Phytomedicines and Cosmetics. Boca Raton, Fla.: CRC Press.

——— & ———. 2000. Sangre de drago (*Croton lechleri* Müell.-Arg.): A phytomedicine for the treatment of diarrhea. Health Notes: Review of Complementary and Integrative Medicine 7(4): 315–320.

———, R. Cooper, S. R. King & E. J. Rozhon. 1997. Modern science and traditional healing. Pp. 84–95 *in* S. Wrigley, Hayes, M., Thomas R., Chrystal E. (eds.), Phytochemical Diversity:

A Source of New Industrial Products. Special Publication 200. Cambridge: Royal Society of Chemistry, Information Services.

———, B. M. Foula, J. A. Chinnock, S. R. King, G. Abdourahmaue, B. M. Sannoussy, A. Bah, S. A. Cisse, M. Camara & R. K. Richter. 2001. Case study on medicinal plant research in Guinea: Prior informed consent, focused benefit sharing, and compliance with the Convention on Biological Diversity. Economic Botany 55(4): 478–491.

Carvalho, L. H. & A. U. Krettli. 1991. Antimalarial chemotherapy with natural products and chemically defined molecules. Memorias do Instituto Oswaldo Cruz, Rio de Janeiro 86 (Suppl. 11): 181–189.

Cragg, G. M., D. J. Newman & K. M. Snader. 1997. Natural products in drug discovery and development. Journal of Natural Products 60: 52–60.

Dicesare, D., H. L. DuPont, J. J. Mathewson, C. D. Ericsson, D. Ashley, F. G. Martínez-Sandoval, J. E. Pennington & S. B. Porter. 1998. A double blind, randomized, placebo controlled study of SP–303 in the symptomatic treatment of acute diarrhea among travelers to Mexico and Jamaica. Abstract presented at the 36th Annual Meeting of the Infectious Diseases Society of America, Denver.

Duke, J. A. & R. Vásquez. 1994. Amazonian Ethnobotanical Dictionary. Boca Raton, Fla.: CRC Press.

Farnsworth, N. R., O. Akerele, A. S. Bingel, D. D. Soejarto & Z. Guo. 1985. Medicinal plants in therapy. Bulletin of the World Health Organization 63(6): 965–981.

Fort, D. M., R. P. Ubillis, C. D. Méndez, S. D. Jolad, W. D. Inman, J. R. Carney, J. L. Chen, T. Ianiro, C. Hasbun, R. C. Bruening, J. Luo, M. J. Reed, M. Iwu, T. J. Carlson, S. R. King, D. E. Bierer & R. Cooper. 2000. Novel antihyperglycemic terpinoid-quinones from *Pycnanthus angolensis*. Journal of Organic Chemistry 65(20): 6534–6539.

Gabriel, S. E., S. E. Davenport, R. J. Steagall, V. Vimal, T. Carlson & E. J. Rozhon. 1999. A novel plant-derived inhibitor of cAMP–mediated fluid and chloride secretion. American Journal of Physiology 276(39): G58–G63.

Gilbert, B. E., P. R. Wyde, P. Z. Wilson & L. Meyerson. 1993. SP–303 small particle aerosol treatment of influenza A virus infection in mice and respiratory syncytial infection in cotton rats. Antiviral Research 21: 37–45.

Guerrant, R. L. 1995. Principles and syndromes of enteric infection. Pp. 2: 945–962 *in* G. L. Mandell, J. E. Bennett & R. Dolin (eds.), Mandell, Douglas and Bennett's Principles and Practice of Infectious Diseases. Ed. 4. New York: Churchill Livingstone.

———, J. M. Hughes, N. L. Lima & J. Crane. 1990. Diarrhea in developed and developing countries: Magnitude, special settings, and etiologies. Reviews of Infectious Diseases 12 (Suppl. 1): S41–S50.

Gupta, M. P., A. Monge, G. A. Karikas, A. López de Ceraín, P. N. Solís, E. de León, M. Trujillo, O. Suárez, F. Wilson, G. Montenegro, Y. Noriega, A. I. Santana, M. Correa & C. Sánchez. 1996. Screening of Panamanian medicinal plants for brine shrimp toxicity, crown gall tumor inhibition, cytotoxicity and DNA intercalation. International Journal of Pharmacognosy: A Journal of Crude Drug Research 34 (1): 19–27.

Hernández, J. & G. Delgado. 1992. Terpenoids from aerial parts of *Croton draco*. Fitoterapia 63(4): 377–378.

Holodniy, M., J. Koch, M. Mistal, J. M. Schmidt, A. Khandwala, J. E. Pennington & S. B. Porter. 1999. A double blind, randomized, placebo-controlled, Phase II study to assess the safety and efficacy of orally administered SP–303 for the symptomatic treatment of diarrhea in patients with AIDS. American Journal of Gastroenterology 94(11): 3267–3273.

Illek, B., T. E. Machen, J. H. Widdicombe, T. Carlson & J. Chow. 2000. A novel botanical extract, DS4/44F2, from the bark latex of *Croton lechleri* inhibits cAMP–mediated chloride secretion in human colonic epithelial cells; the extraction process of DS4/44F2 was later optimized to produce the botanical extract for SP–Normal Stool Formula™. Poster presented at the 17th Annual AIDS Investigators' Meeting and the 3rd Annual Conference on AIDS Research in California, February 25, South San Francisco, Calif.

———, H. Fisher, T. E. Machen, J. H. Widdicombe, T. Carlson & J. Chow. In prep. A novel formulation from the bark latex of *Croton lechleri* inhibits cAMP–mediated chloride secretion in human colonic epithelial cells.

Koch, J. 2000. A phase III, double-blind, randomized, placebo-controlled multicenter study of SP–303 (Provir™) in the symptomatic treatment of diarrhea in patients with acquired immunodeficiency syndrome (AIDS). Poster presented at the 13th International AIDS Conference, July 14, Durban, South Africa.

———, J. Tuveson, T. Carlson & J. M. Schmidt. 1999a. Human studies on SB–300 in HIV–associated diarrhea: A new and effective therapy for HIV–associated diarrhea. Poster presented at the Seventh European Conference on Clinical Aspects and Treatment of HIV–Infection, October 23–27, Lisbon.

———, ———, ——— & ———. 1999b. SB–300: A new and effective therapy for HIV–associated diarrhea improves quality of life. Poster presented at the Seventh European Conference on Clinical Aspects and Treatment of HIV–Infection, October 23–27, Lisbon.

———, ———, ——— & ———. 2000. SB–300: A new and effective therapy for HIV–associated diarrhea. Poster presented at the 13th International AIDS Conference, July 14, Durban, South Africa.

———, ———, ——— & ———. In prep. New and effective therapy for HIV–associated diarrhea: SB–300.

Luo, J., D. Fort, T. J. Carlson, B. K. Noamesi, D. nii-Amon-Kotei, S. K. King, J. Tsai, J. Quan, C. Hobensack, P. Lapresca, N. Waldeck, C. D. Mendez, S. D. Jolad, D. E. Bierer & G. M. Reaven. 1998. Cryptolepine, a potentially useful new antihyperglycemic agent isolated from *Crytolepis sanguinolenta:* An example of the ethnobotanical approach to drug discovery. Diabetic Medicine 15: 367–374.

Miller M. J. S., W. K. MacNaughton & X.-J. Zhang. 2000. Treatment of gastric ulcers and diarrhea with the Amazonian herbal medicine sangre de grado. American Journal of Physiology. Gastrointestinal and Liver Physiology 279: G192–G200.

Orozco-Topete, R., J. Sierra-Madero, C. Caño-Domínguez, J. Kershenovich, G. Ortiz-Pedroza, E. Vázquez-Valls, C. García-Cosío, A. Soria-Córdoba, A. M. Armendariz, X. Terán-Toledo, J. Romo-García, H. Fernández & E. J. Rozhon. 1997. Safety and efficacy of Virend for topical treatment of genital and anal herpes simplex lesions in patients with AIDS. Antiviral Research 35: 91–103.

Peres, M. T. L. P., F. D. Monache, A. B. Cruz, M. G. Pizzolatti & R. A. Yunes. 1997. Chemical composition and antimicrobial activity of *Croton urucurana* Baillon (Euphorbiaceae). Journal of Ethnopharmacology 56 (3): 223–226.

Persinos, G. J., R. Blomster, D. A. Blake & N. R. Farnsworth. 1979. South American plants, II: Taspine, isolation and antiinflammatory activity. Journal of Pharmacological Sciences 68 (1): 124–126.

Pieters, L., T. de Bruyne, G. Mei, G. Lemiere, D. Vanden Berghe & A. J. Vlietinck. 1992. In vitro and in vivo biological activity of South American dragon's blood and its constituents. Planta Medica (Suppl.) 58(1): A582–A583.

———, ———, M. Claeys, A. Vlietnick, M. Calomme & D. Vanden Berghe. 1993. Isolation of a dihydrobenzofuran lignan from South American dragon's blood (*Croton* spp.) as an inhibitor of cell proliferation. Journal of Natural Products 56(6): 899–906.

———, ———, B. Van Poel, R. Vingerhoets, J. Totte, D. Vanden Berghe & A. J. Vlietinck. 1995. In vivo wound healing activity of dragon's blood (*Croton* spp.), a traditional South American drug, and its constituents. Phytomedicine 2(1): 17–22.

Rutter, R. A. 1990. Catálogo de plantas utiles de la Amazonia peruana. Yarinacocha, Peru: Instituto Lingüístico de Verano.

Safrin, S., L. Phan & T. Elbeik. 1993. Evaluation of the in vitro activity of SP–303 against clinical isolates of acyclovir-resistant and foscarnet-resistant herpes simplex virus. Antiviral Research 20 (Suppl. 1): 133.

Spjut, R. W. & R. E. Perdue Jr. 1976. Plant folklore: A tool for predicting sources of antitumor activity? Cancer Treatment Reports 60: 979–985.

Ubillis, R., S. D. Jolad, R. C. Bruening, M. R. Kernan, S. R. King, D. F. Sesin, M. Barrett, C. A. Stoddart, T. Flaster, J. Kuo, F. Ayala, E. Meza, M. Castanel, D. McMeekin, E. Rozhon, M. S. Tempesta, D. Barnard, J. Huffman, D. Smee, R. Sidwell, K. Soike, A. Brazier, S. Safrin, R. Orlando, P. T. M. Kenny, N. Berova & K. Nakanishi. 1994. SP–303, an antiviral oligomeric proanthocyanidin from the latex of *Croton lechleri* (sangre de drago), Phytomedicine 1: 77–106.

Vaisberg, A. J., M. Milla, M. del Carmen Planas, J. L. Córdova, E. Rosas de Agusti, R. Ferreyra, M. del Carmen Mustinga, L. Carlin & G. B. Hammond. 1989. Taspine is the cicatrizant principal in sangre de drago, extracted from *Croton lechleri.* Planta Medica 55: 140–143.

Vlietnick, A. J. & D. A. Vanden Berghe. 1991. Can ethnopharmacology contribute to the development of antiviral agents? Journal of Ethnopharmacology 32(1–3): 141–154.

Wyde, P. R., M. W. Ambrose, L. R. Meyerson & B. E. Gilbert. 1993a. The antiviral activity of SP–303, a natural polyphenolic polymer, against respiratory syncytial and parainfluenza Type 3 viruses in cotton rats. Antiviral Research 28: 145–154.

———, L. R. Meyerson & B. E. Gilbert. 1993b. In vitro evaluation of the antiviral activity of SP–303, an Euphorbiaceae shrub extract, against a panel of respiratory viruses. Drug Development Research 28: 467–472.

Zamora-Martínez, M. C. & C. N. P. Pola. 1992. Medicinal plants used in some rural populations of Oaxaca, Puebla and Veracruz, Mexico. Journal of Ethnopharmacology, 35 (3): 229–257.

Part 3

Ethical Issues in Ethnobiology

10

Ethnobiology and Globalization: Science and Ethics at the Turn of the Century

Miguel N. Alexiades

Introduction

We live in an increasingly interconnected and interdependent world. The use of such terms as "the global environment," "global economy," and "global culture" have become so much a part of our daily vocabulary as to suggest that, beyond the cliché of "globe talk," the world is perceived more than ever as a shared social space, a space molded by transnational ecological, economic, legal, political, and technological forces (Holton 1998). As a field of inquiry, ethnobiology is positioned at the interface of nature and culture, of genetic resources and environmental knowledge, of local and international communities, and of science and society, the very junctures undergoing fundamental transformations through increased globalization. These changes have important implications for ethnobiologists, in terms of both how we think about human-environment relations and how we structure our research questions and relationships. In this chapter I begin to examine some of the economic, political, and social aspects of globalization and their effect on shaping the broader context in which the science and practice of ethnobiology are embedded.

While the processes I will be discussing are by definition transnational, this does not mean that they affect all regions in the same way. Certain of the trends I touch upon, including economic neoliberalism and deregulation, the creation of new spaces for the growth of civil society, the privatization of science and the service sector, and the emergence of cultural knowledge and genetic diversity as contested cultural, political, and economic resources, are much more apparent in some regions, such as the Americas, than in others.

After providing a broad definition and introduction to globalization, I review some of its major economic, political, and cultural dimensions, discussing their relevance to questions regarding genetic resources, ethnobiological knowledge, and social equity. I then focus on four specific processes: the increased commodification of knowledge and ethnobotanical resources, the political economy of globalization and its relationship to science and to human-environment relations, the emergence of powerful international social movements and the politicization of knowledge, and the growing presence of media as an interlocutor between science and society. I conclude with suggestions as to how some of these changes might be productively assimilated in the elaboration of future research and educational agendas within ethnobiology.

Globalization: A Primer

Within academic circles, "globalization" refers to a number of distinct yet interrelated processes, most generally to a growing interdependence among the world's economic, social, and political systems and to a compression of the temporal and spatial aspects of social relations (Harvey 1990; Mittelman 1997a).[1] These processes are not new. The rise of capitalism in the sixteenth century presupposed the creation of a world economic system (Wallerstein 1974), and even before that there were extensive movements of goods, capital, people, technology, and ideas across the surface of the earth (Frank & Gills 1992). A number of scholars (e.g., Held et al. 1999; Mittelman 2000) identify several distinct historical phases of globalization. According to this view, the ongoing neoliberal global expansion corresponds to the contemporary phase of capitalism (Mittelman 2000: 19). In a similar vein, Spybey (1996: 77) draws a distinction between "internationalization," associated with the global colonial expansion of the past 200 years, and modern "globalization," associated with a historically unprecedented degree of "functional integration between internationally dispersed economic activities." The pace of such integration has accelerated dramatically since the cold war ended (Holm & Sørensen 1995; Holton 1998), to the extent that today finance and production are organized in transnational networks that largely escape national and international regulatory powers (Madeuf & Michalet, cited in Cox 1997). Additionally, the scale and intensity of transnational flows have accelerated exponentially in the last few decades (Kearney 1995: 549; Clark 1997: 1). In light of these factors, Held et al. (1999: 16) suggest that contemporary globalization can best be thought of as "a process (or set of processes) which embodies a transformation in the spatial organization of social relations and transactions—assessed in terms of their extensity, intensity, velocity and impact—generating transcontinental or interregional flows and networks of activity, and the exercise of power."

There is considerable debate about the characteristics of modern globalization, as well as its scope, its effects, and its future. Some of the contested specific issues are: the degree to which ongoing globalizing processes are historically unprecedented; the importance of regional, as opposed to global, processes; the causes of and driving forces behind globalization; whether globalization will draw all nations and peoples into a common set of market relations; the role and future of the nation-state; the degree to which globalization is inevitable; and its long-term viability, trajectory, and impact on human well-being and sustainability.[2]

Globalization is best understood as a set or syndrome of related, yet multifaceted and differentiated, political, economic, social, and cultural processes (Holton 1998; Held et al. 1999; Mittelman 2000). In economic terms, the latest wave of globalization is characterized by:

1. Increase in world trade. The volume of world trade more than doubled in the ten-year period between the mid-1980s and the mid-1990s (Neef 1999: 46).
2. Increased international flows of capital. In the same period, the amount of capital flowing through the world's major currency markets increased from $190 million to $1.3 trillion. Particularly dramatic are the total increase in foreign direct investment (FDI) and the proportion of FDI directed to developing economies: FDI flows to developing economies more than doubled between 1991 and 1996, for example (Neef 1999: 52), and accounted for almost 40% of all inflows in 1993 (Mittelman 2000: 20).
3. Growth, expansion, and consolidation of transnational corporations. Regional agreements on tariff reductions have contributed to the globalization of corporate organizations. In particular, a shift begins from the multinational company, based in one country and doing business in others, to the transnational or non-national (Neef 1999: 63) company, which is no longer affiliated with any single country, in terms of ownership, employment, or operation. Four hundred of these multinational and transnational companies are responsible for more than half of the total world output (Need 1999: 13). Of particular relevance to ethnobotany is the increased ability for food, agriculture, and biotechnology multinationals to shape the future production, marketing, and consumption of agricultural and food products (Watts & Goodman 1999).
4. Global reorganization of production, both spatially, as in the international division of labor, and structurally, as in the shift away from a "Fordist" model of centralized mass production to a "post-Fordist," decentralized system of "flexible accumulation." This, in turn, allows companies to adapt to rapid changes in the market and to exploit specialist niche markets (Harvey 1990; Ross 1995). The implications of these changes include deindustrialization of technologically advanced societies, movement of the manufacturing industry to regions or countries that offer cheaper labor and fewer environmental and labor restrictions, and substitution of a labor theory of value with a knowledge theory of value (Bell 1999).
5. Growth of the service and knowledge sector, including telecommunications, media, biotechnology, and agribusiness, and the concomitant development and globalization of intellectual property rights regimes (see below). The increased commodification of biodiversity and of certain forms of traditional environmental knowledge and cultural knowledge raise important ethical, economic, legal, and political questions, which are highly relevant to the subject matter of ethnobiology.

The emergence of new, powerful, and transnational forms of economic and political organization also have important political implications, including:

1. Redefinition and reconstitution of the role and power of the state. The authority of the state has become "increasingly diffused among public and private agencies at the local, national, regional and global levels" (Held et al. 1999: 9). While different states are pursuing distinctive strategies, both as a response to and as an

attempt to engage with the global economy and cross-border issues, several widespread tendencies are associated with the adoption of neoliberal policies, including the demise of the welfare state and the privatization of services, including health, education, and science. The reconstruction and redefinition of state authority have also opened new spaces for the growth of civil society (see below).

2. Race, ethnicity, religion, and gender have begun to supplant labor and the nation as critical markers of identity, which, together with the collapse of the Soviet bloc, have led to challenging state borders and even state sovereignty in many areas (Mittelman 2000). Overall, "the state is being re-formed from above by the tugs of economic globalization and from below by the pull of sub-nationalism" (p. 26). An important dimension of the politics of identity based on ethnicity and religion are claims, at times disputed, to authenticity, ownership, and rights to represent or access past or present cultural institutions. Certain forms of cultural knowledge, for example, have become important political resources in the construction of identity and in articulating claims to tangible and intangible political resources. Not surprisingly, conflicts over ownership, access, and representation have emerged among scientists and other stakeholders, particularly in the contexts of museum collections and anthropological, historical, and archaeological studies (e.g., Wilford 1999; Bray 2001). Bioprospecting, in particular, is an arena in which such conflicts among scientists, the private sector, environmentalists, and local actors have become commonplace (e.g., Rigden 1996; Nuttall 1998; Pollack 1999).
3. Increase in non-national organizations. While some of these, such as the European Union or North American Free Trade Agreement, are distinctly regional in their scope, others, such as the International Monetary Fund and the World Trade Organization, are global in reach. The increase in global communications, the spread of global environmental problems, and the opening of new spaces in civil society (see below) have all contributed to the proliferation and rising prominence of international grass-roots and nongovernmental organizations, particularly in the realms of the environment and human rights. The growth of the nongovernmental organization sector has important implications for development, because it offers new platforms and avenues for political mobilization and for redefining social and ecological relations (Ekins 1992).
4. The widespread, though not universal, preference for democracy in various forms and the growth of civil society.

From a social perspective, manifestations of globalization include massive transfers of populations, goods, and commodities. The proliferation of international media, global communications, and widespread penetration of markets is also associated with cultural homogenization, as well as with increased cultural exchange, hybridization, and, as a response, cultural polarization (Holton 1998). These processes are further mediated by the prevalent tendency to increasingly commodify culture, on one hand, and to resist cultural homogenization and assert identity on the basis of such cultural institutions as religion and ethnicity, on the other hand; that is, culture has become an increasingly important and contested field of economic and political agency.

Having provided a broad overview of the key aspects of globalization, I will now focus on three specific issues, examining their significance for ethnobiology: the rise of the information economy and the commodification of knowledge; the shift in the balance between the public and private sectors; and the growing presence of mass media in human affairs, including science.

Globalization and the Rise of the Information-Based Economy

Beyond the huge impact of such recent technological innovations as the fax machine, the Internet, and satellite television on billions of people throughout the world, the explosive advance in communications technology affects and reflects a critical transformation in the economy and the role of information, particularly within technology-rich societies. The late twentieth century witnessed a broad transition in wealthy nations, from an economic system based on industrial mass production and consumption to an information- and service-based economy, also referred to as a post-industrial or postmodern economy, characterized by the central importance of the "codification of theoretical knowledge and the new relation of science to technology" (Bell 1999: xiv).

This transition of economically powerful nations is indicated by a shift away from manufacturing toward services and by a concomitant rise in the economic importance of human capital and intangible inputs, themselves dependent on knowledge and skills. The service sector, for example, accounts for around 85% and 70% of all employment in the United States and Europe, respectively (Neef 1999: 4, 5). Moreover, whereas 15 years ago material costs accounted for more than 80% of manufacturing costs, today around 80% of the costs associated with production of many goods are in the form of knowledge-rich intangibles, such as product design, marketing, and sales (p. 6).

Increasingly, information has become a cornerstone of economic growth and of corporate and social organization (e.g., Navaretti et al. 1998: 1). In effect, the interdependence of capitalism and industry of the nineteenth century has given way to an interdependence of capitalism and information, whereby information has replaced production as the primary source of wealth and power (Castells 1996). During the 1980s, for example, the value of world telecommunication services grew by 800% (Comor 1994: 2). By the early 1990s, more than two-thirds of the workforce in the United States was engaged in jobs directly involving the production, processing, or distribution of information (Comor 1994: 2). U.S. trade estimates that the percentage of internationally traded goods with high intellectual property content increased from 10% to 30% between the 1940s and the 1990s (Mooney 1997). In the decade between 1983 and 1993, trade in information technology goods grew more rapidly than did all other sectors in world merchandise exports (GATT 1994, cited in James 1999: 7). The central role of information in today's corporate world is exemplified by the cutting-edge business concept of knowledge management, which seeks to optimize how knowledge is produced, stored, and managed, allowing companies to "behave with strategic agility, sensing and responding quickly and innovatively to global market opportunities" (Neef 1999: 13), thus maintaining a competitive edge in an increasingly unpredictable and dynamic business environment (Malhotra 1998).

Science, Technology, and the Commodification of Knowledge

The emergence and rising prominence of such information-centered approaches as chaos and complexity theory, fuzzy logic, computer modeling, robotics, molecular biology, bioinformatics, and genomics all attest to an unprecedented ability to record, store, and manipulate environmental information, suggesting that now, more than ever, information has become a central organizing concept in science and technology, as well as in business and in corporate organization (Bell 1999). I suggest that these technologies are having a profound impact on how we perceive the world around us. Advances in biotechnology and genetic engineering, for example, suggest a view of living organisms as carriers of vast amounts of potentially decipherable information that can potentially be manipulated to create new commodities and to maximize productivity of managed environments and organisms (cf. Kelly 1994). An organism such as a cow or a goat is no longer just an economic resource because it produces dairy products, leather, and meat; it is also an economic resource because we have developed the technological means to manipulate its unique set of genetic data to synthesize useful drugs and hormones (Friend 1999).

The communications revolution is having a similar impact in that our increased ability to gather, store, manipulate, and communicate data has qualitatively changed the meaning, use, and value of information. In these new economies, then, knowledge and information are not simply a means of converting natural resources into commodities. Information itself is a commodity, sold or exchanged according to formal rules conferring specific ownership, rights, and access to that information and to the benefits that arise from its shared use.

Through this "infocentric" perspective, centered on knowledge and information, cultural and biological diversity adopt an entirely new and very significant meaning: the notion that they contain coded information with considerable potential for economic and humanitarian well-being (Alexiades in prep. b). Indeed, this argument is frequently used as an additional justification to encourage the safeguarding of cultural and biological diversity and to highlight their interconnectedness (e.g., Nazarea 1998). As a relatively novel concept (Wilson 1988), biodiversity commonly expresses three late-twentieth- and early-twenty-first-century themes discussed in this chapter: an emphasis on the global (cf. Ingold 1993), the rise of an information-based society and economy, and the increased commodification of genetic resources and information (Alexiades in prep. b).

Globalization and the Flow of Knowledge

The flow of genetic resources and cultural knowledge between different spheres of the global economy inevitably leads to their transformation. I have already suggested that commodification is one such type of transformation (see Alexiades in prep. a). Kloppenburg (1988) describes how the process of industrialization has been sustained by such a transformation of genetic resources, from public goods subject to free access to commodities with restricted access. This process of commodification has accelerated in recent years, following the growth of biotechnology and knowledge-based industries and the dramatic improvements in communications and data-processing technologies.

As certain types of information and genetic resources become increasingly valued as commodities, issues regarding proprietary rights acquire mounting importance. One critical aspect of information is that it is typically very costly to produce but very cheap to reproduce. Trade secrets, copyrights, and patents are some of the institutionalized mechanisms used to privatize information, thus attempting to guarantee economic returns and ensure its continued production (Shapiro & Varian 1998). It is no coincidence that the growth of the information sector and the globalization of information-based technologies and products have been accompanied by the development of international property rights regimes and institutions, such as Trade-Related Aspects of Intellectual Property Rights and the International Union for the Protection of New Varieties of Plants (Holton 1998). The rapid increase in the number of patents filed in the U.S. Patent Office is another indication of the growing economic importance of knowledge-based technology. The granting of patents for biological products since the 1980s has generated a heated public debate regarding the ethical, legal, economic, environmental, and political implications not only of U.S.-based intellectual property rights regimes but also, and more generally, of the emerging "life-sciences" industry and the role of science and technology (e.g., Marshall 1997; Mooney 1997; Shiva 1997; Schwartz 1999).

Questions relating to property rights and cultural and genetic resources are immensely complex and rapidly evolving, and a review of the different issues and positions deserves separate treatment (see, for example, Crucible Group 1994; Brush & Stabinsky 1996; Posey 1996b; Posey & Dutfield 1996; Brown 1999; Crucible Group II 2000; Laird 2002). Suffice to note here that these issues are of central importance to ethnobiology and that the controversies surrounding these issues signal the polarization of the social and political environment through which genetic resources flow as they move through the global economy. Conflicts between different actors are heightened, given that it is often difficult to ascribe clear geographical or cultural boundaries to the distribution of many of these resources and that different stakeholders may have competing claims over these resources, even within the same nation-state. Article 15 of the 1992 Convention on Biological Diversity (CBD), for example, recognizes nation-states as having sovereign rights over genetic resources, suggesting a shift in the status of genetic resources from the "heritage of mankind" model that prevailed until recently (Putterman 1994). On the other hand, while the nation-state is clearly one important stakeholder with proprietary claims on genetic resources, other stakeholders, including local communities and indigenous groups within nations, have claims that to date have not received the same level of legal recognition (Posey 2000: 39).

The commodification and concomitant politicization of genetic resources and cultural knowledge, together with internationalization of human rights and environmental agendas (e.g., Caldwell 1991), have resulted in considerable pressure to develop formal mechanisms for access and benefit sharing, in both academic and corporate research. The CBD provided a legal framework for approaching discussions at national and regional levels, leading to several legislative initiatives among the Andean Pact countries.[3] Numerous professional societies, multilateral development agencies, public and research institutions, and private firms have also been drafting guidelines, codes of practice, recommendations, letters of intent, and contracts, in an attempt to tackle the numerous complex ethical, moral, economic, institutional, and legal aspects of working with genetic resources or cultural knowledge (for a review of these see, for example, Gollin 1999; Laird & Posey 2002; Bannister & Barrett 2004).

To Share or Not to Share: That Is the Question

The ambivalent status of ethnobiological and cultural knowledge as a private and public resource is creating new contradictions and difficulties with respect to how knowledge is used and exchanged and who is accorded rights or access to it. Since the mid-1980s, for example, I have collaborated with several projects and indigenous leaders who see in traditional knowledge a powerful and valuable resource for community development, grassroots change, and health care improvement (e.g., Alexiades & Lacaze 1996). Various programs sought to encourage the exchange and use of ethnobiological information among different ethnic and age groups, through many different vehicles, including workshops, manuals, and brochures (Lacaze & Alexiades 1995). Such initiatives have assisted in the consolidation and reaffirmation of identity and have helped promote and popularize the use of biological and cultural resources, which in many cases were undervalued and underutilized. As such, dissemination and exchange of ethnobiological knowledge have enabled a process often referred to by indigenous leaders in Latin America as *revalorización* (FSI & Kothari 1997), broadly seen to have a desirable political outcome; namely, a greater capacity for self-determination. At the same time, however, there is a growing concern among some indigenous leaders that promoting the diffusion of traditional knowledge can lead to its misappropriation and eventual commodification by third parties. In effect, then, the same act—for example, publishing a community health manual—can lead to contradictory outcomes or result in conflicting expectations. In some cases this has led to a decision to restrict the dissemination of cultural knowledge within community development programs (e.g., Milliken 2002).

Ethnobiologists have also begun to experience a conflict of interest between their responsibility toward their colleagues—who expect free exchange of knowledge—and their responsibility toward their indigenous collaborators—who may insist on limiting the dissemination of knowledge (Bannister & Barrett 2004). Academic publications provides an excellent example: they are critical for the establishment of scientific credibility, consensus, and professional advancement, but they may inadvertently transform the legal status of indigenous knowledge, compromising the proprietary rights and control of the indigenous group over their knowledge (e.g., Laird et al. 2002). A recent survey of a sample of pharmaceutical companies, for example, suggests that firms which use ethnobotanical leads in research on natural products—about half of those interviewed—rely on literature and databases as their primary source of information (ten Kate & Laird 1999). This figure has profound implications, for it suggests that academic research and publications are the primary means through which "traditional knowledge travels from a community to the commercial laboratory" (ten Kate & Laird 1999: 62).

These conflicts of interest have led to productive discussions between researchers and their indigenous counterparts, in some cases bringing about a decision to withhold certain types or aspects of cultural information from publication (e.g., Milliken & Albert 1996; Milliken 1997; Alexiades 1999). Such initiatives, however, have met with considerable resistance from the academic establishment (e.g., Milliken 2002). It is interesting to note that although researchers, often working within university departments and botanical gardens, are frequently either willing or expected to compromise their ethic of free exchange of information in regard to corporate-funded research, such compromises are generally viewed as unfeasible when called for in response to the needs of other interest groups, such as indigenous people. This discrepancy may, in turn, reflect a generalized and structural tendency for institutions in industrialized countries to extend proprietary rights to corporate organizations and scientific institutions but not to indigenous actors, a discrepancy that is creating mounting tension (CIEL 1999; Ruiz 2002). In any event, it is clear that questions concerning the disclosure of cultural knowledge, and more generally rights of access to what is increasingly viewed as cultural property, are likely to become more and more important aspects of discussions among the numerous stakeholders involved in ethnobiological research and are certainly a key element in the process of obtaining prior informed consent (Laird & Noejovich 2002).

Such concerns and conflicts of interest are by no means exclusive to ethnobiologists. At a recent conference organized by the American Association for the Advancement of Science entitled "Secrecy in Science," the chairman of the association's Committee on Scientific Freedom and Responsibility expressed his concern about the effects of the increased privatization of science, which has "resulted in a new regime of secrecy which is very concerning [*sic*] to the scientific community" (Goldberg 1999: 3). In the medical sciences the breakdown of the borders between scientific research and corporate development has proceeded to a much greater extent, raising even deeper and more troubling conflicts of interest (e.g., Eichenwald & Kolata 1999).

Although the concomitant increase in the flow of information creates new opportunities for scientific innovation and for development (e.g., Miller 1993; Burch 1997; Everett 1998), it can also facilitate the appropriation of knowledge by different interest groups and create new forms of exclusion, fueling growing tensions between stakeholders with diverse claims, needs, and expectations.

Globalization and the Production of Ethnobiological Knowledge

The commodification of knowledge raises another important question: What are the effects of these processes on the production of new knowledge and the viability of such knowledge? Here it may be useful to distinguish knowledge from information. The latter refers to data that can be codified in some explicit form; the former "resides in the user's subjective context of action, based on that information" (Malhotra 1998: 58). "Knowledge resides in the ways we speak, the flexibility inherent in linguistic behavior included. . . . [I]t can be stable or in a state of flux[;] . . . it may be available in the form of public beliefs shared by all, and it may reside in special individuals. . . . It may reside in them in the form of general rules that are learned by rote, or as an ability to treat new situations in an imaginative way" (Feyerabend 1987: 106, 109). That is, knowledge is living, dynamic, and socially produced. Moreover, whereas some forms of knowledge can be made explicit and written down, much knowledge is "tacit," not codable, and can only be transferred indirectly, through nonverbal human interaction (cf. Neef 1999: 78).

Reducing indigenous knowledge to a cumulative pool of information objectifies cultures and denies the essential, creative aspect of people continually reinventing themselves and producing new "knowledge" in the face of change (e.g., Shiva 1997). A challenge faced by policymakers and scientists who explore benefit-sharing mechanisms is to consider not only compensation for the use of cultural knowledge but also the impact of their intervention on the viability of the cultural, social, and political systems that produce and manage that knowledge. Indeed, as Maffi (2001: 421) states, different aspects of indigenous heritage need to be protected "as *bodies*—bodies of knowledge, bodies of folklore, bodies of language—and as *living*, constantly developing bodies, not as dead bodies from the past."

This and other fundamental problems have led some to suggest that such market-based mechanisms as intellectual property rights law are not suited to the interests of non-Western systems of ownership, tenure, and access (Dove 1996; Posey 2000; Maffi 2001). Indigenous environmental knowledge is largely produced and managed in community economies that differ in a number of critical respects from the market economic system that gave rise to intellectual property rights. In community economies, innovations are often embedded in some sort of social matrix—kin based or other—and as such form a type of "common property"; in market economies, innovations are seen as the product of individual minds and hence as a form of "individual property." Indeed, the concept of property itself has markedly different meanings in community and market systems. Consequently, as Gudeman (1996: 103) points out, "the use of intellectual property rights at the global level raises issues concerning power and the formulation of practices as well as equity and respect between economies." More recent concepts, such as traditional resource rights, recognize the importance of the social, cultural, and political context in which knowledge is produced and managed and require that these be taken into account when implementing contracts, covenants, material transfer agreements, and information transfer agreements (Posey & Dutfield 1996). Moreover, the concept of "bundle or rights" put forth by traditional resource rights goes beyond the issue of compensation and addresses a deeper problem: the need to help indigenous communities retain or regain control over their resource base.

In sum, then, the shift toward a commodity-based way of defining knowledge is creating new tensions with respect to how knowledge is disseminated and exchanged within and between different interest groups. Ethnobiologists have a considerable vested interest and important role to play in the construction of new models for environmental knowledge use and exchange within the new global society.

The Political Economy of Globalization

Globalization, with its concomitant spatial reorganization of production, interpenetration of industries across borders, spread of financial markets, widespread diffusion of consumer goods, and massive flows of people within and between regions, has profound implications for international political economy. Deregulation, widespread application of structural adjustment policies, and the collapse of the Soviet bloc have reinforced the movement toward an integrated global market, with the creation of regional economies that increasingly transcend national boundaries (Ross 1995; Holton 1998). Economic cooperation and exchange among capitalist nations and regions is increasingly facilitated and mediated through such international regulatory and institutional means as the North American Free Trade Agreement, the European Community, and the World Trade Organization. These changes are, in turn, associated with a shift in the balance between the private and the public sector, leading to wide-scale privatization and deregulation of many sectors of the economy, including transportation, communications, education, health, and science (Drucker 1997; Mittelman 1997b). One concern is that deregulation has shifted power away from national governments, which may be popularly or even democratically accountable, to international treaty organizations, which remain largely unaccountable to the public (Wallach & Sforza 1999).

Increasing interdependence does not necessarily imply increasing homogeneity or equality between nations and regions. Indeed, evidence suggests that globalization has increased social and economic inequalities (UNDP 1999). World-systems theories suggest that, from its onset, capitalism created a global market and global division of labor characterized by "core," "periphery," and "semiperiphery" zones and that the expansion of capitalism has always depended on the unequal exchange between core and periphery. Capital-intensive goods in the core are exchanged with labor-intensive goods in the periphery, a process that leads to the development of cores and the concomitant underdevelopment of peripheries (Frank 1969; Wallerstein 1974).

Peripheries tend to be politically and economically marginal regions with limited access to the manufactured goods and the services produced at the centers but with high biodiversity and diversified resource management systems. Centers are referred to as such because they accumulate political and economic power. In contrast to peripheries, centers are deficient in certain resources—labor, timber, and other raw materials, for example—and depend on the flow of these materials from the peripheries to sustain their growth. While the center-periphery model certainly has its critics and its limitations (see Appadurai 1994: 328; Hall 1996; Shannon 1996, for reviews; Nash 1981), I believe that it nonetheless offers a useful model to illustrate the important structural differences and interdependence that underscore the relationship between different regions and levels.[4]

The center-periphery model is applicable internationally, regionally, and locally. North America, for example, is a center in relation to the periphery of South America; cities like São Paulo are centers in relation to such peripheries as the Brazilian Amazon. Within Amazonia, urban centers such as Manaus function as centers in relation to the peripheries formed by rural forest communities (Clement 1999). There are three important aspects of the relationship between centers and peripheries (Dove 1996). First, there is a historically consistent pattern in the relationship between centers and peripheries when it comes to resource extraction. As natural resources from peripheries acquire value for centers, these assume control over the resources, displacing the communities that traditionally controlled them, typically overexploiting the resource and eventually—where possible and viable—developing alternative, cheaper, or more convenient areas of production. The history of resource extraction in tropical forests— quinine, rubber, peach palms, and guaraná, for example—typifies this dynamic interaction between centers and peripheries, where local resource economies go through boom-and-bust cycles (e.g., Bunker 1988; Homma 1992; Clement 1999; Laird et al. 2002). Second, the structural asymmetry that exists between peripheries and centers makes it extremely difficult for resources to flow from the center to the periphery in ways that strengthen the periphery, a factor that approaches which embody "trickle-down economics" do not take into account. Finally, because the structural asymmetry between peripheries extends from the local to the international, discussions regarding equity should not be restricted to the international scale (see also Robbins 1999: 5). The inequity and environmental degradation that underlie relations between the more and less developed countries are matched—and sometimes exceeded—by those between the centers and peripheries within many countries (Dove 1996). By extension, one might question the validity of emphasizing how little indigenous communities have benefited from the development of pharmaceuticals by international corporations without also remarking on how little these communities have benefited from the extraction of timber or conversion of forests to pasture by other interest groups in the same countries. Laird et al. (2000), for example, note how Cameroon, the third largest exporter of timber in Africa, pays practically no attention to conservation and equity issues associated with timber extraction, whereas considerable attention is paid to these same issues with regard to genetic resources that, "by and large, are part of a poorly defined and understood, and certainly to date within Cameroon a largely unrealized, market" (p. 348).

Globalization and the Privatization of Science

The structural adjustments of neoliberal policies are helping to blur the boundaries between academic and corporate research. While the links between the scientific and corporate sectors are most apparent in such knowledge-based industries as biotechnology and computing, the growing dependence of science on the private sector is felt in many other disciplines, including ethnobiology.

The growing privatization of science means not only that more research agendas are now defined by corporate needs but also that biological materials and data collected for academic research may eventually be used in other contexts. For example, materials initially collected under the premise of academic research might eventually be accessed by bioprospecting firms developing contractual arrangements with the host institution. Although such an act would be considered unethical and is censored by the CBD, some have cautioned about the difficulties of tracking the large amounts of biological materials currently being collected under the premises of scientific research (Parry 2000). Advances in computing and database technologies have also facilitated the flow of information from academia to the private sector. A number of public-access and private, computerized databases compile published information on medicinal plants and natural products, making this information available to academic and corporate users (e.g., Loub et al. 1985).

Of course, the links between industry and ethnobiology, particularly ethnobotany, are by no means novel. Ethnobotany emerged as a discipline within a colonial context (Clément 1998), and botanists have played a role in the transfer of knowledge and genetic resources from peripheries to centers of industrial development for quite some time (Brockway 1979).[5] The direct interface between ethnobotany and industry has been revived in the past couple of decades, but public perceptions frequently overestimate the degree to which this is the case (ten Kate & Laird 1999). Below I will suggest that the media and some ethnobotanists have played an important role in shaping public expectations with regard to ethnobotany's current role in the development of new commodities, though in some cases the outcome has been different from what had been anticipated.

Globalization, Indigenism, and Transnational Social Movements

Over the last 20 years indigenous peoples, particularly in the Americas and parts of Southeast Asia, have become increasingly successful in articulating their political agendas in national and international forums. The internationalization of the indigenous movement is typified by the emergence of such international nongovernmental organizations as the Coordinating Body of Indigenous Organizations of the Amazon Basin, the World Rainforest Movement, the Indigenous Peoples' Biodiversity Network, and the Mataatua Conference Secretariat (Posey 1996a). In some cases empowerment and internationalization of indigenism have been linked to strategic alliances with environmentalists and to the power of international media and communications (see below). More generally, growth of the international environmental and human rights—including indigenous—organizations signals the emergence of new social movements, themselves a product of and a response to economic and political globalization. Indeed, although these movements often appeal to moral values associated with global citizenship, their actions are primarily directed at resisting processes closely associated with "globalization from above" (Falk 1993; Chew 1995). Not surprisingly, a number of these social movements are explicitly organized around the environmental, political, social, moral, and ethical implications of the new "life industries," including genetic engineering, biotechnology, and agribusiness. The issues surrounding these organized forms of resistance to globalization are highly pertinent to ethnobotanists, for they relate specifically to cultural and genetic resources and to the relationship among science, technology, and economic development. The political mobilization of indigenous societies also reflects a global trend toward the increased importance of the politics of identity, based on ethnicity, locality, or religion. Past cultural knowledge and cultural resources—such as plant cultivars—have, in many cases and places, become not only commodities but also valued and contested political resources through which different groups lay claim to broader tangible and intangible claims.

At the same time, the political empowerment of indigenous peoples and local communities also creates new opportunities and roles for science and ethnobiology, particularly in terms of participatory and collaborative ventures (e.g., Chrisman et al. 1999). The Proyecto de Estudio para el Manejo de Areas Silvestres de Kuna Yala in Panama, for example, reportedly only considers collaborative ventures with scientists in those cases in which research is relevant or useful to the Kuna and the research process itself is supervised by the Kuna. The Inuit Tapirisat have produced a background paper in which they set out terms for negotiating research relationships with external agents (Posey 1996a). In other words, globalization, with its concomitant effects on social, economic, and political integration and fragmentation, inclusion and exclusion, is creating new spaces and opportunities to redefine and establish innovative research relationships, albeit in a rapidly changing and frequently polarized and emotionally charged political climate.

Globalization, Media, and Culture

The process of globalization has important and extensive cultural ramifications. Whether it is Amerindian shamanism in Switzerland, Andean music in the Ukraine, MTV in Zimbabwe, or sushi in Argentina, the world is witnessing unprecedented rates of cultural diffusion, appropriation, hybridization, and commodification (Holton 1998). Images, ideas, and messages are projected globally through mass media, shaping—at times in unanticipated ways—powerful and politically significant currents of public opinion. In the last decade, indigenous groups in Brazil and Canada, for example, have been able to halt large development projects and obtain important concessions by gathering public support through media exposure and through widely publicized political alliances with international environmental and human rights movements. The degree to which indigenous activism and its interactions with the international environmental community are mediated and facilitated by the international media illustrates the emerging importance of media in shaping human events and human-environment relations (e.g., Jackson 1994; Conklin & Graham 1995; Brosius 1997).

The effect of specific media messages on public opinion is often unpredictable in such transnational contexts. During the late 1980s and early 1990s, for example, ethnobotanists, journalists, and filmmakers promoted the image of tropical forests as vast, mysterious, yet endangered repositories of unknown biological and medicinal riches. Sean Connery's character in *Medicine Man* portrays the ethnobotanist as a romantic idealist, a twentieth-century Victorian explorer. The image of an ethnobotanist leading "an Indiana Jones kind of life, unlocking the secrets of ancient plant medicine in the world's jungles and rain forests" (Hardy 2000: 1) has been successfully woven into a number of popular accounts and media representations of ethnobotany (e.g., Plotkin 1993). Such accounts resonate with and capitalize on a number of pervasive themes in the Western imagination, including individualism and scientific heroism, pristine nature, and the myth of the "noble savage."

Although such media messages undoubtedly helped advance the cause of tropical forest conservation and foster recognition of the importance of ethnobotany and indigenous knowledge in postindustrial societies, the same messages were read very differently by audiences in different parts of the world, underscoring the contradictions that emerge between the centralized production of media and its decentralized and fragmented reception (Lash & Urry 1994). Indeed, to a citizen of another country—say, Brazil—with an entirely different perception of world history and of current events, *Medicine Man* may suggest not a plot about scientific bravery and altruistic exploration but about biopiracy and scientific neocolonialism.

Of course, I am not suggesting that *Medicine Man* is responsible for shaping the differences in North-South perceptions and positions with respect to genetic resources or ethnobiology. Rather, I use this example to underscore the powerful and complex interplay between media messages and their articulation with groups with distinct historical experiences and expectations. The media hype of the "ethnobotanist finds cures in the jungle" not only backfired by casting ethnobotanists as serving corporate interests but also confirmed the erroneous perception that pharmaceutical companies are interested in medicinal plants and indigenous knowledge, when in fact this is quite exceptional in the industry (ten Kate & Laird 1999).

The level and scale of public concern with regard to biopiracy in tropical countries and the hostile suspicion expressed toward scientists may be perplexing to some, particularly in light of the fact that the scale and level of bioprospecting are relatively small in comparison with other forms of natural resource exploitation. Laird et al. (2000: 348), for example, note how large-scale timber operations proceed apace in Cameroon with little regard for questions of equity or conservation, while "suspicion of outside researchers has become routine."[6] If the concept of biopiracy is a product of the emerging global economy, with its concomitant commodification of genetic resources and knowledge (e.g., Shiva 1997), then "bioparanoia" can be seen as the culturally and historically mediated, and quite understandable, response to globalization by social actors who, rightly or not, perceive globalization and the new biotechnologies as a threat to their individual, social, or national integrity

As communication and media networks continue to consolidate and expand (Everett 1998; Reis 1998), more people will draw a greater part of their life experience from the virtual world, as opposed to the real world (Westmacott 1999). This means that media will probably continue to play an increasingly important role in shaping public perceptions of human-environment relations and the role of science in society, among other issues central to our concerns as ethnobiologists. Moreover, as we enter a period of revisionism and intensified public scrutiny, in which the authority of science and the work of scientists become either contested or subject to public scrutiny (e.g., Hooper 1999; Tierney 2000), it becomes increasingly important for scientists to effectively communicate with the general public. This process will not be facilitated by the fact that as global media will continue to bring into relief the economic and social differences among peoples and regions, a factor that may well feed social tensions (Westmacott 1999). Indeed, the inequities that permeate the planet today create a difficult backdrop against which scientists must articulate their agendas with respect to different interest groups, locally and internationally, and including the general public, officials, colleagues, and the media.

Ethnobiology in the Twenty-First Century: The Search for New Paradigms

In this necessarily brief overview of globalization I have tried to show several major trends with respect to ethnobiology: first, the increasing commodification and politicization of indigenous cultural and genetic resources; second, the intensified flows of knowledge and goods between economic and political centers and peripheries; third, the increasing privatization of science; fourth, the growing role of media; and fifth, the shifting dynamics among local communities, the private sector, science, and government.

Genetic resources and cultural knowledge have become increasingly valued political resources. Researchers and their institutions, willingly or unwillingly serving as mediators in the flow of such resources, suddenly find themselves in the midst of a politically complex international arena and subject to increasing media-processed public scrutiny. The ethical and moral complexities of fieldwork and ethnobiology are compounded by the reality of widespread poverty and malnutrition and by growing inequalities between the world's rich and poor (e.g., Auguste 1998; Cowell 2000). Despite an increase in per capita food production of 20% in developing countries between 1980 and 1990, for example, roughly 800 million people suffer from hunger, and more than one-third of the world's children suffer from malnutrition (UNDP 1996: 20, cited in Watts & Goodman 1999: 1). Only recently have questions pertaining to ethical behavior entered mainstream ethnobiological discussions, yet they have become of central concern and are likely to continue to grow in importance in this field, as they are in the social sciences (e.g., ASA 1997; AAA 1998). Productive discussions are already under way (e.g., Posey 1996b; ISE 1998; SEB 1999), and ethnobiologists and biologists need to participate actively in them (Laird 2002).

Hopefully, these growing ethical concerns will be reflected in the contents and structures of training programs and curricula and will lead to revisions and new developments in how fieldwork is structured (Greaves 1996). There is already a definite trend toward implementing more participatory models of collaboration between researchers and local peoples (e.g., Hersch-Martínez & González-Chávez 1996; Laird 2002), even if the concept of participation is as likely to be rhetorically misappropriated as the concept of sustainability (c.f. Sullivan 2002). Quite possibly, ethnobiologists in the twenty-first century not only will need to know about biology and anthropology, as well as such related fields as linguistics, ecology, and phytochemistry, but also will have to acquire a basic understanding of political economy, history, international relations, communications, law, and economics. Without attempting or needing to be experts, background knowledge will, at the very least, help us engage in productive discussions with colleagues from other disciplines and develop truly effective multidisciplinary approaches. The conflicts of interest that inevitably ensue as different groups and stakeholders lay claim to increasingly commodified and politicized cultural and genetic resources will need to be addressed at many different levels. A first step in the direction of conflict resolution may be to initiate and participate in productive dialogues between the different stakeholders involved in or affected by ethnobiological research.

Ultimately, it seems quite clear that ethnobiology in the twenty-first century will carry the inevitable burden of developing socially responsible research amid a central paradox of our times: increased globalization and interdependence in a world characterized by dislocation, difference, and inequality.

Acknowledgments

I am grateful to Luisa Maffi and Tom Carlson for extending an invitation to submit a chapter for this volume. The ideas expressed here were first explored in a paper delivered at the 1997 symposium on Models of Working with Indigenous Communities: New Paradigms for Botany at the American Institute of Biological Sciences meeting in Montreal, Canada, and subsequently refined and developed for a second paper, delivered at the symposium on Plant Genetic Resources and Intellectual Property Rights in the 21st Century at the International Botanical Congress in Saint Louis, Missouri in 1999. I am grateful to David Lentz and Brian Boom for their respective invitations and support to attend these meetings. Tom Carlson, Charles Clement, Robin Goodman, Sarah Laird, Luisa Maffi, Daniela Peluso, and Steve Rubinstein provided very helpful comments on an earlier draft.

Notes

1. Global processes and globalization do not affect all parts of the world equally or in the same way. Moreover, the presence, degree, and effect of global phenomena are dynamic, discontinuous, contingent, and often contradictory (Dirlik 1998; Rosenau 1997, cited in Held et al. 1999: 14). For one thing, world-systems typically pulsate, undergoing successive expansions and contractions (Hall 2000); during the 1870–1940 period, for example, the world economy was at least as open and integrated as it is today (Hirst & Thompson 1996, cited in Mittelman 2000: 20).

2. For a review of some of the different positions and issues surrounding globalization, see Held et al. 1999. See also, for example, Holm & Sørensen 1995; McMichael 1995; Mander & Goldsmith 1996; Spybey 1996; Fernández & Mommen 1998; Holton 1998; Petras 1999.

3. For overviews of the CBD, see Glowka et al. 1994; Sánchez & Juma 1994.

4. Kearney (1995) pointed out that the last decades of globalization, transnational migration, global marketing, electronic media, and tourism have led to the "implosion of peripheries into centers" (p. 551), or to the "peripheralizaton of the core" (p. 554). In this way, globalization has brought about a "multidimensional global space with unbounded, often discontinuous and interpenetrating sub-spaces" (p. 549), where "the local and the global thoroughly infiltrate each other" (p. 550). The core-periphery model, however, is still conceptually useful in relational terms, allowing us to describe the power dynamics and flow of resources between and within regions. Through their location somewhere along a core-periphery continuum, most places in the world function as both core and periphery in relation to other places (e.g., Clement 1999; Sassen 2000). Alternatively, the core-periphery hierarchy can be better conceptualized as a social, as opposed to geographical, division of the world economy (Hoogvelt 1997, cited in Held et al. 1999: 8).

5. As a field of study, ethnobotany has drawn from two quite distinct theoretical lineages: one consists primarily of natural scientists and focuses more on utilitarian concerns, economic botany, and the properties of plants themselves; the other consists mostly of anthropologists and is concerned with issues pertaining to

human cognition and perception of the natural world. The historic links between ethnobotany and industry are evident largely among the former, and only in recent years have the boundaries between "economic" and "cognitive" ethnobotany become more permeable (Maffi pers. comm. 2000; Carlson & Maffi 2004).

6. Of course, timber operations, no matter how predatory or unsustainable they may be, do provide local employment and hence some benefits. Appropriation of genetic resources or cultural knowledge, when it occurs, requires no or minimal labor, so, from a local perspective, it does not offer even these minimal returns (Clement pers. comm.).

Literature Cited

AAA. 1998. Code of Ethics of the American Anthropological Association. <http://www.aaanet.org/committees/ethics/ethcode.htm>.

Alexiades, M. N. 1999. Ethnobotany of the Ese Eja: Plants, health and change in an Amazonian society. Ph.D. diss., City University of New York.

———. In prep. a. The economic and cultural globalization of traditional knowledge systems: Implications for innovation, flow and appropriation of knowledge. *In* Innovative Wisdom. Submitted to Earthscan Publications.

———. In prep. b. The new economy of nature: Tropical forests and the construction of nature in the post-industrial society.

——— & D. Lacaze. 1996. FENAMAD's program in traditional medicine: An integrated approach to health care in the Peruvian Amazon. Pp. 341–365 *in* M. J. Balick, E. Elisabetsky & S. A. Laird (eds.), Medicinal Resources of the Tropical Forest: Biodiversity and Its Importance to Human Health. New York: Columbia University Press.

Appadurai, A. 1994. Disjuncture and difference in the global cultural economy. Pp. 324–339 *in* P. Williams & L. Chrisman (eds.), Colonial Discourse and Post-Colonial Theory: A Reader. New York: Columbia University Press.

ASA. 1997. American Sociological Association Code of Ethics. <http://www.asanet.org/members/ecoderev.html>.

Auguste, B. G. 1998. What's so new about globalization? New Perspectives Quarterly 15(1): 16–20.

Bannister, K. & K. Barrett. 2004. Weighing the proverbial "ounce of prevention" against the "pound of cure" in a biocultural context: A role for the precautionary principle in ethnobiological research. Pp. 304–336 *in* T. J. S. Carlson & L. Maffi (eds.), Ethnobotany and Conservation of Biocultural Diversity. Advances in Economic Botany, 15. Bronx: New York Botanical Garden Press.

Bell, D. 1999. The Coming of Post-Industrial Society: A Venture in Social Forecasting. Special Anniversary ed. New York: Basic Books.

Bray, T. L. (ed.). 2001. The Future of the Past: Archaeologists, Native Americans, and Repatriation. New York: Garland Publishers.

Brockway, L. 1979. Science and colonial expansion: The role of the British Royal Botanic Gardens. American Ethnologist 6: 449–465.

Brosius, J. P. 1997. Endangered forest, endangered people: Environmentalist representations of indigenous knowledge. Human Ecology 25(1): 47–69.

Brown, M. F. 1999. Can culture be copyrighted? Current Anthropology 39(2): 193–222.

Brush, S. B. & D. Stabinsky (eds.). 1996. Valuing Local Knowledge: Indigenous People and Intellectual Property Rights. Washington, D.C.: Island Press.

Bunker, S. G. 1988. Underdeveloping the Amazon: Extraction, Unequal Exchange, and the Failure of the Modern State. Chicago: University of Chicago Press.

Burch, S. 1997. ALAI: Networking with a diversity focus. Development 40(4): 46–50.

Caldwell, L. K. 1991. Globalizing environmentalism: Threshold of a new phase in international relations. Society and Natural Resources 4(3): 259–272.
Carlson, T. J. S. & L. Maffi. 2004. Introduction: Ethnobotany and conservation of biocultural diversity. Pp. 1–6 *in* T. J. S. Carlson & L. Maffi (eds.), Ethnobotany and Conservation of Biocultural Diversity. Advances in Economic Botany, 15. Bronx: New York Botanical Garden Press.
Castells, M. 1996. The net and the self: Working notes for a critical theory of informational society. Critique of Anthropology 16(1): 9–38.
Chew, S. C. 1995. Environmental transformations: Accumulation, ecological crisis, and social movements. Pp. 201–215 *in* D. A. Smith & J. Böröcz (eds.). A New World Order? Global Transformations in the Late Twentieth Century. Westport, Conn.: Greenwood Press.
Chrisman, N. J., C. J. Strickland, K. Powell, M. D. Squeochs & M. Yallup. 1999. Community partnership research with the Yakama Indian nation. Human Organization 58(2): 134–141.
CIEL. 1999. Comments on improving identification of prior art: Recommendations on traditional knowledge relating to biological diversity, submitted to the United States Patent and Trademark Office, August 2, 1999. Washington, D.C.: Center for International Environmental Law. <http://www.ciel.org/Biodiversity/BiodiversityIntellectualProperty.html>.
Clark, I. 1997. Globalization and Fragmentation: International Relations in the Twentieth Century. New York: Oxford University Press.
Clement, C. R. 1999. Biodiversidade, conhecimento indígena e desenvolvimento sustentável na Amazônia: Conferência 20. Paper presented at the 50° Congresso Nacional de Botânica, Sociedade Botânica do Brasil, Blumenau, Santa Catarina, 18–23 July.
Clément, D. 1998. The historical foundations of ethnobiology. Journal of Ethnobiology 18(2): 161–187.
Comor, E. A. 1994. Introduction: The global political economy of communication and IPE. Pp. 1–19 *in* E. A. Connor (ed.), The Global Political Economy of Communication: Hegemony, Telecommunication and the Information Economy. New York: St. Martin's Press.
Conklin, B. A. & L. R. Graham. 1995. The shifting middle ground: Amazonian Indians and eco-politics. American Anthropologist 97(4): 695–710.
Cowell, A. 2000. To have and have not: What can the new economy do for places without an old one? New York Times, April 23: §4, 7.
Cox, R. W. 1997. A perspective on globalization. Pp. 21–30 *in* J. H. Mittelman (ed.), Globalization: Critical Reflections. Boulder, Colo., and London: Lynne Rienner.
Crucible Group. 1994. People, Plants and Patents: The Impact of Intellectual Property on Trade, Plant Biodiversity and Rural Society. Ottawa: International Development Research Centre.
Crucible Group II. 2000. Seeding Solutions, vol. 1: Policy Options for Genetic Resources: People, Plants and Patents Revisited. Ottawa: International Development Research Centre; Rome: International Plant Genetic Resources Institute; Uppsala, Sweden: Dag Hammarskjöld Foundation.
Dirlik, A. 1998. Globalism and the politics of place. Development 41(2): 7–13.
Dove, M. R. 1996. Center, periphery, and biodiversity: A paradox of governance and a developmental challenge. Pp. 41–67 *in* S. B. Brush & D. Stabinsky (eds.), Valuing Local Knowledge: Indigenous People and Intellectual Property Rights. Washington, D.C.: Island Press.
Drucker, P. F. 1997. The global economy and the nation state. Foreign Affairs 76: 159–171.
Eichenwald, K. & G. Kolata. 1999. When physicians double as entrepreneurs. New York Times, November 30: §A, 1; §C, 16.
Ekins, P. 1992. A New World Order: Grassroots Movements for Global Change. London and New York: Routledge.

Everett, M. 1998. Latin America on-line: The Internet, development and democratization. Human Organization 57(4): 385–393.
Falk, R. 1993. The making of global citizenship. Pp. 39–50 *in* J. Brecher, J. B. Childs & J. Cutler (eds.), Global Visions: Beyond the New World Order. Boston: South End Press.
Fernández J., A. E. & A. Mommen. 1998. Globalization versus regionalization. Pp. 1–26 *in* A. E. Fernández J. & A. Mommen (eds.). Regionalization and Globalization in the Modern World Economy: Perspectives on the Third World and Transitional Economies. London and New York: Routledge.
Feyerabend, P. 1987. Farewell to Reason. London and New York: Verso.
Frank, A. G. 1969. Capitalism and the Underdevelopment of Latin America. New York: Monthly Review Press.
——— & B. Gills. 1992. The five thousand year old system: An interdisciplinary introduction. Humboldt Journal of Social Relations 18(1): 1–80.
Friend, T. 1999. Cloned goats milked for human protein. USA Today, April 27: §D, 1.
FSI (Fundación Sabiduría Indígena) & B. Kothari. 1997. Rights to the benefits of research: Compensating indigenous peoples for their intellectual contributions. Human Organization 56(2): 127–137.
Glowka, L., F. Burhenne-Guilmin & H. Synge. 1994. A Guide to the Convention on Biological Diversity. Gland, Switzerland, and Cambridge: IUCN.
Goldberg, C. 1999. Urging a freer flow of scientific ideas. New York Times, April 6: §F, 3.
Gollin, M. A. 1999. New rules for natural products research. Nature Biotechnology 17: 921–922.
Greaves, T. 1996. Tribal rights. Pp. 25–38 *in* S. B. Brush & D. Stabinsky (eds.), Valuing Local Knowledge: Indigenous People and Intellectual Property Rights. Washington, D.C.: Island Press.
Gudeman, S. 1996. Sketches, qualms, and other thoughts on intellectual property rights. Pp. 102–121 *in* S. B. Brush & D. Stabinsky (eds.), Valuing Local Knowledge: Indigenous People and Intellectual Property Rights. Washington, D.C.: Island Press.
Hall, T. D. 1996. The world-system perspective: A small sample from a large universe. Sociological Inquiry 66(4): 440–454.
———. 2000. Frontiers, ethnogenesis, and world-systems: Rethinking the theories. Pp. 237–270 *in* T. D. Hall (ed.). A World-Systems Reader: New Perspectives on Gender, Urbanism, Cultures, Indigenous Peoples, and Ecology. Lanham, Md.: Rowman & Littlefield.
Hardy, J. 2000. The Medicine Man: New Orleans native Mark Plotkin has lived an Indiana Jones kind of life, unlocking the secrets of ancient plant medicine in the world's jungles and rain forests, far from his boyhood spent traipsing around the Audubon Zoo. Times-Picayune, May 26: §E, 1.
Harvey, D. 1990. The Condition of Postmodernity: An Enquiry into the Origins of Cultural Change. Cambridge, Mass.: Blackwell.
Held, D., A. McGrew, D. Goldblatt & J. Perraton. 1999. Global Transformations: Politics, Economics and Culture. Stanford, Calif.: Stanford University Press.
Hersch-Martínez, P. & L. González-Chávez. 1996. Investigación participativa en etnobotánica: Algunos procedimientos coadyuvantes en ella. Dimensión Antropológica 3(8): 129–153.
Holm, H. & G. Sørensen. 1995. Whose World Order? Uneven Globalization and the End of the Cold War. Boulder, Colo.: Westview.
Holton, R. J. 1998. Globalization and the Nation-State. New York: St. Martin's Press.
Homma, A. K. O. 1992. The dynamics of extraction in Amazonia: A historical perspective. Pp. 23–31 *in* D. C. Nepstad & S. Schwartzman (eds.), Non-Timber Products from Tropical Forests: Evaluation of a Conservation and Development Strategy. Advances in Economic Botany, 9. Bronx: New York Botanical Garden.

Hooper, E. 1999. The River: A Journey to the Source of HIV and AIDS. Boston: Little, Brown and Co.

Ingold, T. 1993. Globes and spheres: The topology of environmentalism. Pp. 31–42 *in* K. Milton (ed.), Environmentalism: The View from Anthropology. New York: Routledge.

ISE. 1998. International Society of Ethnobiology (ISE) Code of Ethics. <http://users.ox.ac.uk/~wgtrr/isecode.htm>.

Jackson, J. E. 1994. Becoming Indians: the politics of Tukanoan ethnicity. Pp. 384–405 *in* A. C. Roosevelt (ed.), Amazonian Indians from Prehistory to the Present. Tucson: University of Arizona Press.

James, J. 1999. Globalization, Information Technology and Development. New York: St. Martin's Press.

Kearney, M. 1995. The local and the global: The anthropology of globalism and transnationalism. Annual Review of Anthropology 24: 547–565.

Kelly, K. 1994. Out of Control: The New Biology of Machines, Social Systems and the Economic World. Cambridge, Mass.: Perseus Books.

Kloppenburg, J., Jr. 1988. First the Seed: The Political Economy of Plant Biotechnology, 1492–2000. New York: Cambridge University Press.

Lacaze, D. & M. N. Alexiades. 1995. "Salud Para Todos": Plantas medicinales y salud indígena en el Departamento de Madre de Dios, Perú: Un manual práctico. Cuzco, Peru: FENAMAD / Centro de Estudios Rurales Bartolomé de las Casas.

Laird, S. A. (ed.). 2002. Biodiversity and Traditional Knowledge: Equitable Partnerships in Practice. London: Earthscan.

——— & F. Noejovich. 2002. Building equitable research relationships with indigenous people and local communities: Consultation and reaching agreement. Pp. 179–220 *in* S. A. Laird (ed.), Biodiversity and Traditional Knowledge: Equitable Partnerships in Practice. London: Earthscan.

——— & D. A. Posey. 2002. Professional society standards for biodiversity research: Codes of ethics and research guidelines. Pp. 16–38 *in* S. A. Laird (ed.), Biodiversity and Traditional Knowledge: Equitable Partnerships in Practice. London: Earthscan.

———, A. B. Cunningham & E. Lisinge. 2000. One in ten thousand? The Cameroon case of *Ancistrocladus korupensis*. Pp. 345–373 *in* C. Zerner (ed.), People, Plants and Justice: The Politics of Nature Conservation. New York: Columbia University Press.

———, M. N. Alexiades, K. P. Bannister & D. Posey. 2002. The publication of biodiversity research results: Employing the precautionary principle. Pp. 77–101 *in* S. A. Laird (ed.), Biodiversity and Traditional Knowledge: Equitable Partnerships in Practice. London: Earthscan.

Lash, S. & J. Urry. 1994. Economies of Signs and Space. New York: Sage.

Loub, W. D., N. R. Farnsworth, D. D. Soejarto & M. L. Quinn. 1985. NAPRALERT computer handling of natural product research data. Journal of Chemical Information and Computer Science 25: 99–103.

Maffi, L. 2001. Language, knowledge, and indigenous heritage rights. Pp. 412–432 *in* L. Maffi (ed.), On Biocultural Diversity: Linking Language, Knowledge, and the Environment. Washington, D.C.: Smithsonian Institution Press.

Malhotra, Y. 1998. Tools@work: Deciphering the knowledge management hype. Journal for Quality and Participation 21(4): 58–60.

Mander, J. & E. Goldsmith (eds.). 1996. The Case against the Global Economy and for a Turn Toward the Local. San Francisco: Sierra Club Books.

Marshall, E. 1997. Whose DNA is it anyway? Science 278(5338): 564–567.

McMichael, P. 1995. The new colonialism: Global regulation and the restructuring of the interstate system. Pp. 37–55 *in* D. A. Smith & J. Böröcz (eds.). A New World Order? Global Transformations in the Late Twentieth Century. Westport, Conn.: Greenwood Press.

Miller, S. E. 1993. The information age and agricultural entomology. Bulletin of Entomological Research 83: 471–474.

Milliken, W. 1997. Traditional anti-malarial medicine in Roraima, Brazil. Economic Botany 51(3): 212–237.

———. 2002. People, plants, and publishing. Pp. 79–81 *in* S. A. Laird (ed.), Biodiversity and Traditional Knowledge: Equitable Partnerships in Practice. London: Earthscan.

——— & B. Albert. 1996. The use of medicinal plants by the Yanomami Indians of Brazil. Economic Botany 50(1): 10–25.

Mittelman, J. H. 1997a. How does globalization really work? Pp. 229–241 *in* J. H. Mittelman (ed.), Globalization: Critical Reflections. Boulder, Colo., and London: Lynne Rienner.

———. 1997b. The dynamics of globalization. Pp. 1–19 *in* J. H. Mittelman (ed.), Globalization: Critical Reflections. Boulder, Colo., and London: Lynne Rienner.

———. 2000. The Globalization Syndrome: Transformation and Resistance. Princeton, N.J.: Princeton University Press.

Mooney, P. R. 1997. Biopiracy and the life industry. Development 40(2): 27–30.

Nash, J. 1981. Ethnographic aspects of the world capitalist system. Annual Review of Anthropology 10: 393–423.

Navaretti, G. B., P. Dasgupta, K.-G. Mäler & D. Siniscalco. 1998. Production and transmission of knowledge: Institutions and economic incentives. An introduction. Pp. 1–10 *in* G. B. Navaretti, P. Dasgupta, K.-G. Mäler & D. Siniscalco (eds.), Creation and Transfer of Knowledge: Institutions and Incentives. New York: Springer.

Nazarea, V. D. 1998. Cultural Memory and Biodiversity. Tucson: University of Arizona Press.

Neef, D. 1999. A Little Knowledge Is a Dangerous Thing: Understanding Our Global Knowledge Economy. Boston: Butterworth-Heinemann.

Nuttall, N. 1998. University ends fight over fungi. The Times (London), November 26: 9.

Parry, B. 2000. The fate of the collections: Social justice and the annexation of plant genetic resources. Pp. 374–402 *in* C. Zerner (ed.), People, Plants and Justice: The Politics of Nature Conservation. New York: Columbia University Press.

Petras, J. 1999. Globalization: A critical analysis. Journal of Contemporary Asia 29(1): 3–37.

Plotkin, M. 1993. Tales of a Shaman's Apprentice: An Ethnobotanist Searches for New Medicines in the Amazon Rain Forest. New York: Penguin Books.

Pollack, A. 1999. Biological products raise genetic ownership issues. New York Times, November 26: §A, 1; §C, 12.

Posey, D. A. 1996a. Protecting indigenous peoples' rights to biodiversity. Environment 38: 6–9.

——— (ed.). 1996b. Traditional Resource Rights: International Instruments for Protection and Compensation for Indigenous Peoples and Local Communities. Gland, Switzerland, and Cambridge: IUCN.

———. 2000. Ethnobiology and ethnoecology in the context of national laws and international agreements affecting indigenous and local knowledge, traditional resources and intellectual property rights. Pp. 35–54 *in* R. Ellen, P. Parkes & A. Bicker (eds.), Indigenous Environmental Knowledge and Its Transformations: Critical Anthropological Perspectives. Amsterdam: Hardwood Academic.

——— & G. Dutfield. 1996. Beyond Intellectual Property: Toward Traditional Resource Rights for Indigenous Peoples and Local Communities. Ottawa: International Development Research Centre.

Putterman, D. M. 1994. Trade and the biodiversity convention. Nature 371: 553–554.

Reis, R. 1998. The impact of television viewing in the Brazilian Amazon. Human Organization 57(3): 300–306.

Rigden, P. 1996. Biopiracy clouds disease fight. Canadian Business and Current Affairs 22(4): 2.
Robbins, B. 1999. Feeling Global: Internationalism in Distress. New York: New York University Press.
Ross, R. J. S. 1995. The theory of global capitalism: State theory and variants of capitalism on a world scale. Pp. 19–35 *in* D. A. Smith & J. Böröcz (eds.). A New World Order? Global Transformations in the Late Twentieth Century. Westport, Conn.: Greenwood Press.
Ruiz, M. 2002. The International Debate on Traditional Knowledge as Prior Art in the Patent System: Issues and Options for Developing Countries. Washington, D.C.: Center for International Law. <http://www.ciel.org/Publications/pubbaw.html>.
Sánchez, V. & C. Juma (eds.). 1994. Biodiplomacy: Genetic Resources and International Relations. Nairobi: African Center for Technology Studies.
Sassen, S. 2000. Cities in a world economy. Ed. 2. Thousand Oaks, Calif.: Pine Forge Press.
Schwartz, J. 1999. For sale in Iceland: A nation's genetic code; deal with research firm highlights conflicting views of progress, privacy and ethics. Washington Post (Final Edition), January 12: §A, 1.
SEB. 1999. Society for Economic Botany Guidelines of Professional Ethics. <http://www.econbot.org/ethics/professional_ethics.html>.
Shannon, T. R. 1996. An Introduction to the World-System Perspective. Ed. 2. Boulder, Colo.: Westview Press.
Shapiro, C. & H. R. Varian. 1998. Information Rules: A Strategic Guide to the Network Economy. Boston: Harvard Business School Press.
Shiva, V. 1997. Biopiracy: The Plunder of Nature and Knowledge. Boston: South End Press.
Spybey, T. 1996. Globalization and World Society. Cambridge: Polity Press.
Sullivan, S. 2002. How sustainable is the communalizing discourse of "new" conservation? The masking of difference, inequality and aspiration in the fledgling "conservancies" of Namibia. Pp. 158–187 *in* D. Chatty & M. Colchester (eds.), Conservation and Mobile Indigenous Peoples: Displacement, Forced Settlement and Sustainable Development. Oxford: Berghahn.
Ten Kate, K. & S. A. Laird. 1999. The Commercial Use of Biodiversity: Access to Genetic Resources and Benefit-Sharing. London: Earthscan.
Tierney, P. 2000. Darkness in El Dorado: How Scientists and Journalists Devastated the Amazon. New York: Norton.
UNDP. 1999. United Nations Human Development Report. New York and London: Oxford University Press.
Wallach, L. & M. Sforza. 1999. Whose World Trade Organization? Corporate Globalization and the Erosion of Democracy. New York: Public Citizen.
Wallerstein, I. 1974. The Modern World-System. New York: Academic Press.
Watts, M. & D. Goodman. 1999. Agrarian questions: Global appetite, local metabolism: Nature, culture, and industry in fin-de-siècle agro-food systems. Pp. 1–32 *in* D. Goodman & M. Watts (eds.), Globalising Food: Agrarian Questions and Global Restructuring. London and New York: Routledge.
Westmacott, T. I. 1999. Culture, capital and communications. Research-Technology Management 42(1): 48–51.
Wilford, J. N. 1999. Archaeology and ancestry clash over skeleton. New York Times, November 9: §F, 1, 4.
Wilson, E. O. 1988. The current state of biological diversity. Pp. 3–18 *in* E. O. Wilson (ed.), Biodiversity. Washington, D.C.: National Academy Press.

11

Weighing the Proverbial "Ounce of Prevention" against the "Pound of Cure" in a Biocultural Context: A Role for the Precautionary Principle in Ethnobiological Research

Kelly Bannister and Katherine Barrett

> *The institutional conception of science as part of the public domain is linked with the imperative for communication of findings. Secrecy is the antithesis of this norm; full and open communication its enactment. The pressure for diffusion of results is reinforced by the institutional goal of advancing the boundaries of knowledge and by the incentive of recognition, which is, of course, contingent upon publication.*
>
> —Robert K. Merton 1973 [1942]

Introduction

With the establishment of ethnobiology as an academic discipline comes a corresponding expectation that ethnobiological study will comply with norms of academic research. The obligation to share data and conclusions with the scientific and wider communities through publication is consistent with such norms and is considered an integral part of ethical research conduct. Indeed, open publication is a hallmark of the modern scientific ethos; it enables credit, accountability, and reproducibility of findings and ensures that outcomes of publicly funded research are available to the public. In the latter respect, publication may be considered a broadly beneficial "public good."

Notwithstanding such recognized benefits, are there circumstances that may give us cause to reconsider publication as a convention of scientific inquiry? We explore this question in the context of potential biological and cultural "harms" caused by the publication and dissemination of ethnobotanical research on traditional plant medicines of indigenous societies in Canada. More specifically, we examine third-party use of ethnomedicinal publications as a basis for literature-based screening programs designed to identify biologically active phytochemicals of perceived medical or economic value. This example seems particularly cogent given current widespread public and commercial interest in plant-based medicines and in ethnobotanical research as a means of gaining access traditional botanical knowledge of indigenous peoples.[1] As a contribution to current debate on appropriate ethical guidelines for research, we propose that application of the legal and ethical principle known as the "precautionary principle" to the publication-related issues raised here may help to provide a broader view of the intentions and potential outcomes of current ethnobiological and related research.

Indigenous Knowledge in the Public Domain

Ethnobiology involves the study of relationships between people and their biological environment; hence, traditional ecological knowledge held by societies of indigenous peoples is an essential and valuable component of ethnobiological research. Indeed, documentation and use of the traditional knowledge of indigenous peoples is seen as invaluable for conservation of biological resources and essential for protection of the integrity and interests of indigenous cultures that, in turn, depend on biological resources. Any disruption to this cultural and biological interdependence seems to conflict with guidelines set by the 1992 Convention on Biological Diversity (CBD),[2] as well as with the general goals of ethnobiology as a discipline.[3]

A large body of published information about the knowledge, languages, traditional beliefs and practices, customs, and lifeways of indigenous peoples throughout the world has accumulated in the academic literature through decades of ethnographic, linguistic, anthropological, and ethnobiological research. Documenting such knowledge for posterity is considered increasingly urgent because in many indigenous societies very few individuals (many of whom are aged) retain fluency in their language and an integrated understanding of traditional lifeways.[4]

The 1990s also saw an unprecedented applied interest in the traditional technologies, land and resource management strategies, and traditional foods and medicines of indigenous peoples. Although documentation of the cultural knowledge of indigenous peoples may provide many benefits to indigenous and nonindigenous peoples alike, it is also important to acknowledge some potential drawbacks. First, preservation of cultural knowledge does not in itself preserve culture. Archives of information, no matter how detailed, cannot fully embody the dynamic and interactive nature of those learned traditions and customs that govern the beliefs and behavior to which we refer as "culture" (Kottak 1991). Nor can such archives fully convey the interdependence of cultural knowledge and biological resources. Thus documentation does not approximate the degree of protection that is afforded through protecting the living systems themselves. In recognition of this, in situ conservation has been proposed as an important measure for protection of knowledge and resources. For example, Brush (1993: 660) noted that "On-site conservation is now accepted as part of the long-term solution to conserving the store of biodiversity that is threatened by such activities as tropical deforestation and agricultural intensification, and this type of conservation depends utterly on indigenous knowledge." Second, the sharing of any knowledge outside the context in which it originates has inherent risks. The following discussion addresses some current implications of disseminating traditional botanical knowledge of indigenous peoples beyond its original biological and cultural contexts.

Indigenous peoples worldwide are expressing increasing concern about loss of control over their knowledge and related biological resources, as indigenous knowledge becomes a sought-after commodity of research, development, and commercialization efforts by nonindigenous people. Posey and Dutfield (1996) pointed out that ethnobiological information published in the academic literature (in the form of journal articles, databases, and field collections) serves as a major source of information and ideas for researchers and industries with commercial objectives. Often these are third parties that have had no direct contact with the indigenous communities whose knowledge they are appropriating.[5]

A striking example is the unprecedented commercial interest in medicinal plant knowledge of indigenous cultures and the consequent rise in ethnobotanical literature–based plant-screening programs over the last two decades. The authors of original ethnobiological research typically give credit to indigenous sources; indeed, this is considered essential to ensure accuracy of information, as well as appropriate recognition. However, subsequent publications by third parties (e.g., on biological activities, pharmacological properties, and phytochemical analyses of plants) often acknowledge only the academic researchers as authorities and overlook the essential role of the original knowledge holders.[6] Is such protocol consistent with the scholarly conventions of acknowledgment, citation, and authorship?

The current ease of access to traditional botanical knowledge of indigenous societies through the ethnobiological literature may shift the perceived value of ethnobiological research from seeking to understand interactive human-plant coevolutionary processes to simply exploiting the useful products of these interactions. How do the impacts of this perceptual shift reflect on the discipline of ethnobiology and on relationships between ethnobiologists and indigenous peoples? Such a shift not only undermines the goals of the discipline but also may devalue the essential role that traditional knowledge has within indigenous communities and indigenous knowledge systems. When information on plants is obtained from the public domain, such literature-based screening programs can proceed without legal obligation to, and often completely without regard for, the indigenous sources of the information, as discussed below. Furthermore, although these research proposals and programs (including justification for funding) are premised on the value of indigenous knowledge in targeting medically relevant or otherwise beneficial plant species, the important innovations are perceived to take place after the knowledge has left its indigenous source. This perspective does not acknowledge the fundamental importance of indigenous peoples as discoverers and innovators over many previous generations or, indeed, millennia. Thus, while a common justification for literature-based screening programs is the potential medical or other benefits accrued to humankind, one could argue that such cures are not without significant costs to indigenous peoples if cultural knowledge is used only as the most expeditious means.

If some of the most significant rights of indigenous peoples derive from physical control over traditional plant resources and the knowledge relating to these plants (Barton 1994), then publication of plant lists, plant locations, and traditional knowledge related to medicinal preparation and use may significantly decrease this control. Cultural and biological contexts are essential to understand the "comprehensiveness of indigenous knowledge as a knowledge system" (Mann 1997a: 29). It follows that the omission of context—particularly apparent in the recent ethnopharmacological and phytochemical literature—only hastens the loss of connection with (and control by) indigenous peoples. Indeed, it is widely agreed that traditional knowledge held by indigenous communities cannot be divorced from its cultural context without concurrent losses in meaning that, in turn, may lead to culturally inappropriate uses. Brouwer (1998) suggested that a part of this problem lies in the separation of technical knowledge of indigenous cultures from broader indigenous knowledge systems. Salmón (1996: 72) articulated the problem thus: "Western scientists accept plant knowledge as something that first must be elevated to raw data that can then be analyzed and synthesized. The problem is that, removed from its source, the knowledge is incomplete. It is only a whisper of its former self." Furthermore, Salmón commented, "Such knowledge undergoes changes as it is separated from its source. Both indigenous peoples and others have yet to fully comprehend the effects of appropriating knowledge" (p. 70). This problem is compounded when indigenous peoples among whom the knowledge originated are unaware of other uses of the knowledge or related resources, as is often the case when third parties appropriate cultural knowledge.

Thus, given the current third-party interest in traditional botanical and other cultural knowledge of indigenous communities, the ease of access to this type of information through ethnobiological publications, and the potential for appropriation and exploitation of knowledge and traditional resources, renewed consideration of the norms of scientific publication seems warranted. What are the implications of continued publication of ethnobiological inventories of plants, descriptions of traditional preparations, and details of medicinal applications for the integrity of cultural and biological diversity, which are the basis of survival for indigenous communities, as well as common heritages of humanity and central concerns of ethnobiological inquiry?

Potential Harms and Uncertainties Resulting from Publication

Potential harms and persisting uncertainties surrounding publication of indigenous cultural knowledge stem from at least two sources: limited definitions of "harm" and the distancing of knowledge from its original context. Disclosure of known effects and prediction of potentially harmful impacts of all research endeavors are essential steps in obtaining prior informed consent,[7] which is a fundamental principle in the ethical codes of most research institutions and many professional societies.[8] However, adequate definitions of "harm" and "effect" are a prerequisite to their prediction, and understanding what is harmful is a prerequisite to any such definition. Perspective and context, not surprisingly, influence such understandings. Thus a critical question arises: How is harm defined, and by whom? With respect to identification of potential harms related to ethnobiological research, the two most obvious aspects to consider are cultural (i.e., pertaining to the well-being of communities of people and the integrity of cultural knowledge, customs, language systems, traditional beliefs, and lifeways) and biological (i.e., pertaining to the well-being and integrity of plant, animal, insect, and microbial populations).

Current work on the protection of indigenous knowledge and traditional resources may serve as a starting point to broaden the scope of existing concepts of harm and assist in the identification of cultural harms and/or biological harms related to publication of indigenous knowledge. The *Report of the Royal Commission for Aboriginal Peoples* (1996) is of value in this regard because it recognizes concerns of indigenous peoples for protection of their intellectual property and identifies three general areas in need of appropriate protection in Canada: assurance of appropriate use of indigenous knowledge; portrayal of authentic identity of originators of the knowledge; and fair compensation for appropriate commercial use of intellectual and cultural property (as in Mann 1997a). Numerous community-based interests that are related to protection of indigenous knowledge are also noted in this report. As Mann (1997a: 27–28) summarized them, these include:

- preventing the loss of control over indigenous knowledge that could lead to its commercialization or to the identification of sacred sites by those who do not appreciate their significance;
- control over revealing spiritual knowledge to outsiders who can destroy its sacredness or twist the meanings of the teachings;
- inappropriate imitation of indigenous practices which are a misrepresentation of Aboriginal culture and weaken its teachings;
- protection from imitative works;
- control over the integrity of [indigenous knowledge] by exercising control over who has access to it and how it can be used;
- where it is appropriate to use the knowledge in a commercial setting which benefits others, a sharing of the benefits should be required;
- a greater sense of protection of the sacredness associated with the knowledge; and
- proper protection of the collective rights as opposed to individual rights, including [indigenous knowledge] that might otherwise be defined to be in the public domain.

These general concerns for protection of cultural knowledge may have relevance to the prediction of potential harms incurred by distancing traditional botanical knowledge from its originators through publication. Additionally, based on personal observation, we suggest that potential cultural and/or biological harms related to popularization of plant products derived from indigenous knowledge may include: loss of access, or limited access, to traditional resources for indigenous peoples due to high market demand for wild-crafted botanical products (e.g., through government-imposed regulations for plant collection on federal lands that are part of traditional plant-collecting territories); overharvesting or other means of resource depletion (e.g., through lack of regulations for plant collection);[9] limitations on preparation and use of traditional herbal remedies (e.g., as a result of government-imposed quality and safety regulations for herbal remedies);[10] displacement of native plant populations and alteration of ecosystems (e.g., by intensive cultivation of economically valued exotic plants at the expense of local ecosystems and soil health);[11] and inappropriate or disrespectful marketing strategies for the sale of commercial products.

Although many adverse consequences of appropriating knowledge from indigenous communities have been identified and experienced firsthand, few legal documents and declarations concerning indigenous knowledge explicitly define harm. Often concepts are general, vague, or absent. The consequent danger is that harms to indigenous and traditional societies may be subject to interpretation by external parties with overriding economic interests, little contextual understanding, and/or no direct experience in working with indigenous communities.

The larger problem, of course, is that harm cannot always be predicted and that the accuracy of predictions is likely to decrease as knowledge is further removed from its original context; that is, uncertainties and potential harms surrounding initial ethnobiological research are extended and intensified through publication, third-party appropriation, and further distancing. The final outcome will likely entail cumulative, synergistic, long-term, and/or unprecedented effects. Addressing these issues will require not only an understanding and balancing of the interests of indigenous communities with the population as a whole but also consideration of the direct or indirect impacts of publishing on local ecosystems, including native plant, animal, insect, and microbial populations. Therefore, while we do not dispute the enormous potential value of exploring the traditional knowledge of indigenous peoples, we recognize that predicting harmful outcomes of such research is not always straightforward.

The Dilemma

Given the inherent uncertainty regarding the consequences of publication, the dilemma for ethnobiologists becomes one of conflicting moral obligations.[12] Implicit in the goals of ethnobiology as a discipline is a moral imperative to both promote and protect populations of indigenous peoples, plants, and animals. Implicit in the academic and scientific nature of the discipline is a moral duty to report publicly funded research in the public domain. The dilemma arises when promotion and protection conflict; for example, when consequences of publishing data from ethnobiological research (i.e., a form of promotion) interfere with the protection or maintenance of cultural or biological diversity (for a case study, see Bannister in press). In this regard, Posey's (1998: 242) caution strikes an uncomfortable chord: "Anthropological research is often a vehicle for the appropriation—not protection—of indigenous knowledge."

In Canada, there are no adequate means to address this dilemma. Problems precipitated by appropriation of cultural knowledge and exploitation of traditional resource are gaining recognition, however. Some proposed mechanisms for guiding appropriate documentation and use of indigenous cultural knowledge include international law and policy, contracts and agreements between researchers and indigenous communities, legal systems of intellectual property rights (IPR) or related alternatives, international initiatives based on moral imperatives, guidelines of professional societies, and institutional research and ethics policies. Still, as discussed below, these mechanisms offer little protection for knowledge after it has been released into the public domain.

Legal and Moral Imperatives Regarding Indigenous Cultural Knowledge

International Law and Policy

In the 1990s issues involving the protection of cultural knowledge and traditional resources came to the forefront of both international law and Canadian law and policy. The heightened interest in and awareness of these issues may be attributed in part to the 1992 United Nations Conference on Environment and Development (the Rio Earth Summit), of which the CBD was one of several outcomes. The objectives of the CBD are stated in Article 1 as follows: "the conservation of biological diversity, the sustainable use of its components and the fair and equitable sharing of the benefits arising out of the utilization of genetic resources, including by appropriate access to genetic resources and by appropriate transfer of relevant technologies, taking into account all rights over those resources and to technologies, and by appropriate funding" (Convention on Biological Diversity 1992).

Canada, as a signatory to the CBD, is legally and morally bound to implementation of all 42 articles in the CBD, but two of these articles—Article 8, "*In-situ* Conservation," and Article 10, "Sustainable Use of Components of Biological Diversity"—are particularly relevant to the protection of indigenous cultural heritage. Article 8(j) acknowledges the importance of: respecting, preserving, and maintaining the knowledge, innovations, and practices of indigenous and local communities embodying traditional lifestyles for the conservation and sustainable use of biological diversity; promoting their wider application with the approval and involvement of the holders of such knowledge, innovations, and practices; and encouraging the equitable sharing of benefits arising from the utilization of such knowledge, innovations, and practices. Article 10(c) acknowledges the importance of protecting and encouraging customary use of biological resources in accordance with traditional cultural practices that are compatible with conservation or sustainable use requirements.[13]

Implementation of these articles has been fraught with problems, partly linked to lack of some fundamental definitions. The tasks of clarification and implementation have largely been left to subsequent CBD working groups and other initiatives.[14] The Canadian Working Group on Article 8(j) found it necessary to address definitions of indigenous knowledge before attempting to identify potential protection strategies in Canada (Augustine 1997; Lambrou 1997; Mann 1997a: 1; Ford 1998).[15] Many hurdles to protection are recognized, especially in wider applications of knowledge use by third parties. Indeed, Mann (1997a: 28) suggested that "the prevention of abuses by others of indigenous knowledge or culture . . . can be seen as a broader public interest issue, requiring more defined legislation directed towards all citizens to prevent imitation, rip-offs, misuses or misidentification of indigenous knowledge or culture."

Contractual Approaches

Contractual approaches explicitly recognizing and protecting rights to knowledge and resources are standard in ethnobiological research involving indigenous communities. Contracts often take the form of research agreements or memoranda of understanding that stipulate the agreed conditions under which research will take place and establish that research will occur with the prior informed consent of the indigenous participants. Although research agreements are considered an essential component of current ethical research conduct, there are several problems that indicate the inadequacy of a contractual approach alone for protection of cultural knowledge. First, How is informed consent established, and who has the authority to give it? Nicholas and Andrews (1997: 8) suggested that "who has the authority to speak for whom, and of what" can be associated with control of power and social posturing as a public exhibition of this control. Each community is likely to have a local protocol to be followed in order to receive permission for conducting research, and local protocols may or may not be consistent with those required by a research institution or funding agency.

Second, few researchers understand or acknowledge that informed consent is a continuing process rather than sealed approval for research given at the onset. The current Tri-Council Policy Statement: Ethical Conduct for Research for Research Involving Humans (n.d.), which governs all such research at Canadian institutions, states: "Free and informed consent lies at the heart of ethical research involving human subjects. It encompasses a process that begins with the initial contact and carries through to the end of the involvement of research subjects in the project."[16]

Third, a significant problem with employing contracts for protection of cultural knowledge is that they are legally binding only between the contracting parties. After research is made public, without formal links to legal mechanisms of intellectual property rights protection (e.g., trade secrets, licensing, patents, and copyright law), the rights of ownership of local communities involved in the research are left open to definition by third parties (Mann 1997b). Even if such links are deemed appropriate, important elements are overlooked by current IPR laws (as indicated below), costs can be prohibitive, and the extent of legal obligations may be limited outside the country of origin (Mann 1997b).

Intellectual Property Rights

The term "intellectual property" refers to precisely defined kinds of knowledge that can be protected by law. The expression of an idea, or the embodiment of an idea in a process or product, can be protected, but the idea itself cannot. Intellectual property laws are seen as investments in innovation; the main purpose of these laws is to promote creativity and the exchange of ideas. Creators and innovators are encouraged and rewarded through protection of ownership and commercial investment, but the rights of creators and innovators are weighed against those of consumers and other stakeholders (Cassidy & Longford n.d.).

Considerable effort has gone into assessing the utility of intellectual property rights tools for protecting cultural knowledge and resources related to traditional plant medicines. Patents are a common means of intellectual property protection for novel, biologically active chemical compounds isolated from plants, but patents are costly and have little utility for protecting either the typically undefined chemical mixtures that comprise most medicinal plant preparations or the traditional preparation methods involved.

Alternatively, trade secrets have been proposed as a means of protecting formulas, methods, techniques, and other commercially valuable know-how related to medicinal plant knowledge (Barton 1994; Posey & Dutfield 1996; Peteru 1997; Stephenson 1999). Confidentiality is a key component in obtaining trade secret protection, however, which implies that most knowledge shared among researchers and community members within the context of an ethnobiological research project would likely be disqualified from this sort of protection. Peteru (1997: 98) commented that "establishing a trade secret may be elusive in a culture that emphasizes common over private property." Indeed, it might appear counterproductive or even antagonistic in an environment where reviving, rebuilding, and celebrating language, culture, and traditional lifeways are priorities.

Licensing agreements have been suggested as potential means of employing IPR protection for medicinal plant knowledge (Barton 1994; Stephenson 1999) because licensing effectively grants rights without relinquishing ownership. Stephenson (1998, 1999) suggests that the adaptability and low cost of licensing agreements may make these IPR instruments the most compatible with protection of indigenous cultural knowledge, provided that they are "grounded in a thorough understanding of the pertinent ethnoecological context" (Stephenson 1999: 241). Brown (1998: 205) also lends support to the concept of "joint stewardship" and to the potential merits offered by "creative licensing partnerships between native communities and corporate interests." Furthermore, Stephenson (1999) reminds us that the constant evolution of IPR tools to keep pace with protection of new technologies may, coincidentally, lead to more suitable IPR tools for indigenous peoples in the future. However, licensing agreements may be of limited utility in practice, because establishing a form of intellectual property protection (e.g., patent or trade secret) is obviously a prerequisite for licensing such rights. Apparently the application of licensing arrangements in this context remains largely theoretical (David Stephenson Jr., pers. comm. 2000).

Copyright offers protection for subject matter in a physical form, so it is applicable to the physical expression of ideas but not to the protection of knowledge itself. This makes copyright largely irrelevant to the laboratory products of medicinal plant research, but it is of importance to the publication of research results.

Interestingly, publication itself is seen as a strategy for protecting knowledge. Public disclosure negates the potential of obtaining a monopoly on inventions derived from cultural knowledge and resources through patent protection because the invention no longer can be considered novel. This strategy has been called "defensive publication" (Crucible Group 1994; Posey & Dutfield 1996). However, in a defensive publication the published material must include an in-depth description of the invention that would enable it to be repeated. Thus, although the published information itself is no longer patentable, defensive publication still may serve the interests of third parties by providing useful leads for development of new products (e.g., modifications of the published information that are patentable), because detailed information is still put into the public domain. This point is supported by estimates that approximately 90% of patents are for improvements to existing patented inventions rather than for new inventions (WWLIA 1995). Thus, public disclosure of data, be it as a requirement of patent protection, through defensive publication to block patent protection, or by other means, still serves as a valuable resource for exploitation by third parties.

There is a general consensus that existing IPR systems are ill equipped to protect the cultural knowledge of indigenous societies from potentially harmful effects of appropriation and exploitation (see, e.g., Brush 1993; Posey & Dutfield 1996; Mann 1997a; Dutfield 2000).[17] The whole premise of applying mechanisms of intellectual property rights to protection of cultural knowledge has sparked international debate. Even if intellectual property protection is employed, Brush (1993) suggests, it is unlikely that its benefits for cultural knowledge and traditional resources would reach indigenous peoples. Hence the feasibility of employing intellectual property protection to indigenous medicinal plant knowledge and resources is a related but nevertheless separate issue from whether these means are culturally appropriate in the first place. Teams of professional lawyers, researchers, and analysts can address the former to some degree, but they cannot determine the latter, which ultimately must be resolved by indigenous communities themselves.

Traditional Resource Rights and Alternatives to Intellectual Property Rights

The conceptual and logistical limitations in existing IPR regimes compelled Posey and Dutfield (1996) to advocate the establishment of a more inclusive set of protective rights, which they called "traditional resource rights." This term builds on the concept of intellectual property protection and compensation but recognizes that traditional resources are both tangible and intangible and that the latter may be "*inalienable* or belonging to no human being" (p. 3).[18]

Application of the term "property" to the traditional resources of indigenous communities is described as largely "inappropriate" in current CBD documents, because the concept of ownership and the ability to transfer ownership in this context are contrary to the worldviews of many indigenous peoples. Hence the traditional resource rights concept has been embraced by CBD working groups as a unifying conceptual approach "that more accurately reflects the views and concerns of indigenous and local communities and yet is entirely compatible with the requirements of the Convention on Biological Diversity, the International Undertaking on Plant Genetic Resources and the [World Trade Organization] Agreement on Trade-Related Aspects of Intellectual Property Rights" (Executive Secretary . . . 2000: ¶41, 50).

A number of provisions in the CBD directly or indirectly involve IPR,[19] and the potential biological and cultural impacts of their implementation were discussed at the third meeting of the Conference of Parties to the CBD (1996).[20] Dutfield (2000: 34) provides a useful summary and discussion of IPR–related aspects of the CBD, noting that "there is no *explicit* reference to who such right holders should be." Current discussions of plant-variety protection under the Trade Related Aspects of Intellectual Property Rights (TRIPs) agreement of the World Trade Organization provide another opportunity to reexamine issues related to indigenous knowledge. The general goals of the TRIPs agreement may conflict with those of the CBD, but Article 27.3(b) of the TRIPs agreement, which allows for protection of plant varieties through sui generis mechanisms, could bridge this divide to some degree. Some sui generis systems of protection currently under discussion include collective systems for traditional resource rights and community and farmers' rights. In this area, Louwaars (1998) suggests that a sui generis system that is compatible with in situ conservation can be achieved by "linking IPR laws with . . . access and remuneration systems for plant genetic resources."[21]

Efforts are also being directed toward alternatives to existing IPR systems. For example, Nijar (1998: 11) outlines a collective rights system in the form of a "Community Intellectual Property Rights Act" and suggests that this could "bridge the gap between the Convention on Biological Diversity and the TRIPs agreement." For several years now, the Organization of African Unity has been working on a model law on community rights and access to biological resources that calls for a complete ban on patenting of plants, animals, and other life-forms (Declaration and Draft Model . . . 1998). It remains to be seen whether such mechanisms will gain widespread support and, if so, how they would affect publication and third-party appropriation of knowledge. An effective system may explicitly grant stewardship and control of traditional knowledge to the indigenous communities where the knowledge originated, require direct consent for use of this knowledge, and specify that the terms of compensation be negotiated according to local custom (Nijar 1998).

Posey (1999) highlights two points that are of utmost importance to the debate on use of IPR mechanisms to protect cultural knowledge of indigenous peoples. First, the property regimes of indigenous peoples themselves will be key in finding solutions to problems of IPR protection. Second, although in some situations commercialization of traditional botanical knowledge and resources may be desired by indigenous peoples, an effective IPR regime must also support the right to not have knowledge and resources privatized and commercialized (i.e., a negative right).

Brown (1998) offers a skeptical view of employing legal means to control cultural appropriation and suggests that there are much broader issues to consider in the development of legal regimes that "impose new limitations on the free exchange of information in the name of protecting ethnic minorities" (p. 195). He explains that while it may be possible to protect culture to some extent, culture cannot be bounded, defined, and protected from contamination, or commodified, without harming it. Indigenous societies borrow from other indigenous societies and from the majority societies in which they live. The concept of heritage protection as the recovery of some unique authenticity cannot be implemented without coming into conflict with such equally important human rights as "the right to innovate, communicate and share"; nor is it easy to reconcile with the realities of indigenous engagement in majority society (Michael Brown pers. comm. 2000).

International Declarations

Due to the inadequacies in existing legal mechanisms for the protection of cultural knowledge and resources of indigenous peoples, the 1990s saw heavy reliance on moral imperatives and ethical standards. Numerous statements and international declarations by indigenous peoples and/or concerned professionals highlight the depth and breadth of concerns relating to the protection of cultural knowledge, resources, and indigenous lifeways.[22]

Each of these statements and declarations makes unique contributions to local and international debates on indigenous intellectual and cultural property rights, yet there are many commonalities among them. Affirmation of basic human rights and recognition of past injustices are seen as essential starting points, and rights to self-determination are seen as fundamental to more equitable interactions. Recognition and respect for cultural knowledge, language, values, spirituality, and all aspects of traditional lifeways and protection of these from exploitation are viewed to be of paramount importance. Willingness exists to share cultural and intellectual properties with all humanity, for "the common good," provided that the past, present, and potential contributions of indigenous peoples are acknowledged and that the fundamental rights of indigenous peoples to define and control these properties are recognized and protected by the international community. An overriding responsibility to both ancestors and descendants and to continued relationships with the land and its inhabitants is expressed, and customary practices and teachings are seen as the most effective means of fulfilling these responsibilities. The role of indigenous peoples as guardians, managers, and perpetuators of biological diversity over millennia is promoted, and the exploitation of biological diversity for commercial or other purposes without the knowledge and consent of indigenous peoples is condemned.

Apparent in these individual documents and in their collective overlap is the "rights-oriented" approach that generates a unifying vocabulary for voicing indigenous concerns. Saunders (1994: 21) pointed out that indigenous peoples' focus on rights has been criticized "as an inadequate framework for advancing the actual situation of particular groups." However, he also argued that "it is compelling and it is available. The dominant society values rights. The gaining of rights could be followed by other gains."

Collectively, such statements and declarations indicate that ethnobiological research involving the cultural knowledge and traditional resources of indigenous peoples is poised on the edge of an historic period of transition. Precisely how future ethnobiological research perspectives and practices will be altered formally is not yet clear. In the meantime, many researchers are left with uncertainty about how to proceed. It is becoming increasingly obvious that some of the standard ways of "doing business" are no longer acceptable, but there is little in the way of precedent to comfortably guide change—and those who attempt to meet these new challenges do so with significant risk of criticism (albeit for different reasons) from the various stakeholders in the research, including their institutions, their sponsors, their professions, and (local and international) indigenous communities. Nicholas and Andrews (1997: 226) describe a parallel situation in archaeology in British Columbia: "Being on the 'leading edge' of anything may carry some status (real or imagined). However, as many of us have discovered, it can also be a very uncomfortable if not dangerous place to be, and there is the constant danger of falling off, of doing the wrong thing."

Professional Societies and Academic Policies

To assist in resolving these complex issues and ethical dilemmas, many nonprofit organizations, professional societies, and research institutes (including botanical gardens[23]) have begun to develop (or revise) codes of conduct and/or codes of ethics. Such codes reflect the earnest need for standards of practice that promote the conservation of biological and cultural diversity and are acceptable and sensitive to the current concerns of indigenous societies, as well as to other stakeholders in the research (for an overview and comparison, see Laird & Posey 2002).

Embodying these objectives in a written form that is acceptable to all those involved in the research process, however, is not necessarily an easy or straightforward task. The 10-year negotiation process involving hundreds of people in developing and ratifying the International Society of Ethnobiology Code of Ethics (1998) attests to this fact. Furthermore, the very concept of developing an ethical code has met with resistance by individuals in some professional societies—apparently some researchers claim not to see the relevance of ethics to their practical fieldwork, while others are opposed to the imposition of authoritarian or inflexible rules or limits on their academic freedom (as detailed in Laird & Posey 2002).

Still others have suggested that the process of discussing, developing, and revising acceptable research guidelines and professional codes of ethics is just as important for individual researchers and for the evolution of academic disciplines as is the resulting end product (Cassell & Jacobs 1987; Laird & Posey 2002). Involvement in this type of process is also invaluable from the perspective of compliance, because enforcement of most (if not all) professional codes of ethics is logistically impossible due to legal implications and excessive costs. Thus, the objectives of most professional codes of ethics are education oriented, and implementation relies largely on the awareness and voluntary compliance of individuals, while drawing also on the power of moral suasion. Here again, however, compliance can be complicated by the multidisciplinary nature of ethnobiology, for most researchers in the field belong to multiple professional societies, each with a code of ethics that may not lead to the same research decisions. Ultimately, decision making is largely left to individuals, who may be required to reconcile obligations to their sponsoring institutions and funding agencies with obligations to the discipline and to local communities involved in the research.

Academic norms such as ethics review procedures, codes of ethics, advancement criteria, and general university policies often are inadequate for acknowledging and addressing complex issues surrounding the incorporation of indigenous cultural knowledge in academic research publications. General university policies on publication and academic freedom, at least in Canada, maintain that the rights to publication of research results belong to the researchers (Mann 1997b). Furthermore, university policies may assign to the university first rights to protection or licensing of a discovery or invention.[24] Current cuts in funding and changes in Canadian university priorities may also inadvertently encourage exploitation of indigenous knowledge by promoting and rewarding ties between academia and industry and by encouraging applied, product-oriented, and/or patentable research.[25] As Posey (1999: 225) points out, such industry-funded research leaves "questions of who controls the resulting data unanswered, even unasked." These observations point to fundamental problems in academic policy and reward structures (see below) and to an important opportunity for professional societies to assume a higher profile in development and enforcement of new or existing codes of conduct and standards of practice in academic research and publication.

Decisions about appropriate publication of knowledge often hinge on moral imperatives rather than on formal legal or institutional rules. Toward this end, Posey and Dutfield (1996) suggested that ethnobiological publications incorporating indigenous knowledge could contain a statement informing readers and users of their moral obligations to indigenous peoples. While such statements make explicit the position of authors regarding appropriate use of their publications and increase awareness among the readership of relevant ethical considerations, nonetheless moral imperatives leave significant potential for misuse and offer little legal protection after publication.[26] Placement of certain ethnobiological research data in the public domain therefore enables third parties (knowingly or unknowingly) to sidestep the most fundamental elements of ethical ethnobiological research; that is, protection of human, plant, and animal communities from adverse effects of the research.

Appropriate mechanisms are needed both to address the uncertainties surrounding publications that incorporate cultural knowledge of indigenous peoples and to balance the promotion and protection of such knowledge. There is a need to identify and document existing harm. If potential harms can be demonstrated, and where uncertainty persists, there must be guidelines to resolve questions of conflicting rights and interest and to guide fair decision-making processes. With the above in mind, we now shift our focus to a possible framework to address these issues: the precautionary principle.

History and Current Status of the Precautionary Principle

The precautionary principle is a broad legal and ethical framework that advocates taking anticipatory, preventative steps to avoid adverse effects of human activities. Rather than a strict code of conduct or specific set of laws, the precautionary principle may be thought of as a set of guidelines to aid decision making in times of uncertainty (Cameron & Abouchar 1991). In general, the principle states that when there is reason to believe that our actions will result in significant harm, we should take active measures to prevent such harm, even if cause-and-effect relationships have not been scientifically proved. In other words, the principle may be likened to the old adage, "An ounce of prevention is worth a pound of cure."

Precaution first arose as a formal normative principle for environmental policy in Germany in the 1970s. There it was known as the principle of *Vorsorge* (foresight or careful forward planning). The German policy context was complex (see Boehmer-Christiansen 1994), but the general impetus was to take anticipatory action to prevent adverse environmental effects of industrial practices, even when conclusive scientific proof of such effects was lacking.

The precautionary principle has been invoked in many international laws, treaties, and declarations.[27] With the incorporation of precaution into the Rio Declaration on Environment and Development (1992) and more recently the Cartagena Protocol on Biosafety (2000) and the Stockholm Convention on Persistent Organic Pollutants (2001), there is growing consensus that this principle has reached the status of international customary law[28] (McIntrye & Mosedale 1997; Backes & Verschuuren 1998). Most applications to date have aimed at preventing or controlling emissions of industrial pollutants. For example, precaution has featured strongly in agreements about and legislation on climate change, marine dumping of pollutants, and general efforts toward sustainability. However, there is increasing awareness of, and support for, wider applications of the precautionary principle in areas such as wildlife and habitat conservation (see Backes & Verschuuren 1998), ecological restoration (Noss 1998), agricultural biotechnology (Barrett 1999), and ethnobotany (Bannister 2000). Of direct relevance to this discussion, the International Society of Ethnobiology (ISE) has incorporated the principle of precaution into its Code of Ethics as part of a professional Code of Conduct.[29] This provides professional impetus for implementation of a precautionary approach in the ethnobiological context described herein.

General Elements of the Precautionary Principle

Current formulations of the precautionary principle range from weak to strong. Weak versions are more flexible and usually tied to cost-benefit analyses; much stronger versions are associated with deep ecological or "green" perspectives (O'Riordan & Jordon 1995). Despite this variability, a number of common themes or elements can be identified as the core of precautionary action, as outlined below. In the final section of this chapter, we consider several of these common elements as means of addressing issues of publication and third-party appropriation of indigenous knowledge.

First, the precautionary principle is primarily an environmental concept, the overall aim of which is to protect or safeguard the earth and its inhabitants from irreversible, persistent, bioaccumulative, or otherwise "serious"[30] effects of human activities. A related goal, stressed in strong versions of the principle, is an appreciation for the intrinsic, rather than merely instrumental, value of the environment. This emphasis suggests that trading costs and benefits is sometimes an inappropriate, if not unworkable, decision-making framework (see below).

Second, the precautionary principle draws our attention to the fact that environmental and health problems are not only scientific but also inextricably bound to social, political, and/or ethical issues. As such, the outcome of interactions among complex ecosystems, cultural traditions, and volatile political circumstances cannot be wholly assessed or predicted through scientific experimentation alone. By highlighting this complexity, we are compelled to identify uncertainties and areas of ignorance that may affect our understanding and assessment of environmental health issues. The precautionary principle asks us to recognize these limitations and to seek additional, more inclusive ways of knowing.

Third, fundamental to the precautionary principle are proactive measures taken before harm occurs, rather than reliance on reactive measures, compensation, or substitution of spent "resources." In many cases, such anticipatory steps must be taken prior to conclusive scientific proof of cause-and-effect relationships. Precaution is an action-oriented principle that, through more extensive and participatory research, seeks alternative solutions to potentially serious problems. A precautionary approach therefore entails a shift in decision-making methods and focus, for example, to include broader spatial and temporal scales (i.e., scales over which harms are likely to occur and thus are able to be assessed), a broader epistemological basis (such as local, experiential, indigenous, and traditional knowledge), and open, interactive dialogue (Barrett & Raffensperger 1999).

Fourth, "burden of proof" is a legal term that refers to an onus to prove a disputed issue based on some level of evidence. In environmental policy, this often means that the public—that is, the people facing harm—carry the burden of proving that an activity or technology is harmful in order to have it halted. A common element of the precautionary principle is shifting this burden of proof to those who propose a potentially damaging activity. That is, rather than placing the burden on communities or the general public to prove harm, researchers, industries, and other proponents of the questionable activity are required up front to demonstrate that technologies are reasonably safe. Prospective developers, therefore, assume a higher degree of liability and responsibility than they currently shoulder.

Fifth, some versions of the precautionary principle require that preventative measures be balanced by cost effectiveness (see, e.g., Rio Declaration ... 1998). For example, weaker versions may hold that precautionary measures should be taken only to the point that these measures do not entail excessive costs or place undue burdens on local economies. However, stronger versions dispute both the technical accuracy and the moral premise of trading ecological costs for economic benefits and tend to reject qualifying clauses such as "cost effectiveness" (see, e.g., Backes & Verschuuren 1998). Although detailed consideration of this point is beyond the scope of this chapter, what may be at issue here are appropriate definitions and time frames for costs and benefits.

The above list suggests that implementing the precautionary principle will entail reconsideration of currently dominant values and priorities, as well as many norms of academic and scientific research. More specifically, a precautionary approach includes a vital role for alternative traditions of science, urges us to acknowledge our limitations, and shifts the burden of proof to those who are proposing a potentially harmful activity. It is this ability of the precautionary principle to question fundamental assumptions and "take a step back" in order to gain a broader view of our intentions and their possible outcomes that lends itself to the issues raised in this chapter.

Applications of the Precautionary Principle to Ethnobiological Research

As discussed above, publication of ethnobiological data relating to traditional knowledge of indigenous peoples can lead to conflicting moral obligations between academic expectations and incentives to share the data and a professional mandate to both respect the interests of indigenous peoples and protect biological and cultural diversity (including cultural knowledge). To address this dilemma, we propose: extending current interpretations of the precautionary principle to include "cultural harms" (i.e., in addition to environmental harms), thereby affirming the irreducible links between ecological and cultural well-being; and applying elements of a precautionary approach (as outlined above) to issues of publication of indigenous knowledge and third-party use. In the following sections we adapt specific elements of the precautionary principle to three problems identified in the first part of our chapter.

Problem 1: Concepts of Harm Are Not Well Defined; Uncertainty of Outcomes Is Not Acknowledged

Too often, what may constitute harmful outcomes of publishing are projected by researchers who are not members of the affected community. Definitions are consequently restricted to foreseeable or precedented harms, rooted in particular cultural perspectives, or not recognized and articulated at all. The multidisciplinary nature of ethnobiology likely accentuates this problem, for researchers from different academic backgrounds have varying degrees of awareness of relevant biological and cultural issues.

The "uncertainty" element of the precautionary principle may address these shortcomings in two ways. First, we suggest that researchers in ethnobiology and related fields work to acknowledge the necessarily complex interactions among cultural, political, and ecological systems; i.e., that science and knowledge are rooted in a sociopolitical context. The resulting difficulty of predicting effects of research and publication must be acknowledged. This is particularly important for third-party users of indigenous knowledge, for the "data" are further displaced from their original context of meaning and values.

Second, because definitions of harm are socially and culturally influenced, potential harm due to publication should be codefined by the threatened communities, researchers, and knowledgeable others. Consistent with this prescription, Backes and Verschuuren (1998), among others, describe an essential role for both public debate within local communities (i.e., to determine the social necessity of an activity) and impact assessment (i.e., to assess potential adverse effects) in implementing the precautionary principle. Thus defined, assessments of harm will more likely encompass elements of traditional and customary lifeways related to indigenous knowledge and reflect community-based values for protection. For example, in the area of environmental impact assessment, Stevenson (1996: 283) stressed the importance of having the indigenous communities that are "most directly affected by proposed developments identify what is important to them and why." He went on to say that "By documenting their concerns, aboriginal people participate directly and effectively in impact prediction and assessment. . . . The knowledge and insights of aboriginal people who have already experienced the impacts of mining and other industrial activities may be especially useful in predicting, assessing, and managing specific and cumulative impacts" (pp. 283–284).

We suggest that including indigenous experiences and perspectives on harm, as described above, is warranted from both a technical and an ethical point of view. Participatory and inclusive processes may help to unearth and/or reduce uncertainties that derive from singular and/or narrow perspectives. Codefining potential harm also grants to indigenous communities a strong positive right and an active responsibility to participate in the research process. Such an undertaking requires adequate time, tools, and resources in order to be fully embraced. Success will therefore require that the restrictive nature of externally imposed research agendas and time lines are recognized and adjusted and that resources to enable active community participation are included in research budgets.

It is noteworthy that this proposal is consistent with the ecosystem approach to biological management that was recommended in a 1998 CBD–initiated workshop.[31] Elements of the ecosystem approach that have direct relevance to the issues raised here include a system of accountability, precautionary actions, consideration of effects (actual or potential) of activities on adjacent or other ecosystems, and spatial and temporal scales appropriate to the objectives. Furthermore, the workshop participants noted, "the ecosystem approach should consider all forms of relevant information, including scientific and indigenous and local knowledge, innovations and practices" and "should involve all relevant sectors of society and scientific disciplines." Also consistent with this approach, recommendations on scientific research that should be undertaken to implement Article 8(j) of the CBD include assurance of "full integration of traditional knowledge (particularly that of traditional sustainable practices) into the process of developing management plans and programmes."[32]

Problem 2: The Burden Is on the Affected Communities to Adapt or Conform to the Laws of Those Proposing the Project

To date, IPR mechanisms and many strategies for protection of indigenous knowledge in Canada are grounded in Euro-Canadian legal systems. To enter into negotiation, indigenous communities must therefore conform to the rules, conventions, and standards of evidence contained in these laws. Although self-determination of indigenous peoples is recognized as a means to more equitable interactions, the time frame and other enormous hurdles required for achieving this status suggests that we continue to support these goals but, in some cases, not wait for national and international law to recognize self-determination before addressing the shortcomings of existing IPR mechanisms.

The burden of proof element of the precautionary principle might be adapted to this situation. We suggest that, at least initially, the onus to conform should rest on external researchers who are proposing a potentially harmful project. That is, nonindigenous researchers ought to adapt to indigenous protocols, laws and customs, and time frames for decision making.[33] As well, concepts of valid evidence of harm should be extended to include the experiential knowledge of indigenous peoples, in addition to experimental knowledge. The latter proposal is consistent with implications of the precedent-setting *Delgamuukw v. The Queen* decision of the Supreme Court of Canada in 1997. This decision enacted a profound change in rules of evidence as they apply to oral history; that is, that rules against "hearsay" no longer apply in aboriginal legal cases in Canada. Importantly, one result of this change is that oral testimony is now admissible as legal evidence for establishing an aboriginal perspectives on relationships with the land and on other integral aspects of culture. By extension, aboriginal elders and others who can establish that they have authority to speak about their oral history and culture are acknowledged in court as legal experts on their culture.

Problem 3: The Focus Is on Sharing Benefits or Compensation for Harm (Reactive Measures) Rather Than on Prevention of Harm (Proactive Measures)

A frequently cited argument against the precautionary principle is that "richer is safer" (e.g., Wildavsky 1988). By this reasoning, pushing ahead with technologies and development will inevitably yield economic benefits that trickle down to all sectors of society, including those harmed by the activity. This logic suggests that a richer society is better equipped (economically) to mitigate or compensate for harm and, therefore, that a precautionary approach yields only foregone opportunities and economic insecurity, leaving society without the necessary resources to control unforeseen hazards. It would appear that this richer-is-safer rationale (i.e., that economic opportunities can mitigate harm) has been adopted by research programs and IPR mechanisms that are focused on compensation and benefit sharing.[34] However, in view of the uncertainty surrounding long-term effects of research and publication, we suggest that such up-front compensation may prove inadequate or inappropriate and that waiting for harm to occur is clearly irresponsible.

Just as the precautionary principle calls for more comprehensive standards of research in evaluating the environmental effects of pollutants, we suggest that precaution might also prescribe more comprehensive standards of research regarding cultural knowledge of indigenous peoples. The "proactive and anticipatory action" and the "recognition of intrinsic value" elements of the precautionary principle are central to this alternative research ethic and may address the problems of reactionary measures, such as compensation and benefit sharing.[35] Recommendations for scientific research that should be undertaken to implement Article 14 of the CBD, "Impact Assessment and

Minimizing Adverse Impacts," are consistent with the principle of preventative action: "It is also important that the assessments focus more on providing early warning of incipient problems rather than recording damage at a stage where it may be irreversible."[36] We suggest three such proactive approaches to avoid potential harms caused by inappropriate dissemination of indigenous cultural knowledge:

Primary Focus on Two-Way Dialogue between Researchers and Indigenous Communities. Long-term, participatory, and open communication processes must be prerequisites to further research and essential components of conducting ethical ethnoscience. Ideally, the community should initiate research projects, and research procedures should be developed cooperatively, with direct and ongoing interactions among all parties—including third parties—so that all stakeholders can stay informed about research outcomes and their implications. Both expected and potential outcomes of research, including concepts of harm and areas of uncertainty, should be discussed prior to initiating research, be based on current concerns and issues, and be revisited and updated throughout the research. That is, when problematic issues arise after the research has been initiated, it is not acceptable to continue the research without joint review of the new circumstances and appropriate redress.

A greater effort is required to identify what the 1996 ISE Draft Code of Conduct called "the extent of reasonably foreseeable harm" to affected communities in order to balance current efforts to identify "the full range of potential benefits." Indigenous communities that have felt adverse effects of research, development, or other impositions may offer unique and valuable insights and guidance based on their direct experiences and may help to redress this current imbalance.

One step in this direction could be accomplished at the editorial level of academic publications. Editorial boards could require mandatory full disclosure of the source(s) of any cultural information cited in articles considered for publication, whether or not the information was obtained firsthand (e.g., through ethnographic research) or was based on previous publications (e.g., derived from ethnobotanical literature)[37]—unless, of course, it is clear that permission for research and publication are contingent on confidentiality. This would help to ensure that full credit is given to indigenous peoples and their communities—i.e., a level of recognition currently granted to all academic research partners—and would facilitate contact with those communities who wish to be consulted if further use of the data or resources is sought. Likewise, such credit should be extended to field collections and literature-based databases.[38]

Changes in Academic Reward Structures and Ethical Guidelines. Institutional requirements that research be disseminated through defined academic routes are a serious but not insurmountable problem for research involving cultural knowledge of indigenous peoples. We propose that the current emphasis on peer-reviewed publications as criteria for promotion, tenure, and doctoral degrees be expanded to include direct contributions to communities, participation in long-term projects, and teaching outside the university, particularly in collaborative education and community-based research programs with indigenous peoples. These contributions may or may not be documented through normal academic avenues.[39] The process of peer review ought to include members of the communities involved[40] and entail review of research grant proposals as well as publications.[41] University policies and ethical guidelines ought to be expanded to acknowledge the added complexities of research involving cultural knowledge and, perhaps, to encourage equitable, mutually beneficial partnerships with indigenous communities. This task requires that policymakers be enlightened as to the concerns and complexities of this area and thus puts an onus on ethnobiologists and other academics to assume a more advocatory role.

Willingness to Forego Some Research Projects or Publications. A fundamental tenet of the precautionary principle is that uncertainty or lack of knowledge about outcomes "should not be used as a reason for postponing measures to prevent environmental [and cultural] degradation" (Bergen Declaration 1990). Therefore, on the basis of discussions at the local community and academic levels, it may be determined that a particular research project should not proceed as planned, that it should not be conducted within the university setting, or that its results should not be released into the public domain through typical academic avenues. Here we should reemphasize that the precautionary principle is action oriented and does not necessarily prohibit research or publication. Rather, the principle suggests that, when faced with uncertainty and the likelihood of potentially harmful effects, we should rethink our overall goals and/or seek alternative ways to meet them. Willingness to abide by this decision reflects a moral commitment to the precautionary principle, respect for the intrinsic value of traditional knowledge and indigenous communities, and willingness to conduct publicly funded research that is truly inclusive of the public interest.

Conclusions: Reconsidering the Ounce of Prevention

Before concluding, it is worth reflecting on some of the broader implications of the above position. We began this chapter with a quotation which states that "secrecy is the antithesis" of the scientific norm of publication (Merton 1973 [1942]: 274). Considered in this light, does our precautionary approach to ethnobiological publication constitute a form of intellectual or cultural secrecy and thereby contribute to erosion of the public domain? It may be perceived as such. However, it is necessary to consider these proposals in the context of current relationships between academia and the public domain. In fact, the present status of the public domain is central to the publication issues raised here. Does the ideal of open publication remain in academia, and if so, does it fulfill its original objectives?

If one considers the influence of corporate and industrial interests on current scientific research—such as the marriage of applied academic research to private industry for provision of scarce research funds, and the widespread application of trade secrets, patent law, and other IPR mechanisms for the protection of research discoveries—one might conclude that, in many cases, open publication has been dismissed as economically impractical. Even though such funding strategies may benefit humankind through the pursuit of knowledge and scientific discoveries, proprietary claims to cultural knowledge may also cause direct harm to individuals and communities. The Bellagio Declaration (1993) insisted that intellectual property rights "can no longer be constructed without reference to their ecological, cultural and scientific effects" and underscored the need for an "increased recognition and protection of the public domain." Likewise, Brown (1998: 196–197) highlights concerns for "the increasingly expansive claims of ownership made by corporate interests—claims that threaten the doctrine of fair use and limit the scope of the public domain."

Parallel concerns are emerging in discussions of rights and access to genetic resources for farmers and plant breeders. For example, Castillo (1997: 21) comments that "It is when scientists working on publicly funded research projects channel research results to the commercial sector for corporate gains that the ethics of science become compromised. Confidentiality and exclusivity take the place of the doctrine that science is public and the time-honored canons of science break down." Furthermore, Menon (1997: 101) insists that "What is at stake today is a question of the relationship between the public and private domains of knowledge. This appropriation of the public domain of knowledge by monopoly capital is what needs to be questioned."

Thus, it is clear that various forms of "intellectual secrecy" presently exist within academia to protect sensitive information and to serve several diverse interests. As Coombe (1998: 207) points out, all IPR mechanisms are "premised on a social bargain" that aims to strike a balance between granting rights and imposing responsibilities in exercising these rights. However, with respect to the "cultural secrecy" of indigenous communities, we make a plea for consistency: Can we expect indigenous communities to freely share knowledge for the benefit of the commons if this requirement is not extended equally to academia as well as the broader public and private sectors, especially if existing IPR mechanisms may then prevent indigenous people from accessing information or benefits derived from their own knowledge? Here we second Brown's appeal (1998: 206) for "earnest reflection on the future of the imperiled intellectual and artistic commons called the public domain, whose survival is of vital significance to us all."

A second critique of precautionary approaches concerns the so-called Law of Unintended Consequences; that is, "that there will always be unforeseen effects, both good and bad, when information enters the public domain" (Brown 1998: 200). Even the most conscientious and well-intentioned of researchers cannot possibly anticipate every impact of their work, much less indirect and downstream effects. However, as van den Daele (1992: 330) argued, "Absolute knowledge, complete control, or totally comprehensive planning cannot be the point in uncertainty arguments." More important for the present case, researchers ought to acknowledge the contribution of ethnobiological literature to the appropriation and industrial commodification of traditional botanical knowledge and take active measures to avoid reasonably foreseeable harmful consequences by, for example, devising and advocating alternatives and pursuing proactive measures, such as described above. Uncertainty, therefore, need not stall research and should not serve as an excuse for avoiding accountability. In current ethnobiological research, ethical issues are necessarily linked to researchers' accountability to local communities (Godbole & Eyzaguirre 1997). As we interpret it, the precautionary principle is not only a means of empowerment but also a tool for mutual engagement and long-term benefit. Its application to ethnobiological and related research may be useful in addressing important issues through facilitation of meaningful dialogue and more equitable, mutually respectful decision-making processes. Recognition of proprietary claims to knowledge, "not as exclusivity of possession but as bundles of multiple rights and relationships still to be delineated" (Coombe 1998: 208), seems prerequisite to an equal voice for indigenous peoples in these discussions.

The complex dilemma outlined in this chapter places researchers in an often uncomfortable and compromising position as they struggle to balance moral obligations to all research partners with pressures of publication, advancement, and scholarship. Much of the problem we have described stems from a lack of interaction between those who originally hold the knowledge and those who apply it downstream. Lack of direct interaction perpetuates the assumption that third-party users have no direct responsibility to indigenous communities. In essence, publications incorporating indigenous cultural knowledge enable third-party users to overlook some significant consequences of their own actions. In our view, respect and acknowledgement are key to a solution, and obviously these cannot be enforced by legislation. However, we suggest that mechanisms to bring third parties into direct engagement with indigenous communities, wherein "each potential user is a new borrower, who must undertake the same process of receiving the consent and supervision of the community" (Mann 1997a: 31), may be of considerable value.

There are no easy solutions, and although we suggest that the precautionary principle can help to establish a framework for addressing the innumerable uncertainties, legal questions, and ethical conflicts surrounding publication of indigenous cultural knowledge, we recognize that it offers no specific rules.[42] The challenge ahead is to create a system that requires and encourages interaction. Specific solutions will necessarily be rooted in specific cases and subject to negotiation among those people involved, but they will also emerge from the willingness to attempt, evaluate, and improve upon new approaches. In the meantime, however, we suggest that researchers, rather than continuing to overlook problems that have no current solutions, may find that the precautionary principle is a helpful tool for redress. In short, we suggest that the issues described herein go beyond requiring simply an ounce of prevention. More appropriately, for the integrity of cultural and biological diversity, they justify an entire pound of precaution.

Acknowledgments

We acknowledge the support and helpful comments of I. E. P. Taylor, T. W. Pearson, G. P. Nicholas, and M. F. Brown on previous versions of this chapter. We thank H. Mann for kindly providing his reports and for assistance in obtaining additional documents of the Canadian Working Group on Article 8(j). An earlier version of this chapter was presented at the 6th International Congress of Ethnobiology (held in November 1998 in Aotearoa / New Zealand) of the International Society of Ethnobiology. We express our appreciation to the participants in the ISE pre-congress workshop, Research Protocols,

Codes of Ethics and Standards of Conduct for Ethnobiologists, for their valuable discussion of this topic. We give thanks also to our gracious Maori hosts of the workshop, at Te Kiri Marae. We thank the Secwepemc Cultural Education Society of Kamloops, British Columbia, and members of the Secwepemc Nation for providing ongoing opportunities to attempt, evaluate, and hopefully improve upon our own efforts to engage in meaningful dialogue. This chapter is dedicated to the memory of Darrell Addison Posey (1947–2001), who was a mentor and friend to Bannister.

Notes

1. That is, "indigenous and local communities embodying traditional lifestyles," as described in the 1992 Convention on Biological Diversity and as employed by Posey and Dutfield (1996). We acknowledge the challenges inherent in choosing appropriate terminology that not only stands the test of time but also crosses cultural and disciplinary boundaries. We believe that appropriate terminology is of utmost importance; however, a more in-depth discussion of this topic is beyond the scope of this chapter. The terms chosen here reflect current common usage in the discipline of ethnobiology. We acknowledge also that the issues raised here may have application beyond the context of indigenous cultures.

2. In particular (and as we discuss subsequently), we refer to the maintenance and protection of knowledge and practices of indigenous peoples that are relevant to conservation of biological diversity and sustainable use of its components, as noted in the Preamble, Article 8(j), and Article 10(c) of the CBD. CBD documents are available at <http://www.biodiv.org>.

3. The Declaration of Belém (1988) states that major concerns of ethnobiology include "preservation and strengthening of indigenous communities and their traditional knowledge," and the International Society of Ethnobiology Draft Code of Conduct (1996) states that there is "a need to protect and enhance the unique relationship between indigenous peoples and their environment in order to better protect and enhance biological and cultural diversity." ISE documents are available at <http://guallart.dac.uga.edu/ISE>.

4. For example, the fourth meeting of the Conference of the Parties to the Convention on Biological Diversity (1998) acknowledged this urgency in the "Recommendations on Scientific Research to be Undertaken for Implementation of Article 8(j)": "Efforts should be made to identify those areas in which loss of traditional knowledge is occurring rapidly, so that appropriate steps can be taken to prevent this loss" (Document UNEP/CBD/COP/4/Inf.18).

5. Posey and Dutfield (1996: 29) also noted that "usually, access to information is not conditional upon recognition of the [intellectual property rights] of indigenous peoples."

6. For various examples, see the *Journal of Ethnopharmacology, Phytochemistry,* and *Planta Medica,* among others.

7. Also known as "informed choice," which is defined in Canada by the National Council on Bioethics in Human Research (1996) as occurring when "a morally valid choice concerning research participation is made: (1) by a competent person; (2) on the basis of adequate information concerning the nature and foreseeable consequences of the research (as these are known at the time the request is made) and all available alternatives; and (3) without controlling influences such as 'force, fraud, deceit, duress, overreaching, or other ulterior forms of constraint or coercion'" (Nuremberg Code).

8. For example, the 1998 ISE Code of Ethics includes the "Principle of Prior Informed Consent and Veto," which recognizes that informed consent must be obtained before research is undertaken and states that "providing prior informed consent presumes that all potentially affected communities will be provided complete information regarding the purpose

and nature of the research activities and the probable results, including all reasonably foreseeable benefits and risks of harm (be they tangible or intangible) to the affected communities."

9. In British Columbia, Canada, no regulations for harvesting of non-timber forest products are yet in place; however, the need for such regulations has been identified (e.g., De Geus 1995; Volpe 1998; Wills & Lipsey 1999), due largely to "the tremendous interest in and demand for [natural health products] and the need to accommodate both the industry and consumers in this area" (Volpe 1998: 3). Moreover, the Shuswap Nation Tribal Council (SNTC) has intervened with the government of British Columbia for acknowledgment of aboriginal rights to control and manage these resources (SNTC Natural Resources Coordinator pers. comm. 1999).

10. In Canada, the *Report of the Standing Committee on Health* (Volpe 1998) recommended that a regulatory framework for natural health products should "not apply to products prepared by Aboriginal Healers where the product was prepared for an individual patient" (p. 66). However, the legislation would apply in any attempt at commercialization, and situations falling between the extremes of single patient treatment and commercialization were not addressed.

11. An example is the intensive commercial American ginseng (*Panax quinquefolium* L.) industry, in the southern interior region of British Columbia, Canada, which (along with urbanization and cattle grazing) displaces an endangered lowlands grassland ecosystem that includes culturally salient plants along the Thompson River systems (see, e.g., Bandringa 1999) and requires prolonged use of pesticides for marketable roots (Province of British Columbia Ministry of Agriculture, Fisheries and Food 1993).

12. Nicholas and Andrews (1997: 2) discuss a parallel "schizophrenia" in the "dual role" of anthropologists.

13. See Posey and Dutfield (1996) for an in-depth analysis of provisions contained in various international agreements, including the CBD, that are relevant to protection of the knowledge and resources of indigenous peoples.

14. For example, recommendations in the *Report of the First Meeting of the Ad Hoc Open-Ended Inter-Sessional Working Group on the Implementation of Article 8(j) and Related Provisions of the CBD* (Document UNEP/CBD/COP/5/5) include the following calls on other CBD parties, governments, international organizations, and organizations representing indigenous peoples: to examine ways and means to establish guidelines for equitable sharing of benefits arising from the utilization of knowledge, innovations, and practices of indigenous and local communities; to provide opportunities for indigenous and local communities to identify their capacity needs; to include funding to build the capacity of communications in proposals and plans for projects carried out in indigenous and local communities; to strengthen and build the capacity for communication among and between indigenous peoples, local communities, and governments at local, national, regional, and international levels; to use clearinghouse mechanisms with direct participation and responsibility of indigenous and local communities; to increase the use of local languages for communication; and to provide case studies of methods and approaches for recording the knowledge, innovations, and practices of indigenous and local communities and for controlling such records.

15. Mann (1997a: 1) suggests that "indigenous knowledge as a concept concerns information, understanding and knowledge that reflects symbiotic relationships between individuals, communities, generations, the physical environment and other living creatures, and the spiritual relationships of a people. [Indigenous knowledge] evolves as ecosystem and other factors change, but remains grounded in the more enduring aspects of identity, culture, generations and spirituality." Cassidy and Longford (n.d: 30) define traditional knowledge as "the expression of the human soul in all its aspects, as well as the foundation for economic, social and spiritual growth."

16. The antiquated term "subjects" is not necessarily deemed appropriate in referring to current ethnobiological research in which community members who engage in the research are typically considered participants or collaborators.

17. However, Mann (1997a) does propose a useful taxonomy of legal mechanisms that may merit case-specific consideration for protection of indigenous knowledge.

18. For more information on traditional resource rights, see <http://users.ox.ac.uk/~wgtrr/decin.htm>.

19. For example: Article 16(5), "Access to and Transfer of Technology," specifies that patents and other IPR tools should be supportive of and not run counter to the objectives of the CBD; Article 17(2), "Exchange of Information," states that information relevant to the conservation of biological diversity should be facilitated from all publicly available sources, should include (among others) indigenous and traditional knowledge, and, where feasible, should include repatriation of information; and Article 18(4), "Technical and Scientific Cooperation," indicates that methods of cooperation for the development and use of technologies, including indigenous and traditional technologies, should be encouraged and developed.

20. Third Meeting of the Conference of Parties to the CBD (1996) on Item 14.1 of the Provisional Agenda: The Impact of Intellectual Property Rights Systems on the Conservation and Sustainable Use of Biological Diversity and on Equitable Sharing of Benefits from Its Use (Document UNEP/CBD/COP/3/22).

21. See Dutfield (2000) for further discussion on bridging TRIPS and the CBD through sui generis options.

22. Examples of international statements and declarations on protection of the cultural knowledge of indigenous peoples include:

- 1988 Declaration of Belém by the International Society of Ethnobiology;
- 1992 Kari-Oca Declaration and the Indigenous Peoples Earth Charter;
- 1992 Charter of the Indigenous-Tribal Peoples of the Tropical Forests;
- 1993 United Nations Draft Declaration on the Rights of Indigenous Peoples;
- 1993 Bellagio Declaration on Cultural Agency / Cultural Policy;
- 1993 Mataatua Declaration on Cultural and Intellectual Property Rights of Indigenous Peoples;
- 1993 Recommendations of the Voices of the Earth Congress;
- 1993 Julayinbul Statement on Indigenous Intellectual Property Rights;
- 1995 Sabah Statement from the United Nations Development Programme Asian Consultation Workshop on the Protection and Conservation of Indigenous Knowledge;
- 1995 Suva Statement from the United Nations Development Programme on Indigenous Peoples' Knowledge and Intellectual Property Rights; and
- 2000 Statement of the Union of British Columbia Indian Chiefs, from the conference on Protecting Knowledge: Traditional Resource Rights in the New Millennium.

23. As many botanical gardens have established bioprospecting collaborations, policies have become necessary, in order to establish the proprietary rights of countries that provided the botanical materials (Dove 1998).

24. For example, the University of British Columbia's Policy #88 ("Patents and Licensing," <http://www.universitycounsel.ubc.ca/policies/policy88.html>) states that "if any member of faculty or staff, any student, or anyone connected with the University proposes to protect or license an invention or discovery in which the university facilities or funds administered by the University were used, a disclosure must be made to the University and the rights assigned to the University" (University of British Columbia Mission Statement 1989).

25. For example, the fifth objective of the mission statement of the University of British Columbia (issued by President Strangway in 1989) states that "the University will foster the transfer and application of knowledge for the benefit of society by encouraging active participation in the intellectual, business and professional life of the community and, when appropriate, by undertaking joint research programs with industry, government and community organizations, and by licensing university intellectual property." See also Policies #87 ("Research," <http://www.universitycounsel.ubc.ca/policies/policy87.html>) and #88 ("Patents and Licensing," <http://www.universitycounsel.ubc.ca/policies/policy88.html>) in University of British Columbia Mission Statement (1989).

26. There is, of course, significant power in moral suasion, especially if moral decisions are recognized and reinforced by professional societies and other bodies that govern research conduct.

27. Examples of international applications of the precautionary principle include the following (see also Hickey & Walker 1995):

- Protocol on Substances That Deplete the Ozone Layer / Montreal Protocol (1987);
- Nordic Council's International Conference on Pollution of the Seas (Nordic Council 1989);
- Bergen Declaration on Sustainable Development (Bergen Declaration 1990);
- Rio Declaration on Environment and Development (1992);
- United Nations Framework Convention on Climate Change (1992); and
- Treaty on the European Union (Maastricht Treaty) (1994).

The CBD (1992) does not use the term; however, the Preamble does state that "where there is a threat of significant reduction or loss of biological diversity, lack of full scientific certainty should not be used as a reason for postponing measures to avoid or minimize such a threat."

28. Customary law is defined as "evidence of a general practice accepted as law" (International Court of Justice, cited in Cameron & Abouchar 1991: 14).

29. Adopted by resolution at the Annual General Meeting of the International Society of Ethnobiology held in Whakatane, Aotearoa/New Zealand on November 28, 1998 (International Society of Ethnobiology Code of Ethics 1998).

30. Reference to "serious" harm is made in several declarations, including the Bergen Declaration (1990) and the Rio Declaration on Environment and Development (1992).

31. *Report on the CBD–Workshop on the Ecosystem Approach. Lilongwe, Malawi, January 1998* (Document UNEP/CBD/COP/4/INF.9).

32. Scientific review of Article 8 of the CBD was carried out by Diversitas (a program established to promote the science of boidiversity) at the request of the Executive Secretary of the CBD Secretariat, Mexico City, 1998 (Document UNEP/CBD/COP/4/Inf.18).

33. This is not to imply that ethnobiologists and indigenous people are mutually exclusive. In fact, if ethnobiology were initiated and implemented by indigenous peoples within their own communities ("an indigenous botany," as termed by Salmón 1996: 72), presumably many of these issues would not arise.

34. In our opinion, these would include the widely cited examples of Shaman Pharmaceuticals (as in King & Tempesta 1994; King et al. 1996; Carlson et al. 2001), the U.S. National Cancer Institute (as in Cragg et al. 1994), and the Merck/InBio agreement (as in Crucible Group 1994).

35. This is not to take away from the importance of benefit sharing; that is, benefit sharing is one of several important considerations.

36. Scientific review of Article 14 of the CBD was carried out by Diversitas at the request of the Executive Secretary of the CBD Secretariat, Mexico City, 1998 (Document UNEP/CBD/COP/4/Inf.18).

37. Whether credit would be most appropriate at an individual or more collective level would likely depend on the case; our point here is that credit for the source of information attributed to "Native Americans" or to "First Nations" (as is common in publications by third parties who use ethnobotanical literature as a basis for bioprospecting) is as inadequate as is credit for scientific discoveries attributed generically to "scientists" rather than to individual discoverers.

38. See also Laird et al. (2002) for a discussion of "community registries" of biodiversity-related knowledge.

39. For example, two forthcoming volumes on Secwepemc (Shuswap) ethnobotany (Bannister & Turner in prep; Turner & Ignace in prep.) have been approved by the Shuswap Nation Tribal Council for copublication and joint copyright by the Secwepemc Cultural Education Society (Kamloops, British Columbia).

40. This suggestion is also consistent with the 1997 decision by the Supreme Court of Canada in *Delgamuukw v. The Queen*; but see Nicholas and Andrews (1997) for a discussion of associated problems with censorship and revisionism.

41. In Canada, the Secwepemc Ethnobotany Project (1990–1998), referred to in note 39 and described briefly in Bannister & Barrett (2001) and Laird et al. (2002), serves as one example. As specified in research agreements with the Secwepemc Cultural Education Society, all grant proposals, field notes, and other records, as well as publications, including dissertations, were subject to a review and approval process.

42. In fact, some commentators on the precautionary principle suggest that secondary legislation is required for its implementation (Cameron 1994).

Literature Cited

Augustine, S. J. 1997. Traditional aboriginal knowledge and science versus occidental science. Report submitted to the Biodiversity Convention Office of Environment Canada.

Backes, C. W. & J. M. Verschuuren. 1998. The precautionary principle in international European and Dutch wildlife law. Colorado Journal of International Environmental Law and Policy 9(1): 43–70.

Bandringa, R. W. 1999. The ethnobotany and descriptive ecology of bitterroot, *Lewisia rediviva* Pursh (Portulacaceae), in the Lower Thompson River Valley, British Columbia: A salient root food of the Nlaka'pamux First Nation. M.S. thesis, University of British Columbia.

Bannister, K. P. 2001. Chemistry rooted in cultural knowledge: Unearthing the links between antimicrobial properties and traditional knowledge in food and medicinal plant resources of the Secwepemc (Shuswap) Aboriginal Nation. Ph.D. diss., University of British Columbia.

———. In press. When promotion and protection conflict: Indigenous knowledge and traditional plant resources of the Secwepemc First Nation. *In* M. Riley (ed.), Towards the Protection of Indigenous Intellectual Property Rights: Facing Legal Obstacles, Developing Innovative Solutions. Walnut Creek, Calif.: AltaMira Press.

——— & K. Barrett. 2001. Challenging the status quo in ethnobotany: A new paradigm for publication may protect cultural knowledge and traditional resources. Cultural Survival Quarterly 24(4): 10–13.

——— & N. J. Turner (eds.). In prep. Secwepemc People and Plants: Research Papers in Shuswap Ethnobotany. Kamloops: Secwepemc Cultural Education Society.

Barrett, K. 1999. Canadian agricultural biotechnology: Risk assessment and the precautionary principle. Ph.D. diss., University of British Columbia.

——— & C. Raffensperger. 1999. Precautionary science. Pp. 106–122 *in* C. Raffensperger & J. A. Tickner (eds.), Protecting Public Health and the Environment: Implementing the Precautionary Principle. Washington, D.C.: Island Press.

Barton, J. H. 1994. Ethnobotany and intellectual property rights. Pp. 214–221 *in* G. T. Prance, D. J. Chadwick & J. Marsh (eds.), Ethnobotany and the Search for New Drugs. CIBA Foundation Symposium 185. Chichester, U.K., and New York: Wiley.

Bellagio Declaration. 1993. Statement of the Bellagio Conference. Cultural Agency / Cultural Policy: Policies and Poetics of Intellectual Property in the Post-Colonial Era. <http://users.ox.ac.uk/~wgtrr/decin.htm>.

Bergen Declaration. 1990. Ministerial Declaration on Sustainable Development in the ECE Region. UN DOC. A/CONF.151/PC/10.

Boehmer-Christiansen, S. 1994. The precautionary principle in Germany: Enabling government. Pp. 31–60 *in* T. O'Riordan & J. Cameron (eds.), Interpreting the Precautionary Principle. London: Earthscan.

Brouwer, J. 1998. Comment on "The development of indigenous knowledge." Current Anthropology 39(3): 351.

Brown, M. F. 1998. Can culture be copyrighted? Current Anthropology 39(2): 193–222.

Brush, S. B. 1993. Indigenous knowledge of biological resources and intellectual property rights: The role of anthropology. American Anthropologist 95: 653–671.

Cameron, J. 1994. The status of the precautionary principle in international law. Pp. 262–283 *in* T. O'Riordan & J. Cameron (eds.), Interpreting the Precautionary Principle. London: Earthscan.

——— & J. Abouchar. 1991. The precautionary principle: A fundamental principle of law and policy for the protection of the global environment. Boston College International and Comparative Law Review 24(1): 1–27.

Carlson, T. J., B. M. Foula, J. A. Chinnock, S. R. King, G. Abdourahmaue, B. M. Sannoussy, A. Bah, S. A. Cisse, M. Camara & R. K. Richter. 2001. Case study on medicinal plant research in Guinea: Prior informed consent, focused benefit sharing, and compliance with the Convention on Biological Diversity. Economic Botany 55(4): 478–491.

Cartagena Protocol on Biosafety. 2000. <http://www.biodiv.org>.

Cassell, J. & S.-E. Jacobs (eds.). 1987. Handbook on Ethical Issues in Anthropology. Washington, D.C.: American Anthropological Association.

Cassidy, M. & J. Longford (eds.). n.d. Intellectual property and aboriginal people: A working paper. Hull, Que.: Department of Indian Affairs and Northern Development, Research Analysis Directorate.

Castillo, G. T. 1997. Whose ethics and whose equity? Issues in the conservation and use of genetic resources for sustainable food security. Pp. 19–31 *in* Ethics and Equity in Conservation and Use of Genetic Resources for Sustainable Food Security. Rome: International Plant Genetic Resources Institute.

Charter of the Indigenous-Tribal Peoples of the Tropical Forests. 1992. Penang, Malaysia. *Reprint:* Pp. 198–204 *in* D. A. Posey & G. DutWeld (eds.), Beyond Intellectual Property: Toward Traditional Resource Rights for Indigenous Peoples and Local Communities. Ottawa: International Development Research Centre, 1996.

Convention on Biological Diversity. 1992. United Nations Conference on Environment and Development, Rio de Janeiro, Brazil. <http://www.biodiv.org>.

Coombe, R. J. 1998. Comment on Can culture be copyrighted? Current Anthropology 39(2): 207–208.

Cragg, G. M., M. R. Boyd, J. H. Cardellina II, D. J. Newman, K. M. Snader & T. G. McCloud. 1994. Ethnobotany and drug discovery: The experience of the US National Cancer

Institute. Pp. 178–196 *in* G. T. Prance, D. J. Chadwick & J. Marsh (eds.), Ethnobotany and the Search for New Drugs. CIBA Foundation Symposium 185. Chichester, U.K., and New York: Wiley.

Crucible Group. 1994. People, Plants and Patents: The Impact of Intellectual Property on Trade, Plant Biodiversity and Rural Society. Ottawa: International Development Research Centre.

De Geus, P. M. J. 1995. Botanical Forest Products in British Columbia: An Overview. Victoria, B.C.: Province of British Columbia Ministry of Forests, Integrated Resources Policy Branch.

Declaration and Draft Model Law by the Scientific, Technical and Research Commission of the Organization of African Unity (OAU/STRC) Task Force on Community Rights and Access to Biological Resources. 1998. <http://www.twnside.org.sg/title/oau1-cn.htm>.

Declaration of Belém. 1988. First International Congress of Ethnobiology, Belém, Brazil. <http://guallart.dac.uga.edu/ISE>.

Dove, A. 1998. Botanical gardens cope with bioprospecting loophole. Science 281(5381): 1273.

Dutfield, G. 2000. Intellectual Property Rights, Trade, and Biodiversity: Seeds and Plant Varieties. London: Earthscan.

Executive Secretary of the Ad Hoc Open-Ended Inter-Sessional Working Group on Article 8(j) and Related Provisions of the Convention on Biological Diversity. 2000. Legal and Other Appropriate Forms of Protection for the Knowledge, Innovations and Practices of Indigenous and Local Communities Embodying Traditional Lifestyles Relevant for the Conservation and Sustainable Use of Biological Diversity. Document UNEP/CBD/WG8J/1/2.

Ford, V. 1998. Traditional knowledge and Article 8(j) of the Convention on Biological Diversity. Report submitted to the Biodiversity Convention Office of Environment Canada.

Godbole, A. & P. Eyzaguirre. 1997. Ethical issues in conservation and ethnobotanical research with examples from indigenous and traditional communities in India. Pp. 83–90 *in* Ethics and Equity in Conservation and Use of Genetic Resources for Sustainable Food Security. Rome: International Plant Genetic Resources Institute.

Hickey, J. E., Jr. & J. R. Walker. 1995. Refining the precautionary principle in international law. Virginia Environmental Law Journal 14: 523–454.

International Society of Ethnobiology Code of Ethics. 1998. Whakatane. <http://guallart.dac.uga.edu/ISE>.

International Society of Ethnobiology Draft Code of Conduct. 1996. Nairobi. <http://guallart.dac.uga.edu/ISE>.

Julayinbul Statement on Indigenous Intellectual Property Rights. 1993. Julayinbul Conference on Intellectual and Cultural Property Rights, 27 November, Jingarrba, Australia. *Reprint:* Pp. 141–144 *in* Proceedings of the Indigenous Peoples Knowledge and Intellectual Property Rights Consultation. Suva, Fiji: Pacific Concerns Resource Centre, 1997.

Kari-Oca Declaration and the Indigenous Peoples Earth Charter. 1992. The World Conference of Indigenous Peoples on Territory, Environment and Development, 25–30 May, Kari-Oca, Brazil. *Reprint:* Pp. 134–140 *in* Proceedings of the Indigenous Peoples Knowledge and Intellectual Property Rights Consultation. Suva, Fiji: Pacific Concerns Resource Centre, 1997.

King, S. R. & M. S. Tempesta. 1994. From shaman to human clinical trials: The role of industry in ethnobotany, conservation and community reciprocity. Pp. 197–213 *in* G. T. Prance, D. J. Chadwick & J. Marsh (eds.), Ethnobotany and the Search for New Drugs. CIBA Foundation Symposium 185. Chichester, U.K., and New York: Wiley.

———, T. J. Carlson & K. Moran. 1996. Biological diversity, indigenous knowledge, drug discovery, and intellectual property rights. Pp. 167–185 *in* S. B. Brush & D. Stabinsky (eds.), Valuing Local Knowledge: Indigenous People and Intellectual Property Rights. Washington, D.C.: Island Press.

Kottak, C. P. 1991. Cultural Anthropology. Ed. 5. New York: McGraw-Hill.

Laird, S. A. & D. A. Posey. 2002. Professional society standards for biodiversity research: Codes of ethics and research guidelines. Pp. 16–38 *in* S. A. Laird (ed.), Biodiversity and Traditional Knowledge: Equitable Partnerships in Practice. London: Earthscan.

———, M. N. Alexiades, K. P. Bannister & D. Posey. 2002. The publication of biodiversity research results: Employing the precautionary principle. Pp. 77–101 *in* S. A. Laird (ed.), Biodiversity and Traditional Knowledge: Equitable Partnerships in Practice. London: Earthscan.

Lambrou, Y. 1997. Control and access to indigenous knowledge and biological resources. Report submitted to the Biodiversity Convention Office of Environment Canada.

Louwaars, N. P. 1998. *Sui generis* rights: From opposing to complementary approaches. Biotechnology and Development Monitor 36: 13–16.

Mann, H. 1997a. Intellectual property rights, biodiversity and indigenous knowledge: A critical analysis in the Canadian context. Report submitted to the Canadian Working Group on Article 8(j) of the Convention on Biological Diversity.

———. 1997b. Indigenous peoples and the use of intellectual property rights in Canada: Case studies relating to intellectual property rights and the protection of biodiversity. Report submitted to Industry Canada, Intellectual Property Policy Directorate, and to the Canadian Working Group on Article 8(j) of the Convention on Biological Diversity.

Mataatua Declaration on Cultural and Intellectual Property Rights of Indigenous Peoples. 1993. Whakatane. Aotearoa/New Zealand. <http://users.ox.ac.uk/~wgtrr/mataatua.htm>.

McIntrye, O. & T. Mosedale. 1997. The precautionary principle as a norm of customary international law. Journal of Environmental Law 9(2): 221–241.

Menon, U. 1997. Designing a regime of access to genetic resources: Beyond the popular logic of farmers' rights and breeders' rights. Pp. 98–106 *in* Ethics and Equity in Conservation and Use of Genetic Resources for Sustainable Food Security. Rome: International Plant Genetic Resources Institute.

Merton, R. K. 1973 [1942]. The normative structure of science. Pp. 267–278 *in* The Sociology of Science: Theoretical and Empirical Investigations. Chicago: University of Chicago Press.

National Council on Bioethics in Human Research. 1996. Facilitating ethical research: Promoting informed choice. Discussion Document 7(2), December.

Nicholas, G. P. & T. D. Andrews. 1997. Indigenous archaeology in the postmodern world. Pp. 1–18 *in* G. P. Nicholas & T. D. Andrews (eds.), At a Crossroads: Archaeology and First Peoples in Canada. Burnaby, B.C.: Archaeology Press.

Nijar, G. S. 1998. Community intellectual rights protect indigenous knowledge. Biotechnology and Development Monitor 36: 11–12.

Nordic Council. 1990. Nordic Council's International Conference on Pollution of the Seas, 1989. P. 99 *in* Nordic Action Plan on Pollution of the Seas. Copenhagen: Nordic Council of Ministers.

Noss, R. 1998. A broad context for restoring ecosystems. Keynote address at the Helping the Land Heal Conference on Ecological Restoration in British Columbia, Victoria. November 5–8.

O'Riordan, T. & A. Jordon. 1995. The precautionary principle in contemporary environmental politics. Environmental Values 4: 191–212.

Peteru, C. 1997. Indigenous peoples knowledge and intellectual property rights working paper. Pp. 85–133 *in* Proceedings of the Indigenous Peoples Knowledge and Intellectual Property Rights Consultation. Suva, Fiji: Pacific Concerns Resource Centre.

Posey, D. A. 1998. Comment on "The development of indigenous knowledge." Current Anthropology 39(2): 242.

———. 1999. Safeguarding traditional resource rights of indigenous peoples. Pp. 217–229 *in* V. D. Nazarea (ed.), Ethnoecology: Situated Knowledge / Located Lives. Tucson: University of Arizona Press.

——— & G. Dutfield. 1996. Beyond Intellectual Property: Toward Traditional Resource Rights for Indigenous Peoples and Local Communities. Ottawa: International Development Research Centre.

Protocol on Substances That Deplete the Ozone Layer / Montreal Protocol. 1987. <http://www.unep.org/ozone/mont_t.htm>.

Province of British Columbia Ministry of Agriculture, Fisheries and Food. 1993. Ginseng Production Factsheet. Kamloops, B.C.: Province of British Columbia Ministry of Agriculture, Fisheries and Food.

Recommendations of the Voices of the Earth Congress. 1993. Amsterdam. *Reprint:* Pp. 209–214 *in* D. A. Posey & G. Dutfield (eds.), Beyond Intellectual Property: Toward Traditional Resource Rights for Indigenous Peoples and Local Communities. Ottawa: International Development Research Centre, 1996.

Report of the Royal Commission for Aboriginal Peoples. 1996. Ottawa: Canadian Communication Group-Publishing.

Rio Declaration on Environment and Development. 1992. <http://www.igc.org/habitat/agenda21/rio-dec.html>.

Sabah Statement from the United Nations Development Programme Asian Consultation Workshop on the Protection and Conservation of Indigenous Knowledge. 1995. 24–27 February. Sabah, East Malaysia. <http://users.ox.ac.uk/~wgtrr/sabah.htm>. *Reprint:* Pp. 145–148 *in* Proceedings of the Indigenous Peoples Knowledge and Intellectual Property Rights Consultation. Suva, Fiji: Pacific Concerns Resource Centre, 1997.

Salmón, E. 1996. Decolonizing our voices. Winds of Change 11(3): 70–72.

Saunders, D. 1994. Developing a modern international law on the rights of indigenous peoples. Research report for the Royal Commission on Aboriginal Peoples. Ottawa: Canadian Communication Group-Publishing.

Statement of the Union of British Columbia Indian Chiefs, from the conference on Protecting Knowledge: Traditional Resource Rights in the New Millennium. 2000. <http://www.ubcic.bc.ca/protect.htm>.

Stephenson, D. J. 1998. Forging a meaningful partnership between international human rights law and ethical bioprospecting. Paper presented at the 6th International Congress of Ethnobiology, Whakatane, Aotearoa / New Zealand, November 24, 1998.

———. 1999. A practical primer on intellectual property rights in a contemporary ethnoecological context. Pp. 230–248 *in* V. D. Nazarea (ed.), Ethnoecology: Situated Knowledge / Located Lives. Tucson: University of Arizona Press.

Stevenson, M. G. 1996. Indigenous knowledge in environmental assessment. Arctic 49(3): 278–291.

Stockholm Convention on Persistent Organic Pollutants. 2001. UNEP/POPS/CONF/4. <http://irptc.unep.ch/pops/POPs_Inc/dipcon/meetingdocs/25june2001/conf4_finalact/en/FINALACT-English.PDF>.

Suva Statement from the United Nations Development Programme on Indigenous Peoples' Knowledge and Intellectual Property Rights. 1995. <http://users.ox.ac.uk/~wgtrr/suva.htm>. *Reprint:* Pp. 181–183 *in* Proceedings of the Indigenous Peoples Knowledge and Intellectual Property Rights Consultation. Suva, Fiji: Pacific Concerns Resource Centre, 1997.

Treaty on the European Union (Maastricht Treaty). 1994. <http://europa.eu.int/en/record/mt/top.html>.

Tri-Council Policy Statement: Ethical Conduct for Research Involving Humans. n.d. <http://www.nserc.ca/programs/ethics/english/policy.htm>.

Turner, N. J. & M. B. Ignace (eds.). In prep. Secwepemc People and Plants: Research Papers in Shuswap Ethnobotany. For submission to Advances in Economic Botany, New York Botanical Garden Press, Bronx.

United Nations Draft Declaration on the Rights of Indigenous Peoples. 1993. <http://www.unhchr.ch/udhr/lang/eng.htm>.

United Nations Framework Convention on Climate Change. 1992. <http://unfccc.int/resource/conv/conv. html>.

University of British Columbia Mission Statement. 1989. U.B.C. Policy and Procedure Handbook (issued by President Strangway). <http://www.universitycounsel.ubc.ca/policies/>.

Van den Daele, W. 1992. Scientific evidence and regulation of technical risks: Twenty years of demythologizing the experts. Pp. 323–340 *in* N. Stehr & R. Ericson (eds.), The Culture and Power of Knowledge: Inquiries into Contemporary Societies. New York: Walter de Gruyter.

Volpe, J. 1998. Natural Health Products: A New Vision. Report of the Standing Committee on Health, House of Commons. Ottawa: Public Works and Government Services Canada-Publishing.

Wildavsky, A. 1988. Searching for Safety. New Brunswick, N.J.: Transaction Books.

Wills, R. & R. G. Lipsey. 1999. An Economic Strategy to Develop Non-Timber Forest Products and Services in British Columbia. Forest Renewal BC Project No. PA97538–ORE. Victoria: British Columbia Ministry of Forests.

WWLIA (World Wide Legal Information Association). 1995. Canadian Patent Law. Canadian Intellectual Property Office. Hull, Que.: Industry Canada.

List of Contributors

Miguel N. Alexiades, Research Fellow, Department of Anthropology, University of Kent, Canterbury, Kent CT2 7NS, England. <m.n.alexiades@ukc .ac.uk>.

Kelly Bannister, Research Associate, POLIS Project on Ecological Governance, Faculty of Law and School of Environmental Studies, University of Victoria, Box 3060, Victoria, B.C., Canada V8W 3R4. <kel@uvic.ca>.

Katherine Barrett, Research Associate, POLIS Project on Ecological Governance, Faculty of Law and School of Environmental Studies, University of Victoria, Box 3060, Victoria, B.C., Canada V8W 3R4. <kbarrett@uvic.ca>.

Thomas J. S. Carlson, Associate Adjunct Professor, University and Jepson Herbaria, 1001 Valley Life Sciences Building, #2465, University of California, Berkeley, Calif. 94720-2465. <tcarlson@socrates.Berkeley.edu>.

S. K. Choge, Research Officer, Kenya Forestry Research Institute., P O Box 20412, Naironi, Kenya. <skchoge2002@yahoo.com>.

A. B. Cunningham, Programme Officer, WWF/UNESCO/Kew People and Plants Initiative, 84 Watkins Street, White Gum Valley, Fremantle 6162, Australia. <peopleplants@bigpond.com>.

Manuel Lizarralde, Assistant Professor of Ethnobotany, Connecticut College, 270 Mohegan Avenue, New London, Conn. 06320. <mliz@conncoll.edu>.

Luisa Maffi, President, Terralingua: Partnerships for Linguistic and Biological Diversity, 1630 Connecticut Ave. N.W., Suite 300, Washington, D.C. 20009. <maffi @terralingua.org>.

Bruce W. Nelson, Dr, Instituto Nacional de Pesquisas da Amazônia (INPA), Al. Cosme Ferreira 2936, Manaus, Amazonas 69083-000, Brazil. <bnelson@internext.com.br>.

Glenn H. Shepard Jr., Post-Doctoral Fellow, Instituto Nacional de Pesquisas da Amazônia, Conj. Villar Camara, R. 3 C. 105, Manaus, Amazonas 69083-000, Brazil. <GShepardJr@aol.com>.

Jennifer C. Sowerwine, Ph.D. Candidate, Department of Environmental Science, Policy and Management, 135 Giannini Hall #3312, University of California, Berkeley, Calif. 94720-3312. <jsowerwi@nature.berkeley.edu>.

Terry C. H. Sunderland, Research Fellow, Dept. of Anthropology, University College, London, Gower Street, London WC1E 6BT, UK. <TCHSunderland@aol.com>.

Douglas W. Yu, Lecturer, School of Biological Sciences, University of East Anglia, Norwich, NR4 7TJ, England. <douglas.yu@uea.ac.uk>.

Egleé López Zent, Postdoctorante, Departamento de Antropología, Laboratorio de Etnología, Instituto Venezolano de Investigaciones Científicas, Apartado Postal 21827, Caracas 1020-A, Venezuela. <elopez@ivic.ve>.

Stanford Zent, Investigador Asociado Titular, Departamento de Antropología, Laboratorio de Etnología, Instituto Venezolano de Investigaciones Científicas, Apartado 21827, Caracas 1020-A, Venezuela. <szent@ivic.ve>.

Index

Page numbers in *italics* refer to tables; page numbers in **boldface** refer to figures.

A

Acacia auriculiformis, 250
Acacia mangium, 250
academic policies, 319–320
  reward structures and, 326–327
acculturation
  of Barí of Venezuela, 115, 124
  of Hotï of Venezuela, 40–41
    Caño Iguana region, 44
    Caño Majagua region, 43
    San José de Kayamá region, 45
Aechmea spp., *145*
African rattans. *See* rattans
Africanized bees, 98
Afzelia quanzensis, 220, 222
age and ethnobotanical knowledge, 39
  among Barí of Venezuela, 117–118, **117–120**
  among Hotï of Venezuela, 53–54, **55–56,** 57, **58–59,** 60–61, **63–64, 67–68**
Agrawal, A., 20–21, 26
agriculture/horticulture
  of Dao women of Vietnam, 237, 238, 241–243, 259
  economic development and, 241
  of Hotï of Venezuela, 85, 95–96, 101–102
    Caño Iguana region, 44
    Caño Majagua region, 43
    Caño Mosquito region, 41
    San José de Kayamá region, 45
  swidden agriculture, 237, 238, 259
Aguaruna ethnic group of Peru, 39
Alibertia pilosa, 145, 161
alternative resources for craft trade, 223–226, **225, 226**
Amazon region
  as anthropogenic ecosystem, 79–82
  craft trade in, 217
  crop diversity within, 81
  diversity of, 133–135
  Matsigenka classification system, 135, 141–143, *142–147,* 147–152, **148–149, 155**
Anacardiaceae, 220
Anacardium giganteum, 88
Angola craft trade, 218
animals. *See* endangered species; fishing; hunting; seed dispersal
anthropocentric perspective on conservation, 9–10
anthropogenic ecosystems, 79–82
anthropogenic landscapes, 81
  linguistically, 12
anthropology, ethnographic approach of, 17
anti-pristine myth, 18–19, 26
Apocynaceae, 220
Apuleia leiocarpa, 145, 157, 163
Araliacaea, 220
Arecaceae. *See* Palmae
Arthrostylidium schomburgkii, 41
Astrocaryum gynacasthum, 91
Astrocaryum murumuru, 144, 156
Astronium graveolens, 120
Attalea spp., *91, 92*
Attalea butyracea, 122, 125, 126, *144,* 156
Attalea macrocarpa, 144, 156
Attalea macrolepis, 90, *91, 92*
Attalea maripa, 90–91, *91, 92,* 92–94, 104
Attalea tessmannii, 144, 161, 163
Aulonemia spp., *144,* 152, 161
Azadirachta indica, 224

B

Ba Vi region of Vietnam, 241–243, 244–245, 247–248, 250–253, *252,* 254–255, 256
Bactris concinna, 144, 152
Bactris gasipaes, 91
Bambusaceae, 218
Bantu language, 176, 178
  life-form categories in, 185–186, *187, 190,* 194
  plant kingdom names in, 185, *190*
Barí of Venezuela
  acculturation of, 115, 124
  age differences among, 117–118, **117–120**
  changes in way of life of, 124–125
  commercial timber and, 121, **121, 123,** 124
  conservation and, 125–128, **127**
  construction by, 121, **121,** 123, **123,** 127–128
  deforestation and, 124, 128
  diet of, 113–114, 119–120, **121,** 122–123, **123**
  economic development of, 114, 124
  education of, 114, 116, 117, 118
  ethnobotanical knowledge of, 115–118, **117, 118,** 118–122, **120, 121,** 125–128, **127**
  ethnographic background of, 113–115, **114**
  firewood use by, 120, **121,** 123, **123**
  fishing by, 113
  forest management by, 126–128
  gathering by, 114
  gender differences among, **117,** 117–118, **118, 119, 120**
  hunting by, 113, **121,** 121–122, **123,** 124
  language of, 116, 117
  medicinal plants and, 120, **121,** 123, **123**
  mode of subsistence of, 117
  population of, 115, 124
  residency and, **120**
  settlement patterns of, 124–125
  use of folk-generic taxa by, 116, **119,** 122–124, **123**
base settlements, 85
basketry, 218–219. *See also* weaving
bees, 96–99, *98*
Benin language, *187*
Berchemia discolor, 219, 222
Berlin's model of ethnobiological classification, 183–184, **184,** 185–190, *190*
biocentric perspective on conservation, 9–10
  deep ecologists and, 9
biocultural diversity, 236, *237*
biocultural systems
  approach to diversity and, 10–14
  defined, 1, 12
biodiversity, 81–82. *See also* Convention on Biological Diversity; diversity
  economic development and, 11
  human impact on, 12
biogeographical factors affecting diversity, 13
Botswana
  craft trade in, 216–217, 218
  weaving in, 222, **225**
Brachylaena huillensis, 220, 221, 222, 223, 224, **226**
Brazilian craft trade, 217
Brosimum lactescens, 88
Brosimum utile, 87
Bulnesia arborea, 123

C

Calamus spp., 177, 186, *191*
Calamus deërratus, 177, *179, 182, 188,* 198–200
Calliandra amazonica, 145
Camelia sinensis, 250
Cameroon
  craft trade in, 217, 218
  language of, 178, *187–188,* **189,** *190*
  rattan use in, *182*
  wood carving in, 220
camp settlements, 85
Caño Iguana region of Venezuela, **42,** 43–44
  acculturation in, 44
  agriculture/horticulture in, 44
  Ecclinusa guianensis in, *90*
  economic development in, 44, 72
  education in, 44
  ethnobotanical knowledge in, 49–71, **50, 52, 56, 58, 63, 65, 67**
  floristic variation in, 47–49, *48*
  interethnic contact in, 43–44
  language in, 44
  missionary presence in, 43–44, 72
  palms in, *92*

Caño Majagua region of Venezuela, **42,** 42–43
  acculturation in, 43
  agriculture/horticulture in, 43
  Ecclinusa guianensis in, *90*
  education in, 43
  ethnobotanical knowledge in, 49–71, **50, 51, 55, 59, 64, 66, 68**
  floristic variation in, 47–49, *48*
  interethnic contact in, 43
  language in, 43
  missionary presence in, 43
  palms in, *92*
  population of, 42
Caño Mosquito region of Venezuela, 41, **42**
  agriculture/horticulture in, 41
  Ecclinusa guianensis in, *90*
  economic development in, 41
  ethnobotanical knowledge in, 49–71, **50, 51, 55, 58, 63, 65, 67**
  floristic variation in, 47–49, *48*
  interethnic contact in, 41
  palms in, *92*
  trade products of, 41
Caring for the Earth, 11, 14
Cariniana pyriformis, 116, 119, 120, 121, 123
Cartagena Protocol on Biosafety, 321
Caryocar microcarpum, 88
Cassia siamea, 224
CBD. *See* Convention on Biological Diversity
Cecropia spp., 148, 156
Cecropia membranacea, 145, 153
Cecropia polystachya, 145, 153
Cecropia sciadophylla, 146, 150, 157, 161
Cedrela odorata, 121, 128, *145,* 153
Cedrelinga spp., 151
Cedrelinga catenaeformis, 145, 151, 157
Central African Republic, rattan use in, *182*
Chrysochlamys ulei, 145, 151
Chrysophyllum argenteum, 88
Chrysophyllum sanguinolentum, 87, *88,* 89–90
Chusquea spp., *144*
Cinchona ledgeriana, 250
Cinnamomum camphora, 250
Cinnamomum cassia, 250
classification systems
  Berlin's model of ethnobiological classification, 183–184, **184,** 185–190, *190*
  of the Matsigenka, 135, 141–143, *142–147,* 147–152, **148, 149, 155**
Clavija longifolia, 145
Clusia spp., *146*
Coccoloba fallax, 88
Cocculus hirsutus, 218
Code of Ethics and Draft Standards of Practice of the International Society of Ethnobiology, 2
Combretum schumannii, 221, 223
Combretum zeyheri, 218
commercial timber, 121, **121, 123,** 124
community-based conservation, 17–18
complex civilizations, 11
Conference on Environment and Development, 11, 313
Conferences of the Parties, 15
Congo
  craft trade in, 218
  language in, 178, *187–188*
  rattan use in, *182*
conservation-based community, 17–18
conservation biology, 16–19, 25–26
construction
  by Barí of Venezuela, 121, **121,** 123, **123,** 127–128
  by Hotï of Venezuela, 90, *91,* 93
contractual approaches to ethnobotanical knowledge, 314
Convention on Biological Diversity, 14–16, 289, 308, 316–317
  Conferences of the Parties, 15
Copaifera langsdorffi, 116, 121
Cordia spp., 220
Cordia platythrysa, 220
Cordia sericicalix, 87
Côte d'Ivoire language, *187*
Couma macrocarpa, 88
crafts trade, 219–222, **221,** 230, *231–234*
  alternative resources for, 223–226, **225, 226**
  deforestation due to, 223
  depletion of natural resources and, 217–222, **221**
  ecological footprints of, 222–223
  economic development and, 215–217, **216**
  endangered species and, 223
  of Hotï of Venezuela, 41
creative disturbance, 79
  disturbance defined, 80–81
  disturbance regimes, 141–142, *143*

creative disturbance (continued)
  diversity of disturbance, 81–82
  by Hotï of Venezuela, 80–81, 86, 100–103, 105
Croton draco, 267
Croton lechleri, 263. *See also* SB–300/SP–303
  chemistry of, 268
  cholera treatment and, 270, **271**
  diarrhea treatment and, *266,* 267, *269,* 270, *272,* 272–273, *276*
  ethnomedical uses of, 265–277, *266, 269,* **271,** *272, 274, 275, 276*
  herpes simplex treatment and, 268, *269,* 271–272, *272, 274*
  HIV treatment and, *272,* 272–273
  human clinical studies of, 270–277, **271,** *272, 274, 275, 276*
  methods of studying, 265
  respiratory infection treatment and, *269,* 269–270, *275*
  wound treatment and, 268, *269, 274*
Croton urucurana, 267
cultural diversity, 12–14, 25, 81–82, 236, *237*
cultural taboos, 18
cultural types
  forest foot nomads, 40
  South American Marginal, 40
culture, 296–297
  adapting to changes in, 13–14
  compatibility of cultural practices, 14–16, 25
  cultural landscapes, 12
Cyathea spp., *145*
Cynometra, 223

D

Dacryodes spp., 87, *88*
Dacryodes chimantensis, 88
Dalbergia melanoxylon, 220, 221, 224
Daniella olivieri, 221
Dao women of Vietnam, 235–237, *237,* 255–257. *See also* medicinal plants; Vietnam
  acquisition and transference of knowledge among, 239–240, **240**
  agriculture/horticulture of, 237, 238, 241–243, 259
  of Ba Vi region, 241–243, 244–245, 247–248, 250–253, *252,* 254–255, 256
Dao women of Vietnam (continued)
  of Sapa region, 240–241, **242,** 245, 248, 249–250, *250,* 254, 256
  settlement patterns of, 238
Declaration on the Rights of Indigenous Peoples, 15
deep ecologists, 9
deforestation, 16. *See also* reforestation
  and Barí of Venezuela, 124, 128
  due to craft trade, 223
  in Venezuela, 115
  in Vietnam, 235
demise of native language, 13
Denya language group, **189**
depletion of natural resources, 217–222, **221.** *See also* deforestation
Dialium guianense, 120, 123
diet
  of Barí of Venezuela, 113–114, 119–120, **121,** 122–123, **123**
  of Hotï of Venezuela, 85, 86–90, 90–94, *91, 92,* 94–95, 96–99, *98,* 104–105
  use of palm grubs in, 94–95, 104–105
  wood carving and, 220
Dioscorea alata, 85
Dioscorea trifida, 85
Diospyros lycioides, 219
Diplasia spp., *145*
Dipteryx spp., *88*
disturbance. *See* creative disturbance
diversity. *See also* biodiversity; Convention on Biological Diversity
  approach to, 10–14
  biocultural, 236, *237*
  biogeographical factors affecting, 13
  biological, 81–82
  crop, 81
  cultural, 12–14, 25, 81–82, 236, *237*
  of disturbance, 81–82
  economic development and, 14
  habitat-mosaic hypothesis of, 133–134
  of language, 12–13, 12–14, 25
  in small-scale societies, 13
  theories re, 133–135
DR Congo
  language in, *187–188*
  rattan use in, *182*
dye resources, 219, 222–223, *232*

E

Earth Charter, 23
EBK. *See* ethnobotanical knowledge
Ecclinusa guianensis, 87–89, *88, 90*
ecological footprints
 of craft trade, 222–223
 defined, 215
economic development
 agriculture/horticulture and, 241
 of Barí of Venezuela, 114, 124
 crafts trade and, 215–217, **216**
 diversity and, 14
 globalization and, 287–292, 292–296
 health care reform and, 243–246
 of Hotï of Venezuela
 Caño Iguana region, 44, 72
 Caño Mosquito region, 41
 indigenous and traditional peoples and, 16–17, 25
 market reform and, 253–255
 rattans and, 176
 reforestation and, 246–253, *250, 252*
 resettlement and, 246–253, *250, 252*
 social effects of, 215–217, **216**
 stemming biodiversity crisis and, 11
 in Vietnam, 235–236, 241, 243–255, *250, 252*
 weaving and, 215–217, **216**
 wood carving and, 215–217, **216**
 World Bank and, 15, 25
education
 acquisition and transference of, 239–240, **240**
 of Barí of Venezuela, 114, 116, 117, 118
 of Hotï of Venezuela
  Caño Iguana region, 44
  Caño Majagua region, 43
  San José de Kayamá region, 45
 as indicator of ethnobotanical knowledge, 39
endangered species, 223
Equatorial Guinea
 language in, *187–188*
 rattan use in, *182*
Eremospatha spp., 176, 186, *191*
Eremospatha barendii, 179, 200
Eremospatha cabrae, 179, 182, 200–201
Eremospatha cuspidata, 179, 201
Eremospatha dransfieldii, 179, 201–202
Eremospatha haullevilleana, 179, 182, 188, 189, 202–203
Eremospatha hookeri, 179, 203
Eremospatha laurentii, 180, 204
Eremospatha macrocarpa, 180, 182, 188, *188,* 189, **189,** *190,* 192, 204–205
Eremospatha quinquecostulata, 180, 188, 205–206
Eremospatha tessmanniana, 180, 188, 206
Eremospatha wendlandiana, 180, 182, 188, 189, 206–207
Erythrina spp., *145*
Erythroxylum coca, 146
ethics
 academic policies and, 319–320, 326–327
 Code of Ethics and Draft Standards of Practice of the International Society of Ethnobiology, 2
 culture and, 296–297
 ethnobiology and, 2, 290–291, 298–299
 intellectual property rights and, 314–320
 international declarations and, 318–319
 International Society of Ethnobiology Code of Ethics, 321
 legal issues and, 313–320
 media and, 296–297
 precautionary principle and, 320–327
 and privatization of science, 294–295
 professional societies and, 319–320
 publication and, 310–313, 323–327
 resource rights and, 316–318
 and transnational social movements, 295–296
ethnobiology
 Berlin's model of ethnobiological classification, 183–184, **184,** 185–190, *190*
 Code of Ethics and Draft Standards of Practice of the International Society of Ethnobiology, 2
 ethics and, 2, 290–291, 298–299
ethnobotanical knowledge. *See also* indigenous and traditional peoples; medicinal plants; traditional ecological knowledge
 age as indicator of, 39
 Barí. *See* Barí of Venezuela
 contractual approaches to, 314

ethnobotanical knowledge (continued)
Dao. *See* Dao women of Vietnam
direction of research into, 38
education as indicator of, 39
ethics and, 290–291
ground-truthing and, 147–152, **148, 149**
Hotï. *See* Hotï of Venezuela
intellectual property rights and, 314–320
language as indicator of, 39
Matsigenka. *See* Matsigenka of Peru
nomenclature of indigenous and traditional peoples, 175–176
in the public domain, 308–310
publication of, 310–313, 323–327
quantitative study of, 37–39
residency and, **120**
social standing as indicator of, 39
use of folk-generic taxa by Barí of Venezuela, 116, **119,** 122–124, **123**
use of for conservation, 125–128, **127**
use of for plant collection, 263–265
ethnoecology, 1
ethnographic approach of anthropology, 17
ethnographic background of Barí of Venezuela, 113–115, **114**
Ethnologue, 185
Eucalyptus alba, 250
Euclea divinorum, 219
Euclea natalensis, 219
Euphoria longan, 250
European Union, 293
conservation efforts of, 15
Euterpe spp., 151
Euterpe precatoria, 91, 92, 144

F

fauna. *See* endangered species; fishing; hunting; seed dispersal
Ficus spp., *145*
Ficus insipida, 145, 153
Ficus trigona, 145
firewood, 120
fishing, 113
flora. *See* ethnobotanical knowledge; medicinal plants; *individual plant names*
folk knowledge. *See* ethnobotanical knowledge; indigenous and traditional peoples; nomenclature of indigenous and traditional peoples; traditional ecological knowledge
food. *See* diet
forest foot nomads, 40
forest management
by Barí of Venezuela, 126–128
by Hotï of Venezuela, 91–93
resources, 85–86
Forestry Stewardship Council, 226

G

Gabon
language in, *187*
rattan use in, *182*
game animal trees. *See* hunting
gap cultivation, 95–96, 101–102
Garcinia macrophylla, 88
gathering
by Barí of Venezuela, 114
by Hotï of Venezuela, 85
gender and ethnobotanical knowledge. *See also* Dao women of Vietnam
among Barí of Venezuela, **117,** 117–118, **118, 119, 120**
among Hotï of Venezuela, 50, **51–52,** 53
generic life-form categories and, 186, *187, 190*
Geonoma spp., *144*
Geonoma deversa, 91, 157
Geonoma stricta, 125
German environmental policy, 321
Ghana
craft trade in, 218
language in, *187–188, 191*
rattan use in, *182*
wood carving in, 221
globalization
culture and, 296–297
defined, 284–287
economic development and, 287–292, 292–296
flow of knowledge and, 288–290
media and, 296–297
and privatization of science, 294–295
and production of ethnobiological knowledge, 291–292
and transnational social movements, 295–296

ground-truthing, 22, 134. *See also* satellite imagery
applications of, 163–166
ethnobotanical, 147–152, **148, 149**
hydrological distinctions in, 141, *142,* **154, 159**
methods of, 137–138
topographic distinctions in, 141, *142,* **154, 159**
uses and limitations of, 138–141
Guadua spp., *144,* 156
Guadua angustifolia, 144, 161
Guadua sarcocarpa, 144
Guadua weberbaueri, 144, 148, 156, 163
Guatteria foliosa, 87
Guazuma crinita, 146
Guinea-Bissau, rattan use in, *182*
Gynerium saccharoides, 145
Gynerium sagittatum, 145

H

habitat-directed research, 38
habitat-mosaic hypothesis of diversity, 133–134
habitat types, Matsigenka of Peru, 162–163
harvesting
of edible fruit trees, 86–90
of medicinal plants, 245
health care reform in Vietnam, 243–246
Heliconia metallica, 145, 156
Helicostylis tomentosa, 88
Hevea brasiliensis, 145, 151
historical ecology, 81
Holarrhena floribunda, 220
honey extraction, 96–99, *98*
horticulture. *See* agriculture/horticulture
Hotï of Venezuela
acculturation of, 40–41, 43, 44, 45
age differences among, 53–54, **55–56,** 57, **58–59,** 60–61, **63–64, 67–68**
agriculture/horticulture of, 41, 43, 45, 85, 95–96, 101–102
bees and, 96–99, *98*
of Caño Iguana region, **42,** 43–44
acculturation of, 44
agriculture/horticulture of, 44
Ecclinusa guianensis in, *90*
economic development of, 44, 72
Hotï of Venezuela (continued)
of Caño Iguana region (continued)
education of, 44
ethnobotanical knowledge in, 49–71, **50, 52, 56, 58, 63, 65, 67**
floristic variation and, 47–49, *48*
interethnic contact among, 43–44
language of, 44
missionary presence among, 43–44, 72
palms in, *92*
of Caño Majagua region, **42,** 42–43
acculturation of, 43
agriculture/horticulture of, 43
Ecclinusa guianensis in, *90*
education of, 43
ethnobotanical knowledge in, 49–71, **50, 51, 55, 59, 64, 66, 68**
floristic variation and, 47–49, *48*
interethnic contact among, 43
language of, 43
missionary presence among, 43
palms in, *92*
population of, 42
of Caño Mosquito region, 41, **42**
agriculture/horticulture of, 41
Ecclinusa guianensis in, *90*
economic development and, 41
ethnobotanical knowledge in, 49–71, **50, 51, 55, 58, 63, 65, 67**
floristic variation and, 47–49, *48*
interethnic contact among, 41
palms in, *92*
trade products of, 41
construction by, 90, *91,* 93
creative disturbance by, 80–81, 86, 100–103, 105
cultural type, 40
diet of, 85, 86–90, 90–94, *91, 92,* 94–95, 96–99, *98,* 104–105
economic development and, 41, 44, 72
education of, 43, 44, 45
ethnobotanical knowledge in, 49–71, **50–52, 55–56, 58–59, 63–68**
first contact with, 40, 82
floristic variation and, 47–49, *48*
forest management by, 91–93
gap cultivation by, 95–96, 101–102

Hotï of Venezuela (continued)
  gathering by, 85
  gender differences among, 50, **51–52,** 53
  harvest of edible fruit trees, 86–90
  honey extraction by, 96–99, *98*
  hunting by, 85
  interethnic contact among, 41, 43–44
  language of, 43, 44, 45
  medicinal plants and, 61–62, **63–68, 67–68**
  methods of studying, 46–47
  missionary presence among, 43–44, 44–46, 72
  population of, 42
  quantitative ethnobotany of, 39–41
  of San José de Kayamá region, 42f, 44–46
    acculturation of, 45
    agriculture/horticulture of, 45
    Ecclinusa guianensis in, *90*
    education of, 45
    ethnobotanical knowledge in, 49–71, **50, 52, 56, 59, 64, 66, 68**
    floristic variation and, 47–49, *48*
    language of, 45
    missionary presence among, 44–46
    palms in, *92*
  settlements of, 84–85
  subsistence orientation of, 85–86
  tools of, 86, 90, *91*
  trade products of, 41
  use of palm grubs by, 94–95, 104–105
  use of palms by, 90–94, *91, 92*
Housing. *See* construction
human impact
  on biodiversity, 12
  complex civilizations and, 11
  levels of, 1
  perspectives on, 9–10, 11, 16
  as predators, 17, 25–26
  unsustainable, 11
human rights, 14–16, 25
Humiria balsamifera, 87, *88,* 89–90
hunting
  by Barí of Venezuela, 113, **121,** 121–122, **123,** 124
  by Hotï of Venezuela, 61–62, **67–68,** 85
  by Matsigenka of Peru, 161
hydrological distinctions in ground-truthing, 141, *142,* **154, 159**
Hymenaea courbaril, 88, 121
Hyospathe spp., *144,* 148
Hyphaene, 218, 223–224
Hyphaene compressa, 218
Hyphaene coriacea, 218

I

indigenous and traditional peoples. *See also* ethnobotanical knowledge; traditional ecological knowledge
  Barí. *See* Barí of Venezuela
  changes in, 124–125
  conservation biology and, 16–19, 25–26
  cultural taboos of, 18
  Dao. *See* Dao women of Vietnam
  defined, 24
  economic development and, 16–17, 25
  globalization and, 295–296
  Hotï. *See* Hotï of Venezuela
  human rights and, 14–16, 25
  intellectual property rights and, 314–320
  knowledge of in the public domain, 308–310
  legal issues involving, 313–320
  Matsigenka. *See* Matsigenka of Peru
  nomenclature of, 175–176
  rattan utilization by, 177–178, *179–182*
  relocation of, 17, 25–26
  researchers and, 326
  resource rights and, 316–318
  tribal peoples defined, 24
  use of folk-generic taxa by Barí of Venezuela, 116, **119**
Indigenous Knowledge and Development Monitor, 20–21, 26
Indigofera, 219
Indigofera arrecta, 219
Indigofera tinctoria, 219
Inga spp., *88,* 120, *145*
Inga alba, 88
Inga bourgoni, 88
Inga edulis, 88
intellectual property rights, 314–320
intensive sampling, 134
interethnic contact
  among Hotï of Venezuela
    Caño Iguana region, 43–44
    Caño Majagua region, 43
    Caño Mosquito region, 41

interinformant consensus analysis, 38–39
intermediate life-form categories, 186, *187, 190*
international declarations, 318–319
international law and indigenous peoples, 313–320
International Society of Ethnobiology, 23, 321
Ipomoea batatas, 85
Iriartea deltoidea, 144, 156
ISE. *See* International Society of Ethnobiology

J

Jacaranda mimosifolia, 224

K

Kenya
  craft trade in, 217, 218
  wood carving in, 219, **221,** 221–222, 223, 224, **226**
kingdom names, plant, 185, *190*
Kirkia acuminata, 222

L

Laccosperma spp., 176, 186, *191,* 192
Laccosperma acutiflorum, 180, 192, 207
Laccosperma korupensis, 180
Laccosperma laeve, 177, *180, 182,* 207–208
Laccosperma opacum, 177, *181, 182, 188,* 189, **189,** 208–209
Laccosperma robustum, 181, 182, 188, 189, 192, 209
Laccosperma secundiflorum, 181, 182, 188, 189, 192, 209–211
LANDSAT. *See* ground-truthing; satellite imagery
language
  adapting to changes in, 13–14
  Bantu. *See* Bantu language
  of Barí of Venezuela, 116, 117
  in Benin, *187*
  Berlin's model of ethnobiological classification and, 185–190, *190*
  in Cameroon, 178, *187–188,* **189,** *190*
  in Congo, 178, *187–188*
  in Côte d'Ivoire, *187*
  demise of native, 13
  Denya language group, **189**
  diversity of, 12–13, 12–14, 25
language (continued)
  in DR Congo, *187–188*
  in Equatorial Guinea, *187–188*
  in Gabon, *187*
  generic life-form categories and, 186, *187, 190*
  in Ghana, *187–188, 191*
  of Hotï of Venezuela
    Caño Iguana region, 44
    Caño Majagua region, 43
    San José de Kayamá region, 45
  as indicator of ethnobotanical knowledge, 39
  intermediate life-form categories and, 186, *187, 190*
  life-form categories and, 185–186, *187, 190,* 194
  linguistic ecologies, 12
  linguistically anthropogenic landscapes, 12
  logosphere theory of, 12
  of Namibia, 178
  Niger-Congo. *See* Niger-Congo language family
  in Nigeria, 178, *187–188, 191*
  nomenclature. *See* nomenclature of indigenous and traditional peoples
  overlap of, 13
  plant kingdom names, 185, *190*
  richness of, and species richness, 12–13
  in Senegal, 178
  shift in, 14, 25
  in Sierra Leone, *190, 191*
  in Somalia, 178
  specific life-form categories and, 186–189, *190*
  structure of, *191*
  in Sub-Saharan Africa, 178
  in Sudan, 178
  Summer Institute of Linguistics' *Ethnologue,* 185
  varietal life-form categories and, 189–190, *190*
Lecythis corrugata, 87
legal issues and indigenous peoples, 313–320
Leguminosae, 217, 219
life-form categories, 185–186, *187, 190,* 194
linguistic ecologies, 12
linguistically anthropogenic landscapes, 12
logosphere theory, 12
Luehea seemanni, 120

M

Malawi wood carving, 219, 223
Mali crafts trade, 218
Malvaceae, 217
Manihot esculenta, 85
Marantaceae spp., *145*
Matayba spp., *88*
Matsigenka of Peru
  classification system of, 135, 141–143, *142–147,* **155, 160**
  ethnobotanical ground-truthing using, 147–152, **148, 149**
  habitat types as defined by, 162–163
  hunting by, 161
  of Tayakome region, ground-truthing and, 152–157, **153–155**
  of Yomybato region, ground-truthing and, 157–162, **158–160**
Mauritia flexuosa, 91, 144, 156
Mauritiella spp., *144,* 150, 151
media, 296–297
medicinal plants, 238–239, 255–257. *See also individual plants*
  acquisition and transference of knowledge about, 239–240, **240**
  and Barí of Venezuela, 120, **121,** 123, **123**
  cholera treatment and, 270, **271**
  diarrhea treatment and, *266,* 267, *269,* 270, *272,* 272–273, *276*
  ethnomedical uses of *Croton lechleri,* 265–277, *266, 269,* **271,** *272, 274, 275, 276*
  harvesting of, 245
  herpes simplex treatment and, 268, *269,* 271–272, *272, 274*
  HIV treatment and, *272,* 272–273
  and Hotï of Venezuela, 61–62, **63–68**
  human clinical studies of, 270–277, **271,** *272, 274, 275, 276*
  interinformant consensus analysis of, 38–39
  knowledge of in the public domain, 308–310
  land use affecting, 248–253, *250, 252*
  market reform and, 253–255
  pharmaceutical industry, 245–246
    approach of to, 263–265
  processing and storage of, 256–257
  respiratory infection treatment and, *269,* 269–270, *275*
medicinal plants (continued)
  in Vietnam, 235–237, *237*
  Western medicine and, 244–246
  wood carving and, 220–221
  wound treatment and, 268, *269, 274*
Micropholis egensis, 88
migration, 11
mineral licks. *See* hunting
missionary presence among Hotï of Venezuela
  Caño Iguana region, 43–44, 72
  Caño Majagua region, 43
  San José de Kayamá region, 44–46
Mozambique
  craft trade in, 218
  wood carving in, 221
Musa x *paradisiaca,* 85

N

Namibia
  craft trade in, 218
  language of, 178
  weaving in, 222
  wood carving in, 220
Native Americans
  pristine myth and, 18–19, 26
  relocation of, 17, 25–26
natural resources, 85–86. *See also* deforestation
  depletion of, 217–222, **221**
nest sites of bees, *98*
Ngamiland weaving, 222
Niger-Congo language family, 176, 178
  intermediate life-form categories in, 186, *187, 190*
  life-form categories in, 185–186, *187, 190,* 194
  plant kingdom names in, 185, *190*
Nigeria
  language in, 178, *187–188, 191*
  rattan use in, *182*
nomenclature of indigenous and traditional peoples, 175–176, 198–213
  Berlin's model of ethnobiological classification and, 185–190, *190*
  parity of with Western classification, 191, *191*
non-Bantu language. *See* Niger-Congo language family
North American Free Trade Agreement, 293

O

Oenocarpus spp., 151
Oenocarpus bacaba, 90, 91, *91, 92,* 92–94, 94–95, 104
Oenocarpus bataua, 94, 122, 126, *144,* 158
Oenocarpus mapora, 122, 126
Olea europaea, 221
Oncocalamus spp., 176, 186
Oncocalamus macrospathus, 181, 211
Oncocalamus mannii, 181, 211–212
Oncocalamus tuleyi, 181, 182, 189, **189,** 212
Oncocalamus wrightianus, 181, 182, 188, 213
Organization of American States, 15
overlap of language, 13
Oxandra venezuelensis, 120

P

palm grubs, 94–95, 104–105
Palmae, 119, 122–123, 218
  Astrocaryum gynacasthum, 91
  Astrocaryum murumuru, 144, 156
  Attalea spp., *91, 92*
  Attalea butyracea, 122, 125, 126, *144,* 156
  Attalea macrocarpa, 144, 156
  Attalea macrolepis, 90, *91, 92*
  Attalea maripa, 90–91, *91, 92,* 92–94, 101, 104
  Attalea tessmannii, 144, 161, 163
  Bactris concinna, 144
  Bactris gasipaes, 90, *91,* 96
  Euterpe spp., 151
  Euterpe precatoria, 91, 92, 144
  Geonoma spp., *144*
  Geonoma deversa, 91
  Hyospathe spp., *144,* 148
  Iriartea deltoidea, 144, 156
  life-form categories and, 185–186, *187, 190,* 194
  Mauritia flexuosa, 91, 144, 156, 158
  Mauritiella spp., *144*
  Oenocarpus spp., 151
  Oenocarpus bacaba, 90, 91, *91, 92,* 92–94, 94–95, 104
  Oenocarpus bataua, 94, 122, 126, *144,* 158
  Oenocarpus mapora, 122, 126
  Phytelephas macrocarpa, 144, 161
  Podococcus barteri, 186
  rattans. *See* rattans
Palmae (continued)
  Sclerosperma spp., 186
  Socratea exorrhiza, 91, 92, 144, 151
  Socratea salazarii, 144, 156, 158, 161
  use of by Hotï of Venezuela, 90–94, *91, 92*
  varietal life-form categories and, 189–190, *190*
  Wettinia augusta, 144, 151, 158, 161
Parkia pendula, 121
Peltogyne purpurea, 121
Permanent Forum on Indigenous Issues, 15
Persea americana, 126
Peru
  Aguaruna ethnic group of, 39
  Amazon. *See* Amazon
  Matsigenka. *See* Matsigenka of Peru
  Tambopata, 39
  Tayakome region of, 152–157, **153–155**
  Yomybato region of, 157–162, **158–160**
pharmaceutical industry, 245–246
  approach of to medicinal plants, 263–265
Phenakospermum guyanense, 87
Pheonix reclinata, 223
Phytelephas macrocarpa, 144, 151, 161
Piaroa males in Venezuela, 39
Pinus massoniana, 250
plant fibers used in weaving, *231*
  depletion of, 217–219
plant kingdom names, 185, *190*
Poaceae spp., *145*
Podococcus barteri, 186
Polylepis spp., *146*
population
  of Barí of Venezuela, 115, 124
  of Hotï of Venezuela, 42
Pourouma melinonii, 88
Pouteria anibaefolia, 119
precautionary principle, 320–327
  applications of, 323–327
  criticisms of, 328–329
  elements of, 321–323
predators, humans as, 17, 25–26
pristine ecosystems myth, 18–19, 26
privatization of science, 294–295
processing of medicinal plants, 256–257
professional societies, 319–320
Protium heptaphyllum, 120
Protium opacum, 88
Protium tenuifolium, 88

Pseudolmedia laevigata, 88
Pseudolmedia laevis, 88
Pterocarpus angolensis, 220, 222
Pterocarpus fulva, 220
publication, ethics of, 310–313, 323–327
Puya spp., *146*

Q

Qualea paraensis, 88
quantitative ethnobotany
  defined, 37–38
  of the Hotï, 39–41

R

random plant collections, 263–265
rattans. *See also* Palmae
  Calamus spp., 177, 186, *191*
  Calamus deërratus, 177, *179, 182, 188,* 198–200
  depletion of, 218
  economic development and, 176
  Eremospatha spp., 176, 186, *191*
  Eremospatha barendii, 179, 200
  Eremospatha cabrae, 179, 182, 200–201
  Eremospatha cuspidata, 179, 201
  Eremospatha dransfieldii, 179, 201–202
  Eremospatha haullevilleana, 179, 182, 188, 189, 202–203
  Eremospatha hookeri, 179, 203
  Eremospatha laurentii, 180, 204
  Eremospatha macrocarpa, 180, 182, 188, *188,* 189, **189,** *190,* 192, 204–205
  Eremospatha quinquecostulata, 180, 188, 205–206
  Eremospatha tessmanniana, 180, 188, 206
  Eremospatha wendlandiana, 180, 182, 188, 189, 206–207
  Laccosperma spp., 176, 186, *191,* 192
  Laccosperma acutiflorum, 180, 192, 207
  Laccosperma korupensis, 180
  Laccosperma laeve, 177, *180, 182,* 207–208
  Laccosperma opacum, 177, *181, 182, 188,* 189, **189,** 208–209
  Laccosperma robustum, 181, 182, 188, 189, 192, 209
  Laccosperma secundiflorum, 181, 182, 188, 189, 192, 209–211
  methods of studying, 184–185
rattans (continued)
  Oncocalamus spp., 176, 186
  Oncocalamus macrospathus, 181, 211
  Oncocalamus mannii, 181, 211–212
  Oncocalamus tuleyi, 181, 182, 189, **189,** 212
  Oncocalamus wrightianus, 181, 182, 188, 213
  taxonomy of, 177–178, *179–182*
  uses of, 176
  utilization of, 177–178, *179–182,* 198–213
reforestation. *See also* deforestation
  in Vietnam, 243–244, 246–253, *250, 252*
relocation of indigenous and traditional peoples, 17, 25–26
  in Vietnam, 246–253, *250, 252*
remote sensing. *See* ground-truthing
Renealmia spp., *145,* 150
Report of the Royal Commission for Aboriginal Peoples, 311
research
  habitat-directed, 38
  use-directed, 38
resettlement of indigenous and traditional peoples, 17, 25–26
  in Vietnam, 246–253, *250, 252*
residency and ethnobotanical knowledge, **120**
resource rights, 316–318
Rio Declaration on Environment and Development, 321
Rubia cordifolia, 219

S

Sagotia spp., *145*
Sagotia racemosa, 119, 120, 122, 123
sampling methods, 134
San José de Kayamá region of Venezuela, 42f, 44–46
  acculturation in, 45
  agriculture/horticulture in, 45
  Ecclinusa guianensis in, *90*
  education in, 45
  ethnobotanical knowledge in, 49–71, **50, 52, 56, 59, 64, 66, 68**
  floristic variation in, 47–49, *48*
  language in, 45
  missionary presence in, 44–46
  palms in, *92*
Sansevieria, 217

Sapa region of Vietnam, 240–241, **242,** 245, 248, 249–250, *250,* 254, 256
satellite imagery. *See also* ground-truthing
applications of, 163–166
habitat types and, 162–163
uses and limitations of, 138–141
SB–300/SP–303, 268. *See also Croton lechleri*
human clinical studies of, 270–277, **271,** *272, 274, 275, 276*
Sclerocarya birrea, 217, 220
Sclerosperma spp., 186
seed dispersal, 96, 101, 105, 126
Attalea maripa, 104
Senegal
language of, 178
rattan use in, *182*
settlement patterns
of Barí of Venezuela, 124–125
base settlements defined, 85
camp settlements defined, 85
of Dao women of Vietnam, 238
in Vietnam, 238, 246–253, *250, 252*
Sierra de Maigualida region of Venezuela, 82–84, **83**
Sierra Leone
language in, *190, 191*
rattan use in, *182*
small-scale societies
diversity in, 13
ecological knowledge of, 11
social standing and ethnobotanical knowledge, 39
Socratea exorrhiza, 91, 92, 144, 151
Socratea salazarii, 144, 156, 161
soil types, 141–142, *143*
Somalian language, 178
Sorghum, 219
Sorocea muriculata, 88
South Africa crafts trade, 218
South American Marginal cultural type, 40
species richness
language richness and, 12–13
in Vietnam, 238
specific life-form categories, 186–189, *190*
Sphaghnum spp., *146*
Spirostachys africana, 220, 221
Spondias mombin, 116, 120, 126
Stockholm Convention on Persistent Organic Pollutants, 321
storage of medicinal plants, 256–257
Sub-Saharan Africa, 178. *See also individual countries*
subsistence orientation, 85–86
Sudan language, 178
Summer Institute of Linguistics' *Ethnologue,* 185
swidden agriculture, 237, 238, 259
Swietenia odorata, 121

T

Tabebuia chrysea, 120, 123, 127, 128
taboos, 18
Tambopata, Peru, 39
Tanzania
craft trade in, 218
wood carving in, 223, 224, 226
taxonomy of rattans, 177–178, *179–182*
Tayakome region of Peru, 152–157, **153–155**
TEK. *See* traditional ecological knowledge
Tephrosia spp., *146*
Tessaria integrifolia, 145, 153
Thymelaeaceae, 217
Tiliaceae, 217
Tillandsia spp., *146*
tools, 86, 90, *91*
topographic distinctions in ground-truthing, 141, *142,* **154, 159**
Tovomita weddeliana, 151
traditional botanical medicine. *See* medicinal plants
traditional ecological knowledge, 10, 19–21, 24, 26. *See also* ethnobotanical knowledge; indigenous and traditional peoples
correspondence of to Western biotaxonomy, 115–116
as distinct from Western scientific knowledge, 20–21, 26
loss of, 14, 25, 117–118
nomenclature of, 175–176
in the public domain, 308–310
publication of, 310–313, 323–327
small-scale societies and, 11
traditional peoples. *See* indigenous and traditional peoples
tribal peoples. *See* indigenous and traditional peoples

U

Ugandan craft trade, 218
United Nations
  Conference on Environment and Development, 11, 313
  Declaration on the Rights of Indigenous Peoples, 15
  Permanent Forum on Indigenous Issues, 15
  Working Group on Indigenous Populations, 15
use-directed research, 38

V

varietal life-form categories and, 189–190, *190*
Venezuela
  Barí. *See* Barí of Venezuela
  Caño Iguana region. *See* Caño Iguana region of Venezuela
  Caño Majagua region. *See* Caño Majagua region of Venezuela
  Caño Mosquito region. *See* Caño Mosquito region of Venezuela
  deforestation in, 115
  ethnobotanical knowledge in, 49–71, **50–52, 55–56, 58–59, 63–68**
  floristic variation in, 47–49, *48*
  Hotï. *See* Hotï of Venezuela
  Piaroa males in, 39
  population, 115
  San José de Kayamá region. *See* San José de Kayamá region of Venezuela
  Sierra de Maigualida region of, 82–84, **83**
Vernonia spp., *146*
Vietnam. *See also* Dao women of Vietnam
  Ba Vi region of, 241–243, 244–245, 247–248, 250–253, *252*, 254–255, 256
  deforestation in, 235
  economic development in, 235–236, 241, 243–255, *250, 252*
  health care reform in, 243–246
  land use in, 248–253, *250, 252*
  market reform in, 253–255
  medicinal plants in, 235–237, *237*
  reforestation in, 243–244, 246–253, *250, 252*
  resettlement in, 246–253, *250, 252*
Vietnam (continued)
  Sapa region of, 240–241, **242,** 245, 248, 249–250, *250*, 254, 256
  settlement patterns in, 238, 246–253, *250, 252*
  species richness in, 238

W

weaving. *See also* basketry
  alternative resources for, 223–224, **225**
  depletion of fibers used in, 217–219
  ecological footprint of, 222
  economic development and, 215–217, **216**
  plant fibers used in, *231*
Western biotaxonomy
  correspondence of to traditional ecological knowledge, 115–116
  folk taxonomic systems parity with, 191, *191*
Western medicine
  pharmaceutical industry, 245–246
    approach of to medicinal plants, 263–265
  vs. traditional, 244–246
Western scientific knowledge, 20–21, 26
Westernization. *See* acculturation
Wettinia augusta, 144, 151, 158, 161
wood carving, 219–222, **221,** *233–234*
  alternative resources for, 224, 226, **226**
  ecological footprint of, 222, 223
  economic development and, 215–217, **216**
Working Group on Indigenous Populations, 15
World Bank, 15, 25
World Parks Congress, 11
World Trade Organization, 293

Y

Yomybato region of Peru, 157–162, **158–160**

Z

Zambian craft trade, 218
Zea mays, 85
Zimbabwe
  craft trade in, **216,** 217, 218
  weaving in, 222
  wood carving in, 220, 222